Praise for

THE EXTENDED MIND

"[Paul's] revolutionary thesis challenges us to rethink what we think about thinking . . . Our minds are bigger than our brains, and if we embrace that fact, there's so much more we can accomplish."
— Emily Balcetis, *Washington Post*

"Annie Murphy Paul's new book, *The Extended Mind*, exhorts us to use our entire bodies, our surroundings and our relationships to 'think outside the brain' . . . Paul writes with precision and flair . . . I'm a convert."
— Susan Pinker, *New York Times Book Review*

"An acclaimed science journalist demystifies how our most important thinking often happens outside our heads."
— Adam Grant, author of *Think Again* and *Originals*, on LinkedIn as "The 12 New Leadership Books to Read This Summer"

"In *The Extended Mind*, author Annie Murphy Paul explains why the key to thinking better sometimes lies in using our brains less. By extending our minds through our bodies, physical surroundings, and relationships, we can work more productively and solve problems more creatively. *The Extended Mind* uses stories and science to show us how it's done."
— Charles Duhigg, author of *The Power of Habit* and *Smarter Faster Better*

"Fascinating, sure-footed and wide-ranging."   — *Wall Street Journal*

"In *The Extended Mind*, science writer Annie Murphy Paul shows us how we can 'think outside the brain'—that is, draw the stuff of the world into our trains of thought. We limit ourselves when we think only with our heads. Extending our minds opens up a host of new possibilities, allowing us to become more focused, more productive, more creative—in a word, smarter."
— Susan Cain, author of *Quiet*

"*The Extended Mind* argues that our creativity, our intelligence, and even our memories are embodied not just in the wet matter of our brains, but in the world all around us. This is a profoundly interesting book that invites us to radically change how we think about thinking."

— Joshua Foer, author of *Moonwalking with Einstein*

"Packed with cutting-edge research, compelling real-world examples, and deep insight, *The Extended Mind* provides a revolutionary framework to help us understand how our brains work. It's one of those rare books that I found so interesting that I couldn't put it down, and the minute I finished, I started making changes in my life."

— Gretchen Rubin, author of *The Happiness Project*

"Just when I thought I was stuck with the brain I have, Annie Murphy Paul reveals that I can do better! Much better! This fascinating tour of the latest science reveals all the ways we can get smarter by changing our physical spaces, moving our hands and bodies, and thinking together with other humans. An inspiring guide to living fuller lives by getting outside our own heads."

— Amanda Ripley, author of *High Conflict* and *The Smartest Kids in the World*

"Powerful, actionable, and whip-smart, this book is proof that when you re-think how thinking works, you open the door to a world of new possibilities. I guarantee that Annie Murphy Paul's concepts, case studies, and research-based tips will help you and your group connect, create, and perform in new ways."

— Daniel Coyle, author of *The Talent Code* and *The Culture Code*

"When it comes to your identity, there is no clear line where your brain ends and your body, your environment, and your culture begin. In this tour de force, Annie Murphy Paul unmasks the larger story of who we are."

— David Eagleman, Stanford University neuroscientist and author of *Incognito* and *Livewired*

"The very smartest people know how to draw upon the wisdom of their entire world, including their environment and also their whole body, not just their brain. *The Extended Mind* is the very best guide on how to do this and an entertaining read as well."

— Tyler Cowen, George Mason University economist and coauthor of the blog *Marginal Revolution*

"Anyone who wants to think better, more creatively, or more collaboratively — and who doesn't? — must read this book. It has the potential to revolutionize classroom design, office practices, parenting techniques, and even exercise habits. Annie Murphy Paul has done the hard work of distilling and synthesizing twenty years of research in multiple fields, all so that we can apply brain science to getting outside our brains. A fascinating read!"

— Anne-Marie Slaughter, CEO of New America
and author of *Unfinished Business*

"Annie Murphy Paul has written a very illuminating book; it made me think more capaciously about the mind and brain."

— Howard Gardner, Harvard University professor of education
and author of *Frames of Mind*

"What if we don't just think with our heads? In this delightful book — clear and provocative and filled with vivid examples and wise advice — Annie Murphy Paul explores the radical idea that our mental processes extend outward from our brains and into our bodies and our physical spaces and the people around us. This ambitious work by one of the best science writers alive will change how you think about thinking."

— Paul Bloom, Yale University psychologist
and author of *How Pleasure Works*

"Annie Murphy Paul has written a book that will help teachers and parents to get the most out of school. *The Extended Mind* is not just a fascinating read, firmly grounded in science — it will help you, and your students and your children, to think better."

— Daniel Willingham, University of Virginia psychologist
and author of *Why Don't Students Like School?*

"Genuine revolutions in thought are few and far between, but this is the real deal. Minds reside in bodies, and our bodies do things in a surrounding world, making use of implements that extend and transform our natural capacities. How do we grasp reality? Not, it turns out, according to the story we have been telling about the mind for about four hundred years. In *The Extended Mind*, we learn that modern society is built on some premises that simply don't hold up and are making us miserable. With grace, clarity, and a humane eye for the truly significant, Annie Murphy Paul brings this revolution to bear on the challenge of living well. This is stuff you can use."

— Matthew B. Crawford, author of *Shop Class as Soulcraft*

"Like *Thinking, Fast and Slow,* this riveting book both transforms our understanding of thought and shows us how to think better. It is not just fascinating but thrilling and liberating to realize our minds are not trapped in the dark inside our skulls but always reaching outside for tools and movement and other minds to think with."

— Larissa MacFarquhar, author of *Strangers Drowning*

"After decades of books narrowly focused on the brain, Annie Murphy Paul's book reveals that to be truly alive, we actually need to be fully embodied. *The Extended Mind* is a brilliant, deeply research-backed guide to living life at our greatest potential." — Emma Seppälä, PhD, author of *The Happiness Track*

"In *The Extended Mind,* Annie Murphy Paul makes the fascinating and convincing argument that our minds continue beyond our heads. Our thoughts reside in our brains, but they're influenced by how our bodies move, the surroundings we inhabit, and the relationships we pursue. A beautifully written, insightful book rich with cutting-edge science and fascinating case studies."

— Adam Alter, associate professor of marketing at New York University and author of *Drunk Tank Pink* and *Irresistible*

"Lucid, entertaining, challenging — and very, very important. But be warned: to extract the juice from this book, don't just read with your intellect. You'll need to resonate with your breath and your belly. Then you'll really get it."

— Guy Claxton, author of *Intelligence in the Flesh*

"Annie Murphy masterfully fits together the parts of an exciting new science, technology, and practice of extending the human mind behind the brain. *The Extended Mind* weaves together strong evidence that we use artifacts, from pencils (try multiplying two six-digit numbers without something to write with) to search engines, to multiply our brains' biological capacities. Besides tools, we shape our environments and rely on our social networks to think in ways that our brains alone could not think."

— Howard Rheingold, author of *Smart Mobs* and *Tools for Thought*

"Paul's knack for finding real-world scenarios to illustrate scientific ideas makes this pop and lends much credence to the theory that an isolated mind isn't the sole source of intelligence and creativity. Her fresh approach hits the mark." — *Publishers Weekly*

"Paul does a good job of drawing together the many extensions of mind that surround us . . . It helps to have a brain to think with, but Paul capably shows that there's much more to the process than all that." — *Kirkus Reviews*

"A practical and mind-expanding guide for writers, artists, teachers, and anyone who wants to increase their brain power and help others do the same."
— *Library Journal*

"In this enthralling book, Annie Murphy Paul shows in practical detail just why mind is not 'all in the head.' Stressing the body, and the social and material context in which we act, this book should alter how we think about ourselves and how we value our worlds. Despite being a longtime proponent of the extended mind, I learned new things on every page. Highly recommended — make it part of your mind today!"
— Andy Clark, professor of cognitive philosophy at the University of Sussex and author of *Natural-Born Cyborgs* and *Supersizing the Mind*

THE EXTENDED MIND

The Cult of Personality:
How Personality Tests Are Leading Us to
Miseducate Our Children, Mismanage Our Companies,
and Misunderstand Ourselves

Origins:
How the Nine Months Before Birth
Shape the Rest of Our Lives

THE
EXTENDED
MIND

THE POWER OF
THINKING OUTSIDE
THE BRAIN

Annie Murphy Paul

MARINER BOOKS
Boston   New York

HarperCollins books may be purchased for educational, business, or sales promotional use. For information, please email the Special Markets Department at SPsales@harpercollins.com.

A hardcover edition of this book was published in 2021 by Houghton Mifflin Harcourt

FIRST MARINER BOOKS PAPERBACK EDITION PUBLISHED 2022

Designed by Chrissy Kurpeski

Library of Congress Cataloging-in-Publication Data has been applied for.

ISBN 978-0-358-69527-1

23 24 25 26 27 LBC 10 9 8 7 6

For Sally, Billy, and Frankie

Contents

Prologue

W HEN YOU'RE WRITING a book about how to think well, your sources — the cognitive scientists, psychologists, biologists, neuroscientists, and philosophers who all have something to contribute on the subject — will often seem to be speaking, via their work, directly to you: *Yes, you there, writing a book!* They cajole and insist, they argue and debate, they issue warnings and pass judgment; as you lay out their recommendations for the reader, they inquire pointedly: *Are you taking your own advice?*

I entered into one such intimate exchange when I read, with a jolt of recognition, a passage written more than 130 years ago; it was as if the author were reaching through the pages that lay open on my desk. Making the meeting more intense, the writer in question was a distinctly intimidating character: the German philosopher Friedrich Nietzsche, he of the severe gaze and vaguely sinister mustache.

"How quickly we guess how someone has come by his ideas," Nietzsche slyly observed, "whether it was while sitting in front of his inkwell, with a pinched belly, his head bowed low over the paper — in which case we are quickly finished with his book, too! Cramped intestines betray themselves — you can bet on that — no less than closet air, closet ceilings, closet narrowness."

The room in which I was writing suddenly seemed rather airless and small.

I encountered his words as I was working on a chapter about how bodily movement affects the way we think. The quote from Nietzsche appears in a book titled *A Philosophy of Walking,* by the contemporary French philosopher Frédéric Gros; Gros has his own thoughts to add. Don't think of a book as issuing only from an author's head, he advises. "Think of the scribe's body: his hands, his feet, his shoulders and legs. Think of the book as an expression of physiology. In all too many books the reader can sense the seated body, doubled up, stooped, shriveled in on itself."

My seated body shifted guiltily in its chair, which it had occupied all morning.

Far more conducive to the act of creation, Gros continues, is "the walking body" — which, he says, is "unfolded and tensed like a bow: opened to wide spaces like a flower to the sun." Nietzsche, he reminds us, wrote that we should "sit as little as possible; do not believe any idea that was not born in the open air and of free movement."

The philosophers were ganging up on me; I closed my laptop and went for a walk.

I was acting not only on their say-so, of course; by this point in my research I had read dozens of empirical studies showing that a bout of physical activity sharpens our attention, improves our memory, and enhances our creativity. And in fact, I found that the forward movement of my legs, the flow of images past my eyes, the slight elevation of my heart rate did work some kind of change on my mind. Upon sitting back down at my desk, I wasted no time resolving a knotty conceptual problem that had tormented me all morning. (I can only hope that the prose I produced also "retains and expresses the energy, the springiness of the body," in Gros's formulation.) Could my brain have solved the problem on its own, or did it require the assist provided by my ambulatory limbs?

Our culture insists that the brain is the sole locus of thinking, a cordoned-off space where cognition happens, much as the workings of my laptop are sealed inside its aluminum case. This book argues otherwise: it holds that the mind is something more like the nest-building bird I spotted on my walk, plucking a bit of string here, a twig there, constructing a whole out of available parts. For humans these parts include, most notably, the feelings and movements of our bodies; the physical spaces in which we learn and work; and the other minds with which we interact — our classmates, colleagues, teachers, supervisors, friends. Sometimes all three elements come together in especially felicitous fashion, as they did for the brilliant intellectual team of Amos Tversky and Daniel Kahneman. The two psychologists carried out much of their groundbreaking work on heuristics and biases — the human mind's habitual shortcuts and distortions — by talking and walking together, through the bustling streets of Jerusalem or along the rolling hills of the California coast. "I did the best thinking of my life on leisurely walks with Amos," Kahneman has said.

Many tomes have been written on human cognition, many theories proposed and studies conducted (Tversky and Kahneman's among them). These efforts have produced countless illuminating insights, but they are limited by their assumption that thinking happens only inside the brain. Much less attention has been paid to the ways people use the world to think: the gestures of the hands, the space of a sketchbook, the act of listening to someone tell a story, or the task of teaching someone else. These "extra-neural" inputs change the way we think; it could even be said that they constitute a part of the thinking process itself. But where is the chronicle of *this* mode of cognition? Our scientific journals mostly proceed from the premise that the mental organ is a disembodied, placeless, asocial entity, a "brain in a vat"; our history books spin tales that attribute world-changing breakthroughs to individual men, thinking great thoughts on their own. Yet a parallel narrative has existed in front of us all along — a kind of secret history of thinking outside the brain. Scientists, artists, authors; leaders, inventors, entrepreneurs: they've all used the world as raw material for their trains of thought. This book aims to exhume that hidden saga, reclaiming its rightful place in any full accounting of how the human race has achieved its remarkable feats of intellect and creativity.

We'll learn about how geneticist Barbara McClintock made her Nobel Prize–winning discoveries by imaginatively "embodying" the plant chromosomes she studied, and about how pioneering psychotherapist and social critic Susie Orbach senses what her patients are feeling by tuning in to the internal sensations of her own body (a capacity known as *interoception*). We'll contemplate how biologist James Watson determined the double-helix structure of DNA by physically manipulating cardboard cutouts he'd made himself, and how author Robert Caro plots the lives of his biographical subjects on an intricately detailed wall-sized map. We'll explore how virologist Jonas Salk was inspired to complete his work on a polio vaccine while wandering a thirteenth-century Italian monastery, and how the artist Jackson Pollock set off a revolution in painting by trading his apartment in frenetic downtown Manhattan for a farmhouse on the verdant south fork of Long Island. We'll find out how Pixar director Brad Bird creates modern movie classics like *Ratatouille* and *The Incredibles* by arguing — vehemently — with his longtime producer, and how physicist Carl Wieman, another Nobel Prize winner, figured out that inducing his students to talk with one another was the key to getting them to think like scientists.

Such stories push back against the prevailing assumption that the brain can, or should, do it all on its own; they are vivid testimony to the countervailing notion that we think best when we think with our bodies, our spaces, and our relationships. But as with Friedrich Nietzsche's commendation of the virtues of walking, the evidence supporting the efficacy of thinking outside the brain is far from merely anecdotal. Research emerging from three related areas of investigation has convincingly demonstrated the centrality of extra-neural resources to our thinking processes.

First, there is the study of *embodied cognition,* which explores the role of the body in our thinking: for example, how making hand gestures increases the fluency of our speech and deepens our understanding of abstract concepts. Second, there is the study of *situated cognition,* which examines the influence of place on our thinking: for instance, how environmental cues that convey a sense of belonging, or a sense of personal control, enhance our performance in that space. And third, there is the study of *distributed cognition,* which probes the effects of thinking with others — such as how people working in groups can coordinate their individual areas of expertise (a process called "transactive memory"), and how groups can work together to produce results that exceed their members' individual contributions (a phenomenon known as "collective intelligence").

As a journalist who has covered research in psychology and cognitive science for more than twenty years, I read the findings generated by these fields with growing excitement. Together they seemed to indicate that it's the stuff *outside* our heads that makes us smart — a proposition with enormous implications for what we do in education, in the workplace, and in our everyday lives. The only problem: there was no "together," no overarching framework that organized these multitudinous results into a coherent whole. Researchers working within these three disciplines published in different journals and presented at different conferences, rarely drawing connections among their areas of specialization. Was there some unifying idea that could pull together these deeply intriguing findings?

Once again a philosopher came to my rescue: this time it was Andy Clark, professor of cognitive philosophy at the University of Sussex in England. In 1995 Clark had co-written a paper titled "The Extended Mind," which opened with a deceptively simple question: "Where does the mind stop and the rest of the world begin?" Clark and his coauthor, philosopher David Chalmers, noted that we have traditionally assumed that the mind is contained within the head—but, they argued, "there is nothing sacred about skull and skin." Elements of the world outside may effectively act as mental "extensions," allowing us to think in ways our brains could not manage on their own.

Clark and Chalmers initially focused their analysis on the way *technology* can extend the mind—a proposal that quickly made the leap from risibly preposterous to self-evidently obvious, once their readers acquired smartphones and began offloading large chunks of their memories onto their new devices. (Fellow philosopher Ned Block likes to say that Clark and Chalmers's thesis was false when it was written in 1998 but subsequently became true—perhaps in 2007, when Apple introduced the first iPhone.)

Yet as early as that original paper, Clark hinted that other kinds of extensions were possible. "What about socially extended cognition?" he and Chalmers asked. "Could my mental states be partly constituted by the states of other thinkers? We see no reason why not." In the years that followed, Clark continued to enlarge his conception of the kinds of entities that could serve as extensions of the mind. He observed that our physical movements and gestures play "an important role in an extended neural-bodily cognitive economy"; he noted that humans are inclined to create "designer environments"—carefully appointed spaces "that alter and simplify the computational tasks which our brains must perform in order to solve complex problems." Over the course of many more published papers and books, Clark mounted a broad and persuasive argument against what he called the "brainbound" perspective—the view that thinking happens only inside the brain—and in favor of what he called the "extended" perspective, in which the rich resources of our world can and do enter into our trains of thought.

Consider me a convert. The notion of the extended mind seized my imagination and has not yet released its grip. During my many years of reporting, I had never before encountered an idea that changed so much about how I think, how I work, how I parent, how I navigate everyday life. It became apparent to me that Andy Clark's bold proposal was not (or not only!) the esoteric thought experiment of an ivory tower philosopher; it was a plainly practical invitation to think differently and better. As I began to catalog the dozens of techniques for thinking outside the brain that researchers have tested and verified, I eagerly incorporated them into my own repertoire.

These include methods for sharpening our interoceptive sense, so as to use these internal signals to guide our decisions and manage our mental processes; they encompass guidelines for the use of specific types of gesture, or particular modes of physical activity, to enhance our memory and attention. This research offers instructions on using time in nature to restore our focus and increase our creativity, as well as directions for designing our learning and working spaces for greater productivity and performance. The studies we'll cover describe structured forms of social interaction that allow other people's cognition to augment our own; they also supply guidance on how to offload, externalize, and dynamically interact with our thoughts—a much more effective approach than doing it all "in our heads."

In time I came to recognize that I was acquiring a second education —one that is increasingly essential but almost always overlooked in our focus on educating the brain. Over many years of elementary school, high school, and even college and graduate school, we're never explicitly taught to think outside the brain; we're not shown how to employ our bodies and spaces and relationships in the service of intelligent thought. Yet this instruction is available if we know where to look; our teachers are the artists and scientists and authors who have figured out these methods for themselves, and the researchers who are, at last, making these methods the object of study.

For my own part, I'm convinced that I could not have written this book without the help of the practices detailed within it. That's not to say that I didn't sometimes fall back into our culture's default position. Before Friedrich Nietzsche's fortuitous intervention that morning, I was in full brainbound mode, my "head bowed low" over my keyboard, working my poor brain ever harder instead of looking for opportunities to extend it. I'm grateful for the nudge my research supplied; it's that gentle push in a more productive direction that this book seeks to offer its own readers.

Frédéric Gros, the French philosopher who brought Nietzsche's words to my attention, maintains that thinkers ought to get moving in a "quest for a different light." As he observes, "Libraries are always too dark," and books written among the stacks manifest this dull dimness — while "other books reflect piercing mountain light, or the sea sparkling in sunshine." It's my hope that this book will cast a different light, bring a bracing gust of fresh air to the thinking we do as students and workers, as parents and citizens, as leaders and creators. Our society is facing unprecedented challenges, and we'll need to think well in order to solve them. The brainbound paradigm now so dominant is clearly inadequate to the task; everywhere we look we see problems with attention and memory, with motivation and persistence, with logical reasoning and abstract thinking. Truly original ideas and innovations seem scarce; engagement levels at schools and in companies are low; teams and groups struggle to work together in an effective and satisfying way.

I've come to believe that such difficulties result in large part from a fundamental misunderstanding of how — and *where* — thinking happens. As long as we settle for thinking inside the brain, we'll remain bound by the limits of that organ. But when we reach outside it with intention and skill, our thinking can be transformed. It can become as dynamic as our bodies, as airy as our spaces, as rich as our relationships — as capacious as the whole wide world.

THE EXTENDED MIND

Introduction:

Thinking Outside the Brain

*U*SE YOUR HEAD.

How many times have you heard that phrase? Perhaps you've even urged it on someone else—a son or daughter, a student, an employee. Maybe you've muttered it under your breath while struggling with an especially tricky problem, or when counseling yourself to remain rational: *Use your head!*

The command is a common one, issued in schools, in the workplace, amid the trials of everyday life. Its refrain finds an echo in culture both high and low, from Auguste Rodin's *The Thinker,* chin resting thoughtfully on fist, to the bulbous cartoon depiction of the brain that festoons all manner of products and websites—educational toys, nutritional supplements, cognitive fitness exercises. When we say it, we mean: call on the more than ample powers of your brain, draw on the magnificent lump of tissue inside your skull. We place a lot of faith in that lump; whatever the problem, we believe, the brain can solve it.

But what if our faith is misplaced? What if the directive to "use your head," ubiquitous though it may be, is misguided? A burgeoning body of research suggests that we've got it exactly backwards. As it is, we use our brains entirely too much—to the detriment of our ability to think intelligently. What we need to do is think *outside* the brain.

Thinking outside the brain means skillfully engaging entities external to our heads—the feelings and movements of our bodies, the physical spaces in which we learn and work, and the minds of the other people around us—drawing them into our own mental processes. By reaching beyond the brain to recruit these "extra-neural" resources, we are able to focus more intently, comprehend more deeply, and create more imaginatively—to entertain ideas that would be literally unthinkable by the brain alone. It's true that we're more accustomed to thinking *about* our bodies, our spaces, and our relationships. But we can also think *with* and *through* them—by using the movements of our hands to understand and express abstract concepts, for example, or by arranging our workspace in ways that promote idea generation, or by engaging in social practices like teaching and storytelling that lead to deeper understanding and more accurate memory. Rather than exhorting ourselves and others to use our heads, we should be applying extra-neural resources to the project of thinking *outside* the skull's narrow circumference.

But wait, you may be asking: What's the need? Isn't the brain, on its own, up to the job? Actually, no. We've been led to believe that the human brain is an all-purpose, all-powerful thinking machine. We're deluged with reports of discoveries about the brain's astounding abilities, its lightning quickness and its protean plasticity; we're told that the brain is a fathomless wonder, "the most complex structure in the universe." But when we clear away the hype, we confront the fact that the brain's capacities are actually quite constrained and specific. The less heralded scientific story of the past several decades has been researchers' growing awareness of the brain's *limits*. The human brain is limited in its ability to pay attention, limited in its capacity to remember, limited in its facility with abstract concepts, and limited in its power to persist at a challenging task.

Importantly, these limits apply to *everyone's* brain. It's not a matter of individual differences in intelligence; it's a matter of the character of the organ we all possess, its biological nature and its evolutionary history. The brain *does* do a few things exquisitely well—things like sensing and moving the body, navigating through space, and connecting with other humans. These activities it can manage fluently, almost effortlessly. But accurately recalling complex information? Engaging in rigorous logical reasoning? Grasping abstract or counterintuitive ideas? Not so much.

Here we arrive at a dilemma—one that we all share: The modern world is extraordinarily complex, bursting with information, built around non-intuitive ideas, centered on concepts and symbols. Succeeding in this world requires focused attention, prodigious memory, capacious bandwidth, sustained motivation, logical rigor, and proficiency with abstractions. The gap between what our biological brains are capable of, and what modern life demands, is large and getting larger each day. With every experimental discovery, the divide between the scientific account of the world and our intuitive "folk" understanding grows more pronounced. With every terabyte of data swelling humanity's store of knowledge, our native faculties are further outstripped. With every twist of complexity added to the world's problems, the naked brain becomes more unequal to the task of solving them.

Our response to the cognitive challenges posed by contemporary life has been to double down on what the philosopher Andy Clark calls "brainbound" thinking—those very capacities that are, on their own, so woefully inadequate. We urge ourselves and others to grit it out, bear down, "just do it"—to *think harder*. But, as we often find to our frustration, the brain is made of stubborn and unyielding stuff, its vaunted plasticity notwithstanding. Confronted by its limits, we may conclude that we ourselves (or our children or our students or our employees) are simply not smart enough, or not "gritty" enough. In fact, it's the way we handle our mental shortcomings—which are, remember, endemic to our species—that is the problem. Our approach constitutes an instance of (as the poet William Butler Yeats put it in another context) "the will trying to do the work of the imagination." The smart move is not to lean ever harder on the brain but to learn to reach beyond it.

In *The Middle Class Gentleman*, a comedy written by the seventeenth-century French playwright Molière, the would-be aristocrat Monsieur Jourdain is delighted by a realization that follows upon his learning the difference between prose and verse. "By my faith! For more than forty years I have been speaking prose without knowing anything about it!" he exclaims. Likewise, we may be impressed to learn that we have long been drawing extra-neural resources into our thinking processes—that we *already* think outside the brain.

That's the good news. The bad news is that we often do it haphazardly, without much intention or skill. It's no wonder this is the case. Our efforts at education and training, as well as management and leadership, are aimed almost exclusively at promoting brainbound thinking. Beginning in elementary school, we are taught to sit still, work quietly, think hard—a model for mental activity that will prevail during all the years that follow, through high school and college and into the workplace. The skills we develop and the techniques we are taught are those that involve using our heads: committing information to memory, engaging in internal reasoning and deliberation, endeavoring to self-discipline and self-motivate.

Meanwhile, there is no corresponding cultivation of our ability to think outside the brain — no instruction, for instance, in how to tune in to the body's internal signals, sensations that can profitably guide our choices and decisions. We're not trained to use bodily movements and gestures to understand highly conceptual subjects like science and mathematics, or to come up with novel and original ideas. Schools don't teach students how to restore their depleted attention with exposure to nature and the outdoors, or how to arrange their study spaces so that they extend intelligent thought. Teachers and managers don't demonstrate how abstract ideas can be turned into physical objects that can be manipulated and transformed in order to achieve insights and solve problems. Employees aren't shown how the social practices of imitation and vicarious learning can shortcut the process of acquiring expertise. Classroom groups and workplace teams aren't coached in scientifically validated methods of increasing the collective intelligence of their members. Our ability to think outside the brain has been left almost entirely uneducated and undeveloped.

This oversight is the regrettable result of what has been called our "neurocentric bias" — that is, our idealization and even fetishization of the brain — and our corresponding blind spot for all the ways cognition extends beyond the skull. (As the comedian Emo Philips has remarked: "I used to think that the brain was the most wonderful organ in my body. Then I realized who was telling me this.") Seen from another perspective, however, this near-universal neglect represents an auspicious opportunity — a world of unrealized potential. Until recently, science shared the larger culture's neglect of thinking outside the brain. But this is no longer the case. Psychologists, cognitive scientists, and neuroscientists are now able to provide a clear picture of how extra-neural inputs shape the way we think. Even more promising, they offer practical guidelines for enhancing our thinking through the use of these outside-the-brain resources. Such developments are unfolding against the backdrop of a broader shift in how we view the mind — and, by extension, how we understand ourselves.

But first — to gain a sense of where we've been and where we're headed, it's worth taking several steps back in time, to the moment when our current ideas about the brain were born.

ON FEBRUARY 14, 1946, a breathless bustle filled the halls of the Moore School of Electrical Engineering in Philadelphia. On this day, the school's secret jewel was going to be revealed to the world: the ENIAC. Inside a locked room at Moore hummed the Electronic Numerical Integrator and Computer, the first machine of its kind capable of performing calculations at lightning speed. Weighing thirty tons, the massive ENIAC used around eighteen thousand vacuum tubes, employed about six thousand switches, and encompassed upwards of half a million soldered joints; it had taken more than 200,000 man-hours to build.

The bus-sized contraption was the brainchild of John Mauchly and J. Presper Eckert Jr., two young scientists at the University of Pennsylvania, Moore's parent institution. With funding from the US Army, the ENIAC had been developed for the purpose of computing artillery trajectories for American gunners fighting the war in Europe. Compiling trajectory tables — necessary for the effective use of new weapons being introduced by the military — was a laborious process, requiring the service of teams of human "computers" working in shifts around the clock. A machine that could do their job with speed and accuracy would give the army an invaluable edge.

Now, six months after V-Day, the demands of wartime were giving way to the needs of an expanding economy, and Mauchly and Eckert had called a press conference to introduce their invention to the world. The two men had prepared for the event with deliberate care, and no small amount of stagecraft. As the ENIAC chugged away at a given task, some three hundred neon lights built into the machine's accumulators flickered and flashed. Presper Eckert, known to all as "Pres," judged the effect of these small bulbs insufficiently impressive. On the morning of the press conference, he ran out and purchased an armful of Ping-Pong balls, each of which he cut in half and marked with a number. The plastic domes, glued over the neon bulbs, now cast a dramatic glow — especially once the room's overhead lights were dimmed.

At the appointed hour, the door to the room that held the ENIAC was opened, and a gaggle of officials, academics, and journalists filed in.

Standing in front of the hulking machine, lab member Arthur Burks welcomed the group and sought to impart to them a sense of the moment's magnitude. The ENIAC was engineered to carry out mathematical operations, he explained, and these operations, "if made to take place rapidly enough, might in time solve almost any problem." Burks announced that he would begin the day's demonstration by asking the ENIAC to multiply 97,367 by itself five thousand times. The reporters in the room bent over their notepads. "Watch closely, you may miss it," he warned, and pushed a button; before the newsmen had time to look up, the task was complete, executed on a punch card delivered to Burks's hand.

Next Burks fed the machine a problem like those for which it had been designed: the ENIAC would now calculate the trajectory of a shell taking thirty seconds to travel from the gun to its target. Such a task would take a team of human experts three days to compute; the ENIAC completed the job in twenty seconds, faster than the shell itself could fly. Jean Bartik, one of a group of pioneering female engineers who helped program the ENIAC, was on hand for the demonstration. She recalled, "It was unheard of that a machine could reach such speeds of calculation, and everyone in the room, even the great mathematicians, were in complete wonder and awe at what they had just seen."

The next day, admiring accounts of the ENIAC appeared in newspapers all over the world. "PHILADELPHIA — One of the war's top secrets, an amazing machine which applies electronic speeds for the first time to mathematical tasks hitherto too difficult and cumbersome for solution, was announced here tonight by the War Department," the *New York Times* reported on its front page. The *Times* reporter, T. R. Kennedy Jr., sounded dazzled by what he'd seen. "So clever is the device," he wrote, "that its creators have given up trying to find problems so long that they cannot be solved."

The introduction of the ENIAC was not just a milestone in the history of technology. It was a turning point in the story of *how we understand ourselves*. In its early days, Mauchly and Eckert's invention was frequently compared to a human brain. Newspaper and magazine articles described the ENIAC as a "giant electronic brain," a "robot brain," an "automatic brain," and a "brain machine." But before long, the analogy got turned around. It became a commonplace that the *brain* is like a *computer*. Indeed, the "cognitive revolution" that would sweep through American universities in the 1950s and 1960s was premised on the belief that the brain could be understood as a flesh-and-blood computing machine. The first generation of cognitive scientists "took seriously the idea that the mind is a kind of computer," notes Brown University professor Steven Sloman. "Thinking was assumed to be a kind of computer program that runs in people's brains."

Since those early days at the dawn of the digital age, the brain-computer analogy has become only more pervasive and more powerful, engaged not just by researchers and academics but by the rest of us, the public at large. The metaphor provides us with a model, sometimes conscious but often implicit, of how thinking works. The brain, according to this analogy, is a self-contained information-processing machine, sealed inside the skull as the ENIAC was sequestered in its locked room. From this inference emerges a second: the human brain has attributes, akin to gigabytes of RAM and megahertz of processing speed, that can be easily measured and compared. Following on these is the third and perhaps most significant supposition of all: that some brains, like some computers, are just *better*; they possess the biological equivalent of more memory storage, greater processing power, higher-resolution screens.

To this day, the computer metaphor dominates the way we think and talk about mental activity — but it's not the only one that shapes our notion of the brain. A half-century after the ENIAC was unveiled, another analogy rose to prominence.

"NEW RESEARCH SHOWS That the Brain Can Be Developed Like a Muscle," read the headline of the news article, set in bold type. The year was 2002, and Lisa Blackwell, a graduate student at Columbia University working with psychology professor Carol Dweck, was handing out copies of the article to a classroom full of seventh-graders at a public school in New York City. Dweck and Blackwell were testing a new theory, investigating the possibility that the way we conceptualize the brain can affect how well we think. The study's protocol required Blackwell to guide the students through eight informational sessions; in this, the third session in the sequence, students were to take turns reading the text of the article aloud.

"Many people believe that a person is born either smart, average, or dumb — and stays that way," one student began. "But new research shows that the brain is more like a muscle—it changes and gets stronger when you use it." Another student picked up the thread: "Everyone knows that when you lift weights, your muscles get bigger and you get stronger. A person who can't lift 20 pounds when they start exercising can get strong enough to lift 100 pounds after working out for a long time. That's because the muscles become larger and stronger with exercise. And when you stop exercising, the muscles shrink and you get weaker. That's why people say, 'Use it or lose it!'" A giggle rippled through the room. "But," a third pupil read on, "most people don't know that when they practice and learn new things, parts of their brain change and get larger, a lot like muscles do when they exercise."

Dweck's idea, which she initially called "the incremental theory of intelligence," would eventually become known to the world as the "growth mindset": the belief that concerted mental effort could make people smarter, just as vigorous physical effort could make people stronger. As she and her colleagues wrote in an account of their early research in schools, "The key message was that learning changes the brain by forming new connections, and that students are in charge of this process." From these beginnings, growth mindset became a popular phenomenon—spawning a book, *Mindset,* that has sold millions of copies, and inspiring an untold number of speeches, presentations, and workshops, delivered to corporate and organizational audiences as well as to students and teachers.

At the center of it all is a metaphor: the brain as muscle. The mind, in this analogy, is akin to a biceps or a quadriceps—a physical entity that varies in strength among individuals. The comparison has been incorporated into another hugely popular concept originating in academic psychology: "grit." Angela Duckworth, the University of Pennsylvania psychologist who defines grit as "perseverance and passion for long-term goals," echoes Dweck in her own book. "Like a muscle that gets stronger with use, the brain changes itself when you struggle to master a new challenge," she wrote in the best-selling *Grit,* published in 2016. The emphasis in *Grit* on mustering more of one's own internal resources makes the brain-as-muscle analogy a perfect fit. The comparison is made even more explicitly by purveyors of so-called "cognitive fitness" exercises, which have drawn millions of hopeful users under names like "CogniFit" and "Brain Gym." (So pervasive is the metaphor that some scientists concerned about the spread of "neuromyths" — common misconceptions about the brain — have begun to point out that the brain is not *actually* a muscle but rather an organ made up of specialized cells known as neurons.)

These two metaphors — brain as computer and brain as muscle — share some key assumptions. To wit: the mind is a discrete thing that is sealed in the skull; this discrete thing determines how well people are able to think; this thing has stable properties that can easily be measured, compared, and ranked. Such assumptions feel comfortably familiar; indeed, they weren't particularly novel even at the moment they were first proposed. For centuries, brains had been likened to machines — to whichever appliance of the time appeared most advanced: a hydraulic pump, a mechanical clock, a steam engine, a telegraph machine.

In a lecture delivered in 1984, philosopher John Searle noted: "Because we do not understand the brain very well, we are constantly tempted to use the latest technology as a model for trying to understand it. In my childhood we were always assured that the brain was a telephone switchboard." Teachers, parents, and other adults all proffered the metaphor of brain as switchboard, recounted Searle, for "what else could it be?"

Brains had also long been likened to muscles that could be strengthened with exercise — a theme promulgated, for example, by physicians and health experts in the nineteenth and early twentieth centuries. In his *First Book in Physiology and Hygiene,* published in 1888, doctor John Harvey Kellogg made an argument that sounds very much like Carol Dweck's. "What do we do when we want to strengthen our muscles? We make them work hard every day, do we not?" Kellogg inquired of his intended youthful readership. "The exercise makes them grow large and strong. It is just the same with our brains. If we study hard and learn our lessons well, then our brains grow strong and study becomes easy."

Entrenched historical foundations support these metaphors; they rest upon deep cultural underpinnings as well. The computer and muscle analogies fit neatly with our society's emphasis on individualism — its insistence that we operate as autonomous, self-contained beings, in possession of capacities and competencies that are ours alone. These comparisons also readily conform to our culture's penchant for thinking in terms of good, better, best. Scientist and author Stephen Jay Gould once included in his list of "the oldest issues and errors of our philosophical traditions" our persistent inclination "to order items by ranking them in a linear series of increasing worth." Computers may be slow or fast, muscles may be weak or strong — and so it goes, we assume, with our own and others' minds.

There even appear to be hard-wired *psychological* factors underlying our embrace of these ideas about the brain. The belief that some core quantity of intelligence resides within each of our heads fits with a pattern of thought, apparently universal in humans, that psychologists call "essentialism" — that is, the conviction that each entity we encounter possesses an inner essence that makes it what it is. "Essentialism shows up in every society that has been studied," notes Yale University psychology professor Paul Bloom. "It appears to be a basic component of how we think about the world." We think in terms of enduring essences — rather than shifting responses to external influences — because we find such essences easier to process mentally, as well as more satisfying emotionally. From the essentialist perspective, people simply "are" intelligent or they are not.

Together, the historical, cultural, and psychological bases of our assumptions about the mind — that its properties are individual, inherent, and readily ranked according to quality — give them a powerful punch. Such assumptions have profoundly shaped the views we hold on the nature of mental activity, on the conduct of education and work, and on the value we place on ourselves and others. It's therefore startling to contemplate that the whole lot of it could be misconceived. To grasp the nature of this error, we need to consider another metaphor.

ON THE MORNING of April 18, 2019, computer screens went dark across a swath of Seoul, South Korea's largest city. Lights flickered out in schools and offices across the 234-square-mile metropolis, home to some 10 million people. Stoplights at street intersections blinked off, and electric-powered trains slowed to a halt. The cause of the blackout was as small in scale as its effects were widespread: a power outage caused by magpies, the black-and-white-feathered birds who build their nests on utility poles and transmission towers. Magpies — members of the corvid family, which also includes crows, jays, and ravens — are well known for making their nests out of whatever is available in the environment. The birds have been observed using an astonishing array of materials: not only twigs, string, and moss, but also dental floss, fishing line, and plastic Easter grass; chopsticks, spoons, and drinking straws; shoelaces, eyeglass frames, and croquet wickets. During the American Dust Bowl of the 1930s, which eliminated vegetation from huge swaths of the West, magpies' corvid cousins made nests out of barbed wire.

The densely packed urban neighborhoods of modern-day Seoul feature few trees or bushes, so magpies use what they can find: metal clothes hangers, TV antennas, and lengths of steel wire. These materials conduct electricity — and so, when the birds build their nests on the city's tall electrical transmission towers, the flow of electricity is regularly disrupted. According to KEPCO, the Korea Electric Power Corporation, magpies are responsible for hundreds of power outages annually in areas all across the country. Each year, KEPCO employees work to remove upwards of ten thousand nests, but just as quickly the magpies build them up again.

Magpies may pose a headache for power companies, but their activity supplies a felicitous analogy for the way the mind works. Our brains, it might be said, are like magpies, fashioning their finished products from the materials around them, weaving the bits and pieces they find into their trains of thought. Set beside the brain-as-computer and brain-as-muscle metaphors, it's apparent that the brain as magpie is a very different kind of analogy, with very different implications for how mental processes operate. For one thing: thought happens not only inside the skull but out in the world, too; it's an act of continuous assembly and reassembly that draws on resources external to the brain. For another: the kinds of materials available to "think with" affect the nature and quality of the thought that can be produced. And last: the capacity to think well — that is, to be intelligent — is not a fixed property of the individual but rather a shifting state that is dependent on access to extra-neural resources and the knowledge of how to use them.

This is, admittedly, a radically new way of thinking about thinking. It may not feel easy or natural to adopt. But a growing mass of evidence generated within several scientific disciplines suggests that it's a much more accurate rendering of how human cognition actually works. Moreover, it's a gratifyingly generative conceptualization, because it offers so many practical opportunities for improving how well we think. It has arrived just in time. Recasting our model of how the mind functions has lately become an urgent necessity, as we find ourselves increasingly squeezed by two opposing forces: we need ever more to think outside the brain, even as we have become ever more stubbornly committed to the brainbound approach.

First, as to that growing need to think outside the brain: as many of us can readily recognize—in the accelerated pace of our days and the escalating complexity of our duties at school and work—the demands on our thinking are ratcheting up. There's *more* information we must deal with. The information we have to process is coming at us *faster*. And the *kind* of information we must deal with is increasingly specialized and abstract. This difference in kind is especially significant. The knowledge and skills that we are biologically prepared to learn have been outstripped by the need to acquire a set of competencies that come far less naturally and are acquired with far more difficulty. David Geary, a professor of psychology at the University of Missouri, makes a useful distinction between "biologically primary" and "biologically secondary" abilities. Human beings, he points out, are born ready to learn certain things: how to speak the language of the local community, how to find their way around a familiar landscape, how to negotiate the challenges of small-group living. We are not born to learn the intricacies of calculus or the counterintuitive rules of physics; we did not evolve to understand the workings of the financial markets or the complexities of global climate change. And yet we dwell in a world where such biologically secondary capacities hold the key to advancement, even survival. The demands of the modern environment have now met, and exceeded, the limits of the biological brain.

For a time, it's true, humanity was able to keep up with its own ever-advancing culture, resourcefully finding ways to use the biological brain better. As their everyday environments grew more intellectually demanding, people responded by upping their cognitive game. Continual engagement with the mental rigors of modern life—along with improving nutrition, rising living conditions, and reduced exposure to infectious disease and other pathogens—produced a century-long climb in average IQ score, as measured by intelligence tests taken by people all over the globe. But this upward trajectory is now leveling off. In recent years, IQ scores have stopped rising, or have even begun to drop, in countries like Finland, Norway, Denmark, Germany, France, and Britain. Some researchers suggest that we have now pushed our mental equipment as far as it can go. It may be that "our brains are already working at near-optimal capacity," note Nicholas Fitz and Peter Reiner, writing in the journal *Nature*. Efforts to wrest more intelligence from this organ, they add, "bump up against the hard limits of neurobiology."

As if to protest this unwelcome truth, attempts to subvert such limits have received growing attention in recent years. Commercial brain-training regimens like Cogmed, Lumosity, and BrainHQ have attracted many who desire to improve their memory and increase their focus; Lumosity alone claims 100 million registered users in 195 countries. At the same time, so-called neuroenhancement — innovations like "smart pills" and electrical brain stimulation that claim to make their users more intelligent — have drawn breathless media coverage, as well as extensive investment from pharmaceutical and biotechnology companies.

So far, however, these approaches have yielded little more than disappointment and dashed hopes. A team of scientists who set out to evaluate all the peer-reviewed intervention studies cited on the websites of leading brain-training companies could find "little evidence" within those studies "that training improves everyday cognitive performance." Engaging in brain training does improve users' performance — but only on exercises highly similar to the ones they've been practicing. The effect does not seem to transfer to real-life activities involving attention and memory. A 2019 study of Cogmed concluded that such transfer "is rare, or possibly inexistent." A 2017 study of Lumosity determined that "training appears to have no benefits in healthy young adults"; similarly dismal results have been reported for older individuals. In 2016, Lumosity was forced to pay a $2 million fine for deceptive advertising to the US Federal Trade Commission. Smart pills haven't fared much better; a clinical trial of one "nootropic" drug popular among Silicon Valley tech workers found that a cup of coffee was more effective at boosting memory and attention.

Medications and technologies that might, someday, actually enhance intelligence remain in the early stages of laboratory testing. The best way — and, at least for now, the *only* way — for us to get smarter is to get better at thinking outside the brain. Yet we dismiss or disparage this kind of cognition, to the extent that we consider it at all. Our pronounced bias in favor of brainbound thinking is long-standing and well entrenched — but a bias is all it is, and one that can no longer be supported or sustained. The future lies in thinking outside the brain.

WE CAN BETTER grasp the future of thinking outside the brain by taking a look back at the time when the idea first emerged. In 1997, Andy Clark — then a professor of philosophy at Washington University in St. Louis, Missouri — left his laptop behind on a train. The loss of his usually ever-present computer hit him, he later wrote, "like a sudden and somewhat vicious type of (hopefully transient) brain damage." He was left "dazed, confused, and visibly enfeebled — the victim of the cyborg equivalent of a mild stroke." The experience, distressing as it was, provided fodder for a notion he had been pondering for some time. His computer, he realized, had in a sense become a *part* of his mind, an integral element of his thinking processes. His mental capacities were effectively extended by the use of his laptop, allowing his brain to overachieve — to think more efficiently and effectively, more *intelligently,* than it could without the device. His brain plus his computer equaled his mind, extended.

Two years earlier, Clark and his colleague David Chalmers had co-authored an article that named and described just this phenomenon. Their paper, titled "The Extended Mind," began by posing a question that would seem to have an obvious answer. "Where does the mind stop and the rest of the world begin?" it asked. Clark and Chalmers went on to offer an unconventional response. The mind does *not* stop at the standard "demarcations of skin and skull," they argued. Rather, it is more accurately viewed as "an extended system, a coupling of biological organism and external resources." A recognition of this reality, they acknowledged, "will have significant consequences" — in terms of "philosophical views of the mind," but also "in moral and social domains." The authors were aware that the vision they were setting out would require a thorough reimagining of what people are like and how they function, a reimagining they saw as necessary and right. Once "the hegemony of skin and skull is usurped," they concluded, "we may be able to see ourselves more truly as creatures of the world."

The world, at first, was not so sure. Before being published in *Analysis* in 1998, the paper received rejections from three other journals. Once in print, "The Extended Mind" was greeted with perplexity — and no small amount of derision. But the idea it proposed turned out to have surprising power, within the academy and well beyond it. What at first appeared radical and out-there quickly came to seem less so, as daily life in the digital age provided a continuous proof-of-concept demonstration of people extending their minds with their devices. Initially derided as wacky, the notion of the extended mind came to seem eminently plausible, even prescient.

In the more than twenty years since the publication of "The Extended Mind," the idea it introduced has become an essential umbrella concept under which a variety of scientific sub-fields have gathered. Embodied cognition, situated cognition, distributed cognition: each of these takes up a particular aspect of the extended mind, investigating how our thinking is extended by our bodies, by the spaces in which we learn and work, and by our interactions with other people. Such research has not only produced new insights into the nature of human cognition; it has also generated a corpus of evidence-based methods for extending the mind.

That's where this book comes in: it aims to *operationalize* the extended mind, to turn this philosophical sally into something practically useful. In chapter 1, we'll learn how to tune in to our interoception—the sensations that arise from within the body—and how to use these signals to make sounder decisions. In chapter 2, we'll find out how moving our bodies can nudge our minds toward deeper understanding. Chapter 3 looks at how the gestures we make with our hands can bolster our memory. Chapter 4 examines how time spent in natural spaces can restore our depleted attention. In chapter 5, we'll see how built spaces—the interiors we inhabit at school and at work—can be designed to promote creativity. In chapter 6, we will explore how moving our thoughts out of our heads and into "the space of ideas" can lead us to new insights and discoveries. Chapter 7 probes how we can think with the minds of experts; chapter 8 considers how we can think with classmates, colleagues, and other peers. Finally, in chapter 9, we'll examine how groups thinking together can become more than the sum of their members.

Across these varied instantiations of the extended mind, several common themes are apparent. The first of these concerns the source of Andy Clark's initial inspiration: the role of technology in extending our thinking. Our devices can and do extend our minds, of course—but not always; sometimes they lead us to think *less* intelligently, as anyone who's been distracted by clickbait or misled by a GPS system can tell you. The failure of our technology to consistently enhance our intelligence has to do with a metaphor we encountered earlier in this introduction: the computer as brain. Too often, those who design today's computers and smartphones have forgotten that users inhabit biological bodies, occupy physical spaces, and interact with other human beings. Technology itself is brainbound — but by the same token, technology itself could be extended, broadened to include the extra-neural resources that do so much to enrich the thinking we do in the offline world. In each of the chapters that follow, we'll encounter examples of such "extended technology" — from an online foreign-language-learning platform that encourages its users to make gestures and not just repeat words; to a Waze-like app that plots not the fastest route but the one most filled with nature's greenery; to a video game that induces players to look not at the screen but at one another, synchronizing their movements in pursuit of a shared experience.

A second theme to emerge from a review of research on the extended mind is its distinctive take on the nature of expertise. Traditional notions of what makes an expert are highly brainbound, focused on internal, individual effort (think of the late psychologist Anders Ericsson's famous finding that mastery in any field requires "10,000 hours" of practice). The literature on the extended mind suggests a different view: experts are those who have learned how best to marshal and apply extra-neural resources to the task before them. This alternative perspective has real implications for how we understand and cultivate superior performance. For example: although the conventional take on expertise highlights economy, efficiency, and optimality of action — geniuses and superstars "just do it" — research in the vein of the extended mind finds that experts actually do *more* experimenting, more testing, and more backtracking than beginners. They are more apt than novices to make skillful use of their bodies, of physical space, and of relationships with others. In most scenarios, researchers have found, experts are less likely to "use their heads" and more inclined to extend their minds — a habit that the rest of us can learn to emulate on our way to achieving mastery.

Finally, in surveying the study of the extended mind, there's one more theme that is impossible to ignore: the matter of what we might call "extension inequality." Our schools, our workplaces, the very structure of our society are based on the assumption that some people are able to think more intelligently than others. The reason for such individual differences is taken as self-evident: obviously it's because those people are smarter—because they have more of the stuff called "intelligence" inside their heads. Research on the extended mind points to a different explanation. That is: some people are able to think more intelligently because they are *better able to extend their minds*. They may have more knowledge about how mental extension works, the kind of knowledge that this book aims to make accessible. But it's also indisputable that the extensions that allow us to think well—the freedom to move one's body, say, or the proximity of natural green spaces; control over one's personal workspace, or relationships with informed experts and accomplished peers—are far from equally distributed. When reading the chapters that follow, we should keep in mind the way access, or lack of access, to mental extensions might be shaping the thinking of our students, employees, co-workers, and fellow citizens.

Metaphors are powerful, and none more so than the ones we use to understand our own minds. The value of the approach described in these pages ultimately lies in the novel analogy it offers, an analogy we can apply to our everyday efforts to learn and remember, to solve problems and imagine possibilities. We extend beyond our limits, not by revving our brains like a machine or bulking them up like a muscle—but by strewing our world with rich materials, and by weaving them into our thoughts.

PART I

THINKING WITH OUR BODIES

1

.

Thinking with Sensations

DURING HIS YEARS of working as a financial trader at Goldman Sachs, Merrill Lynch, and Deutsche Bank, John Coates watched it happen again and again. "Using my best analytical efforts, drawing on my education" — Coates has a PhD in economics from the University of Cambridge — "and a wide reading of economic reports and statistics," he would devise a brilliant trade, one that was impeccable in its logic and unassailable in its reasoning. And — it would lose money, every time.

Then there were other occasions, equally puzzling. "I would catch a glimpse with peripheral vision of another possibility, another path into the future. It showed up as a mere blip in my consciousness, a momentary tug on my attention, but it was a flash of insight coupled with a gut feeling that gave it the imprimatur of the highly probable." When he obeyed these "gut feelings," Coates found, he was usually rewarded with a profitable outcome. Against all his assumptions, all his training, Coates was forced to arrive at an unconventional conclusion: "Good judgment may require the ability to listen carefully to feedback from the body."

Further, he observes, "some people may be better at this than others." On any Wall Street trading floor "you will find high-IQ, Ivy League–educated stars who cannot make any money at all, for all their convincing analyses; while across the aisle sits a trader with an undistinguished degree from an unknown university, who cannot keep up with the latest analytics, but who consistently prints money, to the bafflement and irritation of his seemingly more gifted colleagues." It is possible, muses Coates, "though odd to contemplate, that the better judgment of the money-making trader may owe something to his or her ability to produce bodily signals, and equally to listen to them."

Coates shares these reflections in a captivating book, *The Hour Between Dog and Wolf*, which draws on his years as a trader as well as on his surprising second career as an applied physiologist. Over time, the questions generated by his work in finance — "Could we tell whether one person has better gut feelings than another? Could we monitor feedback from their bodies?" — became more compelling than the work itself, and Coates left Wall Street to pursue the answers in scientific research. He presented the fruits of his inquiry in 2016, detailing the results of a collaboration with academic neuroscientists and psychiatrists in the journal *Scientific Reports*.

Coates and his new colleagues examined a group of financial traders working on a London trading floor, asking each one to identify the successive moments when he felt his heart beat — a measure of the individual's sensitivity to bodily signals. The traders, they found, were much better at this task than were an age- and gender-matched group of controls who did not work in finance. What's more, among the traders themselves, those who were the most accurate in detecting the timing of their heartbeats made more money, and tended to have longer tenures in what was a notably volatile line of work. "Our results suggest that signals from the body — the gut feelings of financial lore — contribute to success in the markets," the team concluded. Confirming Coates's informal observations, those who thrived in this milieu were not necessarily people with greater education or intellect, but rather "people with greater sensitivity to interoceptive signals."

Interoception is, simply stated, an awareness of the inner state of the body. Just as we have sensors that take in information from the outside world (retinas, cochleas, taste buds, olfactory bulbs), we have sensors inside our bodies that send our brains a constant flow of data from within. These sensations are generated in places all over the body — in our internal organs, in our muscles, even in our bones — and then travel via multiple pathways to a structure in the brain called the insula. Such internal reports are merged with several other streams of information — our active thoughts and memories, sensory inputs gathered from the external world — and integrated into a single snapshot of our present condition, a sense of "how I feel" in the moment, as well as a sense of the actions we must take to maintain a state of internal balance.

All of us experience these bodily signals — but some of us feel them more keenly than others. To measure interoceptive awareness, scientists apply the heartbeat detection test, the one John Coates used with his group of financial traders: test takers are asked to identify the instant when their heart beats, without placing a hand on the chest or resting a finger on a wrist. Researchers have found a surprisingly wide range in terms of how people score. Some individuals are interoceptive champions, able to determine accurately and consistently when their heartbeats happen. Others are interoceptive duds: they can't feel the rhythm. Few of us are aware that this spectrum of ability even exists, much less where we fall on it — so preoccupied are we with more conventionally brain-bound capacities. We may remember down to the point our SAT scores or our high school GPA, but we haven't given this particular aptitude a moment's thought.

Vivien Ainley recalls a clear demonstration of this common oversight. Ainley, an interoception researcher at Royal Holloway, University of London, was administering the heartbeat detection test to members of the public as part of an exhibit at London's Science Museum. Visitors to the exhibit were instructed to place a finger on a sensor that detected their pulse; the readout of the sensor was visible only to Ainley.

"Please tell me when your heart beats," she would say to each patron who stepped forward. An elderly couple who stopped by the booth had very different reactions to Ainley's request.

"How on earth would I know what my heart is doing?" the woman asked incredulously. Her husband turned and stared at her, equally dumbfounded.

"But of course you know," he exclaimed. "Don't be so stupid, everyone knows what their heartbeat is!"

"He had always been able to hear his heart, and she had never been able to hear hers," Ainley observed in an interview, smiling at the memory. "They had been married for decades, but they had never talked of or even recognized this difference between them."

Though we may not notice such differences, they are real, and even visible to scientists using brain-scanning technology: the size and activity level of the brain's interoceptive hub, the insula, vary among individuals and are correlated with their awareness of interoceptive sensations. How such differences arise in the first place is not yet known. All of us begin life with our interoceptive capacities already operating; interoceptive awareness continues to develop across childhood and adolescence. Differences in sensitivity to internal signals may be influenced by genetic factors, as well as by the environments in which we grow up, including the communications we receive from caregivers about how we should respond to our bodily prompts.

What we do know is that interoceptive awareness can be deliberately cultivated. A series of simple exercises can put us in touch with the messages emanating from within, giving us access to knowledge that we already possess but that is ordinarily excluded from consciousness — knowledge about ourselves, about other people, and about the worlds through which we move. Once we establish contact with this informative internal source, we can make wise use of what it has to tell us: to make sounder decisions, for example; to respond more resiliently to challenges and setbacks; to savor more fully the intensity of our emotions while also managing them more skillfully; and to connect to others with more sensitivity and insight. The heart, and not the head, leads the way.

TO UNDERSTAND HOW interoception can act as such a rich repository, it's important to recognize that the world is full of far more information than our conscious minds can process. Fortunately, we are also able to collect and store the volumes of information we encounter on a *non-conscious* basis. As we proceed through each day, we are continuously apprehending and storing regularities in our experience, tagging them for future reference. Through this information-gathering and pattern-identifying process, we come to *know* things — but we're typically not able to articulate the content of such knowledge or to ascertain just how we came to know it. This trove of data remains mostly under the surface of consciousness, and that's usually a good thing. Its submerged status preserves our limited stores of attention and working memory for other uses.

A study led by cognitive scientist Pawel Lewicki demonstrates this process in microcosm. Participants in Lewicki's experiment were directed to watch a computer screen on which a cross-shaped target would appear, then disappear, then reappear in a new location; periodically they were asked to predict where the target would show up next. Over the course of several hours of exposure to the target's movements, the participants' predictions grew more and more accurate. They had figured out the pattern behind the target's peregrinations. But they could not put this knowledge into words, even when the experimenters offered them money to do so. The subjects were not able to describe "anything even close to the real nature" of the pattern, Lewicki observes. The movements of the target operated according to a pattern too complex for the conscious mind to accommodate — but the capacious realm that lies below consciousness was more than roomy enough to contain it.

"Nonconscious information acquisition," as Lewicki calls it, along with the ensuing application of such information, is happening in our lives all the time. As we navigate a new situation, we're scrolling through our mental archive of stored patterns from the past, checking for ones that apply to our current circumstances. We're not aware that these searches are under way; as Lewicki observes, "The human cognitive system is not equipped to handle such tasks on the consciously controlled level." He adds, "Our conscious thinking needs to rely on notes and flowcharts and lists of 'if-then' statements — or on computers — to do the same job which our non-consciously operating processing algorithms can do without external help, and instantly."

But — if our knowledge of these patterns is not conscious, how then can we make use of it? The answer is that, when a potentially relevant pattern is detected, it's our interoceptive faculty that tips us off: with a shiver or a sigh, a quickening of the breath or a tensing of the muscles. The body is rung like a bell to alert us to this useful and otherwise inaccessible information. Though we typically think of the brain as telling the body what to do, just as much does the body guide the brain with an array of subtle nudges and prods. (One psychologist has called this guide our "somatic rudder.") Researchers have even captured the body in mid-nudge, as it alerts its inhabitant to the appearance of a pattern that she may not have known she was looking for.

Such interoceptive prodding was visible during a gambling game that formed the basis of an experiment led by neuroscientist Antonio Damasio, a professor at the University of Southern California. In the game, presented on a computer screen, players were given a starting purse of two thousand "dollars" and were shown four decks of digital cards. Their task, they were told, was to turn the cards in the decks face-up, choosing which decks to draw from such that they would lose the least amount of money and win the most. As they started clicking to turn over cards, players began encountering rewards—bonuses of $50 here, $100 there—and also penalties, in which small or large amounts of money were taken away. What the experimenters had arranged, but the players were not told, was that decks A and B were "bad"—they held lots of large penalties in store —and decks C and D were "good," bestowing more rewards than penalties over time.

As they played the game, the participants' state of physiological arousal was monitored via electrodes attached to their fingers; these electrodes kept track of their level of "skin conductance." When our nervous systems are stimulated by an awareness of potential threat, we start to perspire in a barely perceptible way. This slight sheen of sweat momentarily turns our skin into a better conductor of electricity. Researchers can thus use skin conductance as a measure of nervous system arousal. Looking over the data collected by the skin sensors, Damasio and his colleagues noticed something interesting: after the participants had been playing for a short while, their skin conductance began to spike when they contemplated clicking on the bad decks of cards. Even more striking, the players started avoiding the bad decks, gravitating increasingly to the good decks. As in the Lewicki study, subjects got better at the task over time, losing less and winning more.

Yet interviews with the participants showed that they had no awareness of why they had begun choosing some decks over others until late in the game, long after their skin conductance had started flaring. By card 10 (about forty-five seconds into the game), measures of skin conductance showed that their bodies were wise to the way the game was rigged. But even ten turns later—on card 20—"all indicated that they did not have a clue about what was going on," the researchers noted. It took until card 50 was turned, and several minutes had elapsed, for all the participants to express a conscious hunch that decks A and B were riskier. Their bodies figured it out long before their brains did. Subsequent studies supplied an additional, and crucial, finding: players who were more interoceptively aware were more apt to make smart choices within the game. For them, the body's wise counsel came through loud and clear.

Damasio's fast-paced game shows us something important. The body not only grants us access to information that is more *complex* than what our conscious minds can accommodate. It also marshals this information at a pace that is far *quicker* than our conscious minds can handle. The benefits of the body's intervention extend well beyond winning a card game; the real world, after all, is full of dynamic and uncertain situations, in which there is no time to ponder all the pros and cons. If we rely on the conscious mind alone, we lose.

HERE, THEN, is a reason to hone our interoceptive sense: people who are more aware of their bodily sensations are better able to make use of their non-conscious knowledge. Mindfulness meditation is one way of enhancing such awareness. The practice has been found to increase sensitivity to internal signals, and even to alter the size and activity of that key brain structure, the insula. One particular component appears to be especially effective; this is the activity that often starts off a meditation session, known as the "body scan." Rooted in the Buddhist traditions of Myanmar, Thailand, and Sri Lanka, the body scan was introduced to Western audiences by mindfulness pioneer Jon Kabat-Zinn, now a professor emeritus at the University of Massachusetts Medical School. "People find the body scan beneficial because it reconnects their conscious mind to the feeling states of their body," says Kabat-Zinn. "By practicing regularly, people usually feel more in touch with sensations in parts of their body they had never felt or thought much about before."

To practice the body scan, he explains, we should first sit or lie down in a comfortable place, allowing our eyes to close gently. He recommends taking a few moments to feel the body as a whole and to sense the rising and falling of the abdomen with each in-breath and out-breath. We then begin a "sweep" of the body, starting with the toes of the left foot. Advises Kabat-Zinn, "As you direct your attention to your toes, see if you can channel your breathing to them as well, so that it feels as if you are breathing in *to* your toes and out *from* your toes." After focusing on the toes for a few breaths, we shift our attention to the sole of our foot, the heel, the ankle, and so on up to the left hip. The same procedure is repeated for the right leg, focusing on each section for the length of a few breaths. The roving spotlight of our attention now travels up through the torso, the abdomen and chest, the back and shoulders, then down each arm to the elbows, wrists, and hands. Finally, the attentional spotlight moves up through the neck and face. If our attention should wander during the exercise, we can gently guide it back to the part of the body that is the object of focus. Kabat-Zinn recommends doing the body scan at least once a day.

The aim of this practice is to bring nonjudgmental awareness to any and all feelings that arise within the body. In the rush of everyday life, we may ignore or dismiss these internal signals; if they do come to our notice, we may react with impatience or self-criticism. The body scan trains us to observe such sensations with interest and equanimity. But tuning in to these feelings is only a first step. The next step is to *name* them. Attaching a label to our interoceptive sensations allows us to begin to regulate them; without such attentive self-regulation, we may find our feelings overwhelming, or we may misinterpret their source. Research shows that the simple act of giving a name to what we're feeling has a profound effect on the nervous system, immediately dialing down the body's stress response.

In an experiment conducted by researchers at the University of California, Los Angeles, study subjects were required to give a series of impromptu speeches in front of an audience (a reliable way to induce anxiety). Half of the participants were then asked to engage in what the researchers call "affect labeling," filling in responses to the prompt "I feel _____," while the other half were asked to complete a neutral shape-matching task. The affect-labeling group showed steep declines in heart rate and skin conductance compared to the control group, whose levels of physiological arousal remained high. Brain-scanning studies offer further evidence of the calming effect of affect labeling: simply naming what is felt reduces activity in the amygdala, the brain structure involved in processing fear and other strong emotions. Meanwhile, thinking in a more involved way about feelings and the experiences that evoked them actually produces *greater* activity in the amygdala.

The practice of affect labeling, like the body scan, is a kind of mental training intended to get us into the habit of noting and naming the sensations that arise in our bodies. Psychologists recommend keeping two things in mind as we try it out. The first is to be as *prolific* as possible: the UCLA scientists reported that study participants who came up with a larger number of terms for what they were feeling subsequently experienced a greater reduction in their physiological arousal. The second is to be as *granular* as possible: that is, to choose words that are precise and specific when describing what we feel. Accurately distinguishing among interoceptive sensations is associated with making sounder decisions, acting less impulsively, and planning ahead more successfully—perhaps because it gives us a clearer sense of what we need and what we want.

Sensing and labeling our internal sensations allows them to function more efficiently as our somatic rudder, steering a nimble course through the many decisions of our days. But does the body really have anything to contribute to our thinking—to processes we usually regard as taking place solely in our heads? It does. In fact, recent research suggests a rather astonishing possibility: the body can be *more rational* than the brain. Recall that, in the study conducted by John Coates, traders with keener interoceptive awareness earned more money: that is, they made more rational choices about buying and selling, as judged by the market, than investors who were less attuned to their bodies. Outcomes like these may result from the fact that the body is not subject to the cognitive biases that so often distort our conscious thought—the glitches that appear to be hardwired into the human brain.

Take, for example, our stubborn tendency to insist on notions of fairness, even at the cost of spiting ourselves. In the "ultimatum game," an experimental paradigm often employed by behavioral economists, participants are paired up with a partner; one of the partners is given a pot of money to divide as she wishes. The other partner may then choose to accept or reject the proposed division. Accepting even a very low offer is more rational than rejecting the offer outright, which leaves the receiving partner with nothing. Yet studies consistently find that many players decline low offers out of a sense of being unjustly wronged—a sense that they *should* have gotten more.

In a study published in 2011, researchers from Virginia Tech scanned the brains of two groups of people as they played the ultimatum game: a group who regularly practiced meditation, and a group of control subjects who did not meditate. The scans revealed that in the meditators, the insula—the brain's interoceptive center—was active during game play, indicating that they were relying on their bodies' signals to make their decisions. The controls exhibited a different pattern: their scans showed activity in the prefrontal cortex, the part of the brain that makes conscious judgments about what's fair and unfair. The two groups also diverged in their behavior, researchers reported. The interoceptively aware medita- tors were more likely to elect the rational option of accepting a low offer over no money at all, while the cogitating controls were more apt to snub a proposed division that was tilted in their partners' favor.

Among social scientists, a character named *Homo economicus* is of- ten invoked; the term describes an idealized agent who always makes the perfectly logical and rational choice. This figure has proved hard to find in the real world — and yet, the Virginia Tech researchers write, "in this study, we identified a population of human beings who play the ul- timatum game more like *Homo economicus*." In a tone of some surprise, they continue, "Experienced meditators were willing to accept even the most asymmetrical offers on more than half of the trials, whereas control members of *Homo sapiens* did so in just over one-quarter of the trials."

The bias shown by the non-meditators in the Virginia Tech study is one of many catalogued by behavioral economists. Others include the *anchoring effect,* in which we rely too heavily as a point of reference on the first piece of information we encounter; the *availability heuristic,* in which we overestimate the likelihood of events that come more readily to mind; and the *self-serving bias,* in which our personal preferences incline our beliefs in an overly optimistic direction. What to do about such biases? The strategy of many economists and psychologists has been to inform people of their existence, then recommend that people monitor their mental activity for signs that their thinking is being swayed. In the terminology popularized by psychologist Daniel Kahneman, we're supposed to use rational, reflective "System 2" thinking to override the bias-riddled responses of the faster "System 1."

Mark Fenton-O'Creevy, a professor of organizational behavior at The Open University in the UK, was once a believer in this highly brainbound approach. Then he conducted a series of interviews with expert traders at six investment banks and found that they almost never proceeded in this fashion. Instead, the traders told him, they relied heavily on the sen- sations they felt stirring within their own bodies. One investor described the process to Fenton-O'Creevy in particularly visceral terms. "You have to trust your instincts, and a lot of the decisions are split second, so you need to know where the edge is and what you are going to do about it," he related. "Having a feeling is like having whiskers, like being a deer;

just hearing something that the human ear can't hear and all of a sudden you're on edge. Something somewhere just gave you a slight shiver, but you're not quite sure what, but it's something to be careful about, something's around."

Successful financiers are exquisitely sensitive to these subtle physiological cues, Fenton-O'Creevy discovered. What's more, they seem to pick up on such signals early on, just as the feelings start to emerge — and act on them in that moment, rather than dismissing them, suppressing them, or holding them off for later inspection. Because this approach proceeds rapidly and with little mental effort, it's much better suited to addressing the complex, fast-paced decisions that many of us are called upon to make, says Fenton-O'Creevy. And going around our cognitive biases in this way is more effective than laboriously trying to correct them. "De-biasing approaches which rely primarily on shifting cognition from System 1 to System 2 are unlikely to succeed," he maintains. "The human capacity for self-monitoring and effortful System 2 cognition is limited and is rapidly depleted. Attempts to reduce biases by learning about biases and engaging in self-monitoring rapidly come up against human cognitive limits."

Fenton-O'Creevy has experimented with techniques intended to increase investors' interoceptive awareness — through the practice of mindfulness, and through the provision of frequent physiological feedback. In his lab, he had participants play a specially designed video game called *Space Investor*; as part of the game, they periodically estimated how fast their hearts were beating. The more accurate their guesses, as gauged by a wireless sensor placed on the chest, the more game points they accrued. Fenton-O'Creevy reports that repeated play appears to produce lasting improvements in participants' interoceptive awareness.

This approach suggests a novel way to support smart decision making: not through the application of painstaking deliberation and analysis, but through the cultivation of what we might call "interoceptive learning." This is a process of learning, first, how to sense, label, and regulate our internal signals — and second, how to draw connections between the particular sensations we feel within and the pattern of events we encounter in the world. When we feel a flutter in the stomach as we embark on a certain course of action, what consequences seem to follow? When we feel our heart leap at the thought of one option before us, and our heart sink at the mention of another, what does that portend for the choice we ultimately make?

We can clarify and codify the body's messages by keeping an "interoceptive journal" — a record of the choices we make, and how we felt when we made them. Each journal entry has three parts. First, a brief account of the decision we're facing. Second, a description — as detailed and precise as possible — of the internal sensations we experience as we contemplate the various options available. An interoceptive journal asks us to consider the paths that lie before us, one by one, and take note of how we feel as we imagine choosing one path over another. The third section of the journal entry is a notation of the choice on which we ultimately settle, and a description of any further sensations that arise upon our making this final selection.

Once you know how a particular decision turned out — Did the investment make money? Did the new hire work out? Was the out-of-town trip a good idea? — you can return to the record of the moment when you made that choice. Over time, you may perceive that these moments arrange themselves into a pattern. Perhaps you'll see in retrospect that you experienced a constriction in your chest when you contemplated a course of action that would, in fact, have led to disappointment — but that you felt something subtly different, a lifting and opening of the ribcage, when you considered an approach that would prove successful. Such distinctions are delicate and fleeting; an interoceptive journal can help us fix them in place long enough to see them clearly.

THE BODY, THEN, can act as a sagacious guide to good decision making — in the words of John Coates, as an *"éminence grise,"* more knowledgeable and judicious than the easily overwhelmed conscious mind. The body and its interoceptive capacities can also play another role: as the coach who pushes us to pursue our goals, to persevere in the face of adversity, to return from setbacks with renewed energy. In a word, an awareness of our interoception can help us become more *resilient*.

This may seem surprising. If there's any human capacity that calls for mind over matter, the mental over the corporeal, it would seem to be resilience. We think of ourselves as *deciding* to grit it out, as *resolving* to exert willpower, often over the protests of an unwilling body. But, in fact, resilience is rooted in our awareness of the sensations that originate in our organs and extremities — and the more alert we are to these inner signals, the more resilient we are able to be in the face of life's hardships.

The reason for this: every action we take requires the expenditure of scarce, precious energy. On a level below awareness, we're constantly keeping tabs on how much energy we have on hand and how much energy we will need to take the actions the world demands of us. Interoception acts as a continually updated gauge of our present status. Its cues let us know when we can push ourselves and when we have to give ourselves a rest. They help us match our effort to the magnitude of the challenge and pace ourselves so that we can see it through to the end. And just as some people are better than others at using bodily sensations to guide their decisions, some people are better than others at using interoceptive signals to monitor and manage their moment-by-moment expenditure of energy.

Martin Paulus is a professor of psychiatry at the University of California, San Diego, who investigates the role of interoception in promoting resilience. In a study he conducted in 2016, Paulus gave participants a list of statements like these, asking them to agree or disagree with each one:

> I can deal with whatever comes my way.
> I tend to bounce back after a hardship or illness.
> I give my best effort, no matter what.
> When things look hopeless, I don't give up.
> Under pressure, I focus and think clearly.
> I am not easily discouraged by failure.

With their replies, his subjects sorted themselves into two distinct groups: high resilience and low resilience. By their own accounts, when faced with adversity or challenge, the high-resilience group was likely to push on through to success, while members of the low-resilience group were more likely to struggle, burn out, or give up. Paulus found an additional difference between the two groups: on average, the low-resilience individuals exhibited poor interoception, as measured by the heartbeat detection test, while the high-resilience people possessed a keen sense of their internal world.

In order to explore such intriguing findings, Paulus has devised a protocol that exposes volunteers to a challenging internal experience as their brains are being scanned. Over the past decade, Paulus has administered this regimen, called the inspiratory breathing load task, to hundreds of people. One of the most famous of his guinea pigs is the champion swimmer Diana Nyad. A world-record holder in distance swimming, Nyad made history in 1975 by becoming the first woman to swim around Manhattan Island. Four decades later, at the age of sixty-four, she set out to swim from Cuba to Florida. Nyad was a model of resilience as she battled fatigue, nausea, and potentially deadly jellyfish stings over the course of the 110-mile swim. She failed four times in the attempt before trying, and succeeding, in August 2013.

Later that same year she became a pioneer in another fashion, arriving at Paulus's lab to be studied. Before climbing into an MRI machine, she was fitted with a nose clip, which prevented her from breathing through her nose, and with a tube that went into her mouth. At the end of the tube was a stopper. When the stopper was removed, it was possible for Nyad to breathe freely through the tube; when the stopper was inserted, only a very small amount of air could pass through.

Once inside the MRI chamber, Nyad was directed to look at a computer screen mounted in front of her eyes. When the screen turned blue, the breathing tube was left open; when the screen turned yellow, there was a 25 percent chance that the tube would be plugged, forcing Nyad to struggle for breath. Observing the activity of Nyad's brain under each condition allowed Paulus and his colleagues to investigate how she *anticipated* a stressor, how she *responded* to a stressor, and how she *recovered* from a stressor. While all this was going on, Nyad was also responding to questions in a test of her cognitive ability. (Writing about the experience in her autobiography, Nyad commented, "Of course I'm competitive, so I wanted to score higher than anyone who's ever had the MRI.")

The scan of Nyad's brain revealed a distinctive response to this uncomfortable experience. Her insula mounted an intense anticipatory response *before* the stressor — when the screen turned yellow — but settled down to a state of relative quiescence *during* and *after* the stressor. As for the cognitive test, Nyad recalls Paulus showing her the results, presented as dots charted on a digitized graph: "Clumped at the bottom are ordinary people, who did very poorly on the test during the periods of oxygen restriction and when they were anticipating the upcoming oxygen restriction," she recounts. "Next were a group significantly above the control group, who did much better; these were the Marines. The next group was a big bump up from that: Navy SEALS. Then Dr. Paulus pointed high up to the right, almost off the computer screen. This, he said, is me."

Nyad is truly an outlier, but Paulus has found the same pattern in elite performers of all stripes. Astonishingly, putting these individuals through an extremely unpleasant interoceptive experience actually *improves* their cognitive performance. These champions have a superior ability to sense their bodies' cues, and are therefore better able to monitor and manage their bodies' resources as they rise to meet a challenge. They are like efficient, well-calibrated motors that don't waste even a bit of power, keeping plenty of energy in reserve.

People with low resilience, by contrast, present a very different profile. When undergoing the inspiratory breathing load task, their brain scans show a pattern that is the opposite of Diana Nyad's: low levels of activity *before* a stressor, and high levels *during* and *after* the stressor. The self-management of these individuals is sloppy, all over the place, like poorly calibrated motors that leak power. They are brought up short by challenges, and then waste energy in the scramble to catch up. They begin to struggle to answer the test questions. Discouraged by their failures, their energy reserves depleted, they lose motivation and give up.

Such differences clearly matter in attempting a physical feat of courage or endurance, but they matter, too, when the pursuit is of a more cerebral sort. Mental activity, like any other we undertake, requires the mobilization and management of energy; indeed, the brain consumes fully 20 percent of the body's energy supply. The ability to allocate our internal resources effectively in tackling mental challenges is a capacity researchers call "cognitive resilience."

For one of Martin Paulus's collaborators, cognitive resilience holds a special importance. Elizabeth Stanley, an associate professor of security studies at Georgetown University, is a member of a storied military family who spent years working as an intelligence officer in the US Army, including postings in Germany, South Korea, and the Balkans. In military service and in civilian life, Stanley drove herself relentlessly; she describes her modus operandi as "dig in, access deep wells of willpower and determination, and power through." For decades, she writes, "I considered my capacity to ignore and override my body and my emotions in this way to be a good thing — a sign of strength, self-discipline, and determination." Eventually, however, she came to recognize that "this default strategy was actually undermining my performance and well-being." (The time she threw up all over her computer keyboard, after months of sixteen-hour days at work on her PhD dissertation, was one hint.)

Seeking a different approach, Stanley found her way to mindfulness meditation, which she practices daily; she also created a program, called Mindfulness-based Mind Fitness Training, designed to bolster the cognitive resilience of service members facing high-stress situations. MMFT, as it is known, places an emphasis on recognizing and regulating the body's internal signals. In collaboration with academic psychologists and neuroscientists, Stanley has tested the efficacy of the program among troops as they prepare to deploy to combat; results show that the training helps participants maintain their attentional focus and preserve their working memory even under the most challenging circumstances. In workshops around the country, Stanley teaches the method not only to members of the military but also to others in high-stress occupations: firefighters, police officers, social workers, health care providers, disaster relief workers.

Like the expert traders interviewed by Mark Fenton-O'Creevy, and like the elite athletes studied by Martin Paulus, Stanley has found that the most cognitively resilient soldiers pay close attention to their bodily sensations *at the early stage* of a challenge, when signs of stress are just beginning to accumulate. She instructs her workshop participants to do the same, using mindfulness techniques similar to the ones described by Jon Kabat-Zinn. By remaining alert to these preliminary signals, she says, we can avoid being taken by surprise and then overreacting, entering a state of physiological arousal from which it is hard to come down. (Stanley notes ruefully that many of us take just the opposite approach, as she once did: pushing aside internal red flags in the hope that we can "power through" and get the job done.)

Stanley also demonstrates for her students a technique she calls "shuttling" — moving one's focus back and forth between what is transpiring internally and what is going on outside the body. Such shifts are useful in ensuring that we are neither too caught up in external events nor too overwhelmed by our internal feelings, but instead occupy a place of balance that incorporates input from both realms. This alternation of attention can be practiced at relaxed moments until it becomes second nature:

a continuously repeated act of checking in that provides a periodic infusion of interoceptive information. The point is to keep in close contact with our internal reality at all times — to train ourselves "to pay attention and notice what's happening while it's happening," as Stanley puts it. The vision of resilience she offers is not a formidable display of will and grit of the kind she once would have embraced; it is, rather, a flexible, moment-by-moment responsiveness to changing conditions — both inside and out.

AN AWARENESS OF our interoceptive signals can assist us in making sounder decisions and in rebounding more readily from stressful situations. It can also allow us to enjoy a richer and more satisfying emotional life. Research finds that people who are more interoceptively attuned feel their emotions more intensely, while also managing their emotions more adeptly. This is so because interoceptive sensations form the building blocks of even our most subtle and nuanced emotions: affection, admiration, gratitude; sorrow, longing, regret; irritation, envy, resentment. People who are more interoceptively aware can interact more intimately and more skillfully with the emotions that interoceptive sensations help construct.

But first — understanding the relationship between interoception and emotion requires correcting a basic misconception most of us hold about how feelings come about. The story we're used to telling goes like this: on the basis of what's happening to us, the brain determines the appropriate emotion (happy, sad, scared), then directs the body to act accordingly (smile, cry, scream). In fact, the causal arrow points in the *opposite* direction. The body produces sensations, the body initiates actions — and only then does the mind assemble these pieces of evidence into the entity we call an emotion.

The pioneering American psychologist William James deduced this more than a century ago. Imagine you meet a bear in the woods, James wrote. Your heart pounds, your palms sweat, your legs break into a run — why? It might seem that it's because your brain generates a feeling of fear, and then tells your body to get moving. But James suggested that it works the other way around: we feel fear *because* our heart is racing, *because* our palms are sweating, *because* our legs are propelling us forward. As he put it: "Common sense says, we lose our fortune, are sorry and weep; we meet a bear, are frightened and run; we are insulted by a rival, are angry and strike." But, he went on, "this order of sequence is incorrect." It would be more accurate, wrote James, to say that "we feel sorry because we cry, angry because we strike, afraid because we tremble."

In recent years, scientists have begun to elaborate on James's theory with the help of modern investigative techniques like brain scanning. Their research has confirmed that the thing we call "emotion" (and experience as a unified whole) is actually constructed from more elemental parts; these parts include the signals generated by the body's interoceptive system, as well as the beliefs of our families and cultures regarding how these signals are to be interpreted. This perspective carries two important implications. First: the greater our awareness of interoceptive sensations, the richer and more intense our experience of emotion can be. And second: equipped with interoceptive awareness, we can get in on the ground floor of emotion construction; we can participate in creating the type of emotion we experience.

Psychologists who study the construction of emotion call this practice "cognitive reappraisal." It involves sensing and labeling an interoceptive sensation, as we've learned to do here, and then "reappraising" it — reinterpreting it in an adaptive way. We can, for example, reappraise "nervousness" as "excitement." Consider the interoceptive sensations that accompany these two emotions: a racing heart, sweaty palms, a fluttering stomach. The *feelings* are almost identical; it's the *meaning* we assign to them that makes them, variously, an ordeal to be dreaded or a thrill to be enjoyed. The one thing we are not, in such moments, is cool or calm — and yet most of us are convinced that when we are in anxiety's grip, what we ought to do is try to *calm down*.

Alison Wood Brooks, an associate professor at Harvard Business School, had a different notion of how to handle nervousness. In a series of three studies, she subjected groups of people to experiences that most everyone would find nerve-racking: completing "a very difficult IQ test" administered "under time pressure"; delivering, on the spot, "a persuasive public speech about 'why you are a good work partner'"; and most excruciating of all, belting out an 80s pop song ("Don't Stop Believin'," by Journey). Before beginning the activity, participants were to direct themselves to stay calm, or to tell themselves that they were excited.

Reappraising nervousness as excitement yielded a noticeable difference in performance. The IQ test takers scored significantly higher. The speech givers came across as more persuasive, competent, and confident. Even the singers performed more passably (as judged by the Nintendo Wii *Karaoke Revolution* program they used). All reported genuinely feeling the pleasurable emotion of excitement — a remarkable shift away from the unpleasant discomfort such activities might be expected to engender.

In a similar fashion, we can choose to reappraise debilitating "stress" as productive "coping." A 2010 study carried out with Boston-area undergraduates looked at what happens when people facing a stressful experience are informed about the *positive* effects of stress on our thinking — that is, the way it can make us more alert and more motivated. Before taking the GRE, the admissions exam for graduate school, one group of students was given the following message to read: "People think that feeling anxious while taking a standardized test will make them do poorly on the test. However, recent research suggests that arousal doesn't hurt performance on these tests and can even help performance. People who feel anxious during a test might actually do better. This means that you shouldn't feel concerned if you do feel anxious while taking today's GRE test. If you find yourself feeling anxious, simply remind yourself that your arousal could be helping you do well." A second group received no such message before taking the exam. Three months later, when the students' GRE scores were released, the students who had been encouraged to reappraise their feelings of stress scored an average of 65 points higher.

Reappraisal research has begun to elucidate the mechanisms by which this technique exerts its effects. In the GRE study, saliva samples were collected from all the participants and analyzed for the presence of a hormone associated with nervous system arousal. Among the students who engaged in reappraisal, the level of this hormone was elevated — suggesting that their bodies had identified the presence of a challenge and were mounting an effective response, enhancing their alertness and sharpening their attention. Another study explored the neural effects of the reappraisal technique on students who struggle with math anxiety. Their brains were scanned twice as they completed a set of math problems inside an fMRI machine. Before the first round, participants were told to use whatever strategies they usually employed. Before the second round, participants were given instructions on how to engage in reappraisal.

When employing the reappraisal approach, the students answered more of the math questions correctly, and the scans showed why: brain areas involved in executing arithmetic were more active under the reappraisal condition. The increased activity in these areas suggests that the act of reappraisal allowed students to redirect the mental resources that previously were consumed by anxiety, applying them to the math problems instead.

Psychologists offer two additional points of interest for those adopting the strategy of reappraisal. The first is that reappraisal works best for those who are interoceptively aware: we have to be able to identify our internal sensations, after all, before we can begin to modify the way we think about them. Second, the sensations we're actually feeling have to be *congruent* with the emotion we're aiming to construct. We're able to reappraise nervousness as excitement because the physiological cues associated with the two emotions are so similar; if what we're feeling is a heavy sense of apathy or lassitude, exclaiming "I'm so excited!" isn't going to work.

Becoming aware of our internal sensations can help us handle our own emotions. Perhaps more surprisingly, the body's interoceptive faculty can also bring us into closer contact with *other* people's emotions. That's because the brain, on its own, has no direct access to the contents of other people's minds, no way to feel what others are feeling. Interpreting others' spoken words and facial expressions may yield only a coolly abstract sense of the emotions that churn within. The body acts as a critical conduit, supplying the brain with the visceral information it lacks. It does so in this way: When interacting with other people, we subtly and unconsciously mimic their facial expressions, gestures, posture, and vocal pitch. Then, via the interoception of our own bodies' signals, we perceive what the other person is feeling *because we feel it in ourselves.* We bring other people's feelings onboard, and the body is the bridge. In an act akin to taking a bite off our partner's plate, or borrowing an earbud to hear the song our friend is listening to, we are sampling their emotions.

When people can't engage in such mimicking, they have a harder time figuring out what others are feeling. A striking example: people injected with the wrinkle reducer Botox, which works by inducing mild paralysis of the muscles used to generate facial expressions, are less accurate in their perceptions of others' emotions — presumably because they can't simulate others' feelings within themselves. On the other end of the spectrum, interoceptively attuned people are more likely to mimic the expressions of others, and more accurate in their interpretation of others' feelings, than are people who are less aware of their bodily sensations. They also tend to have more empathy for others. Via mimicking, all of us "feel" the pain of others: research demonstrates that the areas of the brain involved in sensing our own pain are also activated when we see other people experience physical harm. But when interoceptively attuned people view someone experiencing pain, they rate the other person's pain as more intense.

The interoceptive champions among us may be clinical psychologists, who are professionally trained to read their own bodies' signals for clues to what their patients are feeling — even when their clients are not yet able to verbalize their emotions. "For me it's like using the body as radar, you think of those sort of dishes that collect satellite messages and funnel them down — well, I see the body in that way," observed one clinician in a 2004 study of how therapists use their bodies to understand their patients. In similar fashion, Susie Orbach, a clinical psychologist and the author of groundbreaking books on women and body image, has found that her own body is a sensitive instrument for picking up on what her patients are feeling. Focusing on the sensations that well up inside her during therapy, says Orbach, "has helped me to realize that the body's development is every bit as crucial as the mind's."

Like psychotherapists, we can develop the capacity of the body to enhance our connections with others — a faculty called "social interoception." Research suggests that looking our conversational partner directly in the eye increases our interoceptive attunement, as does a brief touch on the hand or the arm. Studies also show that when interpersonal situations become challenging — when we feel socially rejected or excluded, for example — we tend to shift our focus away from our own internal sensations and toward external events, perhaps in an urgent effort to repair the breach. This shift, however well intentioned, may cut us off from a source of insight about the other person just when we need it most. Better to attempt a flexible back-and-forth movement between attending to others' social cues and to our own interoceptive signals (a process that recalls Elizabeth Stanley's technique of "shuttling"). By drawing on data from both sources, we can feel our way into the other person's emotional world while maintaining a vivid sense of our own.

JOHN COATES, the trader turned scientist we met at the beginning of this chapter, likens our bodies to "sensitive parabolic reflectors, registering a wealth of predictive information." These biological antennae are continually receiving and sending important messages, he notes, yet "these messages are notoriously, frustratingly hard to hear. They fade in and out like a radio picking up a distant station." Coates believes that technology can help — not by replacing gut feelings with data-driven algorithms, but by amplifying the body's own accumulated insight. He has now embarked on a third career, this time as an entrepreneur. His company, Dewline Research, collects data on financial traders' physiological signals via wearable sensors, tracking the relationship between the gyrations of the market and the traders' bodily reactions.

Another group of interoception researchers, meanwhile, have developed a similar device — this one oriented around enhancing the body's role in promoting resilience. Heartrater, a technology introduced by scientists at the Brighton and Sussex Medical School in the UK, allows athletes to monitor their internal climate more closely, permitting a more efficient use of energy and a more rapid recovery from exertion. Perhaps the most intriguing use of such extended technology — that is, digital tools extended by the body — has been pioneered by interoception researcher Manos Tsakiris, a professor of psychology at Royal Holloway, University of London. With a team from the firm Empathic Technologies, he helped develop a device called doppel. It provides users with bodily feedback that is not amplified but rather deliberately distorted; in effect, doppel seeks to trick the user into believing that her heart is beating slower or faster than it actually is.

Recall that — as William James explained so eloquently — our brains take their cues about the emotions we're experiencing from the sensations generated by the body. Tsakiris's device intervenes in this loop by offering the brain a message that is different from the one the body is actually producing. Worn on the wrist like a watch or a Fitbit, doppel generates the persuasive sensation of a slow, relaxed heartbeat — or, on a different setting, of a fast, excited one. When set to slow mode, doppel induces a sense of calm in people who are nervous about having to engage in public speaking; when it's in fast mode, people wearing the device are more alert and more accurate in their performance on a challenging test of sustained attention. The technology allows us, says Tsakiris, "to harvest our natural responses to heartbeat-like rhythms" in the interest of improving our performance.

That such trickery is even possible only serves to underline the robustness of the connection between body and mind, the two-way flow of information that does so much to shape our daily decisions, our everyday efforts, our most intimate relationships. This link may also form the foundation of something even more fundamental: our sense of self. Among all their other functions, the steady flow of internal sensations we experience provides us with a sense of personal continuity. Thinkers have long mused on how it is that we can regard ourselves as unique and persistently existing entities. "What makes me the same person throughout my life, and a different person from you?" wondered the late philosopher Derek Parfit. Their answers have usually had something to do with the brain—with our thoughts or our memories. As the French philosopher René Descartes declared, "I think, therefore I am."

According to neuroanatomist and interoception expert A. D. Craig, it would be more accurate to say, "I *feel*, therefore I am." Craig maintains that interoceptive awareness is the basis of the "material me," the source of our most fundamental knowledge of ourselves. Because our hearts beat, because our lungs expand, because our muscles stretch and our organs rumble—and because all these sensations, unique to us, have carried on without interruption since the day of our birth—we know what it is to be one continuous self, to be ourselves and no other. Interoception, says Craig, is nothing less than "the feeling of being alive."

2

■

Thinking with Movement

D R. JEFF FIDLER is a radiologist at the Mayo Clinic in Rochester, Minnesota. In the course of his professional duties, he regularly reviews fifteen thousand or more images *a day* — and he used to do so sitting down. Not anymore. These days Fidler walks as he looks, having set up a treadmill in front of the large screen that displays the radiological slides he is charged with examining. Within the first year of creating his "walking workstation," Fidler lost twenty-five pounds — and, he felt convinced, he was doing a better job of identifying the abnormalities hidden in the X-ray images he scrutinized.

With a colleague, Fidler designed a study to test his hunch. Radiologists inspected a batch of images while seated, and while walking on a treadmill at one mile per hour. The participating physicians identified a total of 1,582 areas of concern in the slides, and rated 459 of these as posing potentially serious risks to the health of the patient. When they compared the "detection rates" they achieved while sitting and while moving, the results were clear: radiologists who remained seated spotted an average of 85 percent of the irregularities present in the images, while those who walked identified, on average, fully 99 percent of them.

Other evidence supports Fidler's findings. A study conducted at the University of Maryland Medical Center, for example, found that radiologists reviewing images of patients' lungs were more likely to identify potentially problematic nodules if they walked rather than sat as they worked. And a study conducted by physicians in the radiology department of the Naval Medical Center in Portsmouth, Virginia, found that radiologists who used a treadmill workstation did their work faster, with no loss of accuracy.

A different set of research findings helps to explain the radiologists' results. When we're engaged in physical activity, our visual sense is sharpened, especially with regard to stimuli appearing in the periphery of our gaze. This shift, which is also found in non-human animals, makes evolutionary sense: the visual system becomes more sensitive when we are actively exploring our environment. When our bodies are at rest — that is, sitting still in a chair — this heightened acuity is dialed down.

Such activity-induced alterations in the way we process visual information constitute just one example of how moving our bodies changes the way we think. Scientists have long known that overall physical fitness supports cognitive function; people who have fitter bodies generally have keener minds. In recent years, however, researchers have begun to explore an exciting additional possibility: that single bouts of physical activity can enhance our cognition in the short term. By moving our bodies in certain ways, that is, we're immediately able to think more intelligently. Scientists investigating this phenomenon have approached it from two different directions: the *intensity* of the movement and the *type* of movement. As we'll soon see, low-, medium-, and high-intensity physical activity each exerts a distinct effect on our cognition. Later on in the chapter, we'll explore how certain types of movement — congruent movements, novel movements, self-referential movements, and metaphorical movements — can also extend our thinking beyond what is possible when we remain stationary.

The tight connection between thinking and moving is a legacy of our species's evolutionary history. The human brain is approximately three times larger than it "should" be, given the dimensions of the human body; according to fossil evidence, a remarkable expansion in the size of the brain took place about 2 million years ago. Scientists have proposed various reasons for this increase, such as the growing complexity of our forebears' social interactions or the need to adapt to changing ecological conditions. Recently, another explanation has been put forth: "At the same time as brain size began to increase in the human lineage, aerobic activity levels appear to have changed dramatically," notes David Raichlen, a professor of biological sciences at the University of Southern California. "Human ancestors transitioned from a relatively sedentary ape-like existence to a hunting and gathering lifestyle which required an increased amount of physical activity compared to earlier hominins."

Raichlen, who has conducted extensive studies of some of the world's remaining hunter-gatherer tribes, points out that this way of life is both physically *and* cognitively demanding. Foraging requires vigorous, sustained physical activity; it also makes demands on attention, memory, spatial navigation, motor control, and executive functions like planning and decision making. Hunting, too, poses both a mental and a physical challenge: the hunter has to locate the animal and track its unpredictable movements even while mustering the energy to outrun it. Such are the conditions under which the unique human brain evolved. The dual demands of physical challenge and cognitive complexity shaped our special status as *Homo sapiens;* to this day, bodily activity and mental acuity are still intimately intertwined.

Of course, things have changed for those of us who live in modern societies; we are no longer a species on the move. While the Hazda people of East Africa — one of the hunter-gatherer tribes studied by Raichlen — spend an average of 135 minutes a day engaged in moderate to vigorous physical activity, most inhabitants of industrialized countries fail to meet health experts' minimum recommendation of 150 minutes of activity *a week*. Put another way, contemporary hunter-gatherers engage in more than fourteen times as much moderate-to-vigorous physical activity as the typical American. The lack of physical movement in our society is due in large part to the dominance of academic learning and of knowledge work, and to the habits and convictions that have grown up around these endeavors. While we're thinking, we believe, we should be sitting still.

Challenges to this belief may be met with derision. When Jeff Fidler published his findings in the *Journal of the American College of Radiology,* some of his colleagues reacted with mocking ridicule. "I see that I can depend on this publication for comic relief," wrote Robert Feld, a radiologist from Hartford, Connecticut, in a letter to the editor. Fidler's study, he declared, was "a parody of clinical research gone awry"; in Feld's view, making provisions for physicians to move as they work is "a staggering waste of effort and resources."

Such attitudes are widely reflected in the way students and workers use their time. Children spend an average of 50 percent of the school day sitting, a proportion that increases as they enter adolescence. Adults in the workplace move even less, remaining seated for more than two-thirds of the average workday. We inherited a "mind on the hoof," in the phrase of philosopher Andy Clark—but in today's classrooms and offices, the vigorous clatter of hoofs has come to an eerie halt.

THAT'S EMPHATICALLY NOT the case in the classroom helmed by Maureen Zink, a fourth-grade teacher at Vallecito Elementary School in San Rafael, California. Her students don't sit still at their desks; in fact, most of them are not sitting at all. In 2013, the entire school replaced traditional desks and chairs with standing desks, and the school's "activity-permissive" ethos allows pupils to stand upright, perch on stools, sit on the floor, and otherwise move around as they wish. Though some were hesitant about the change, Zink and the other teachers at Vallecito now say it's been a resounding success; students are more alert, more attentive, and more engaged. "I taught at sitting desks for 30 years," says Zink, "and I'll never go back." Tracy Smith, the principal at Vallecito during the switch to standing desks, agrees that students are "more focused, confident, and productive" when given license to move.

The community's initial trepidation is telling. We associate stillness with steadiness, seriousness, and industriousness; we believe there's something virtuous about controlling the impulse to move. At times and places where there's work to be done, physical movement is regarded with disapproval, even suspicion. (Consider the way we associate fidgeting with a certain moral shiftiness.) What this attitude overlooks is that the capacity to regulate our attention and our behavior is a limited resource, and some of it is used up by suppressing the very natural urge to move.

This trade-off is highlighted by the work of Christine Langhanns and Hermann Müller of Justus Liebig University in Germany. For a study published in 2018, they asked groups of volunteers to solve a set of math problems in their heads while staying still, while remaining relaxed "but without substantial movement," or while moving slightly in a rhythmic pattern. All the while, the participants' cognitive load—how hard their brains were working—was being measured with a brain-scanning technology called functional near-infrared spectroscopy (fNIRS). The results were illuminating. Subjects' cognitive load "considerably increased under the instruction 'not to move,'" Langhanns and Müller report. Significantly, the stay-still command increased brain activity in the same area as did the mental calculations: the prefrontal cortex, responsible for carrying out intellectual tasks like arithmetic *and* for keeping our impulses in check. Of the three conditions, the requirement to remain still produced the poorest performance on the math problems; the greater their overall cognitive load as registered by fNIRS, the worse the subjects did on the calculations. "Sitting quietly," the researchers conclude, "is not necessarily the best condition for learning in school."

The continual small movements we make when standing as opposed to sitting — shifting our weight from one leg to another, allowing our arms to move more freely — constitute what researchers call "low-intensity" activity. As slight as these movements appear, they have a marked effect on our physiology: an experiment carried out by researchers at the Mayo Clinic determined that simply by standing rather than sitting, study participants expended 13 percent more energy. The impact on our cognitive functioning is also significant. Research has found that the use of a standing desk is associated with an enhancement in students' executive function — that crucial capacity for planning and decision making — and with an increase in "on-task engagement." In adults, working at a standing desk has been shown to boost productivity.

It's not only that such activity-permissive setups relieve us of the duty to monitor and control our inclination to move; they also allow us to fine-tune our level of physiological arousal. Such variable stimulation may be especially important for young people with attention deficit disorders. The brains of kids with ADHD appear to be chronically *under*-aroused; in order to muster the mental resources needed to tackle a difficult assignment, they may tap their fingers, jiggle their legs, or bounce in their seats. They move as a means of increasing their arousal — not unlike the way adults down a cup of coffee in order feel more alert.

Julie Schweitzer, a professor of psychiatry at the University of California, Davis, led a 2016 study of children aged ten to seventeen who had been diagnosed with ADHD. As the young participants worked on a challenging mental task, their movements were monitored by a sensor, called an actometer, strapped to their ankles. She found that more intense physical movement was associated with better cognitive performance on the task. The more the children moved, in other words, the more effectively they were able to think. Parents and teachers often believe they have to get kids to stop moving around before they can focus and get down to work, Schweitzer notes; a more constructive approach would be to allow kids to move around *so that* they can focus.

Even among those without an ADHD diagnosis, the amount of stimulation required to maintain optimal alertness varies from person to person. Indeed, it may differ for the same individual over the course of a day. We have at our disposal a flexible and sensitive mechanism for making the necessary adjustments: fidgeting. At times we may use small rhythmic movements to calm our anxiety and allow us to focus; at other moments, we may drum our fingers or tap our feet to stave off drowsiness, or toy with an object like a pen or a paperclip as we ponder a difficult concept. All of these activities, and many others, were submitted to researcher Katherine Isbister after she put out a call on social media for descriptions of people's favorite "fidget objects" and how they used them.

Isbister, a professor of computational media at the University of California, Santa Cruz, believes that the social disapproval directed at fidgeting is misplaced. Though we imagine that we can manage our mental activity from within our heads, it's often more effective to employ the movements of our bodies for that purpose — to engage in what she calls "embodied self-regulation." Isbister would reverse the usual chain of command in which the brain tells the body what to do. "Changing what the *body* does," she notes, "can change our feelings, perceptions, and thoughts."

Her research and that of others suggests that fidgeting can extend our minds in several ways beyond simply modulating our arousal. The playful nature of these movements may induce in us a mildly positive mood state, of the kind that has been linked to more flexible and creative thinking. Alternatively, their mindless and repetitive character may occupy just enough mental bandwidth to keep our minds from wandering from the job at hand. One study found that people who were directed to doodle while carrying out a boring listening task remembered 29 percent more information than people who did not doodle, likely because the latter group had let their attention slip away entirely.

Perhaps most intriguing is Isbister's theory that fidgeting can supply us with a range of sensory experiences entirely missing from our arid encounters with screen and keyboard. "Today's digital devices tend to be smooth, hard, and sleek," she writes, while the fidget objects she crowdsourced exhibit "a wide range of textures, from the smoothness of a stone to the roughness of a walnut shell to the tackiness of cellophane tape." The words contributors used to describe their favored objects were vivid: such articles were "crinkly," "squishy," "clicky-clackety"; with them they could "scrunch," "squeeze," "twirl," "roll," and "rub." It's as if we use fidgeting to remind ourselves that we are more than just a brain—that we have a body, too, replete with rich capacities for feeling and acting. Thinking while moving brings the full range of our faculties into play.

ACTIVITY-PERMISSIVE SETTINGS are still the exception in schools and workplaces, but we ought to make them the rule; we might even dispense with that apologetic-sounding name, since low-intensity physical activity clearly belongs in the places where we do our thinking. Meanwhile, medium- and high-intensity activity each exerts its own distinct effect on cognition—as the psychologist Daniel Kahneman has discovered for himself.

Kahneman spends a few months each year in Berkeley, California, and on most days he takes a four-mile walk on a marked path in the hills, with a view of San Francisco Bay. Ever the scientist, Kahneman has subjected the experience to close analysis. "I usually keep track of my time and have learned a fair amount about effort from doing so," he writes. "I have found a speed, about 17 minutes for a mile, which I experience as a stroll. I certainly exert physical effort and burn more calories at that speed than if I sat in a recliner, but I experience no strain, no conflict, and no need to push myself. I am also able to think and work while walking at that rate. Indeed, I suspect that the mild physical arousal of the walk may spill over into greater mental alertness."

He noted, however, that "accelerating beyond my strolling speed completely changes the experience of walking, because the transition to a faster walk brings about a sharp deterioration in my ability to think coherently. As I speed up, my attention is drawn with increasing frequency to the experience of walking and to the deliberate maintenance of the faster pace. My ability to bring a train of thought to a conclusion is impaired accordingly. At the highest speed I can sustain on the hills, about 14 minutes for a mile, I do not even try to think of anything else."

Kahneman's careful self-observations are backed up by empirical research. Moderate-intensity exercise, practiced for a moderate length of time, improves our ability to think both during and immediately after the activity. The positive changes documented by scientists include an increase in the capacity to focus attention and resist distraction; greater verbal fluency and cognitive flexibility; enhanced problem-solving and decision-making abilities; and increased working memory, as well as more durable long-term memory for what is learned. The proposed mechanisms by which these changes occur include heightened arousal (as Kahneman speculated), increased blood flow to the brain, and the release of a number of neurochemicals, which increase the efficiency of information transmission in the brain and which promote the growth of neurons, or brain cells. The beneficial mental effects of moderately intense activity have been shown to last for as long as two hours after exercise ends.

The encouraging implication of this research is that we have it within our power to induce in ourselves a state that is ideal for learning, creating, and engaging in other kinds of complex cognition: by exercising briskly just before we do so. As things stand, however, we don't often take intentional advantage of this opportunity. Our culture conditions us to see mind and body as separate — and so we separate, in turn, our periods of thinking from our bouts of exercise. Consider how many of us make our visits to the gym only *after* work, for example, or on weekends. Instead, we should be figuring out how to incorporate bursts of physical activity into the work day and the school day — which means rethinking how we approach our breaks. Lunch breaks, coffee breaks, downtime between tasks or meetings: all become occasions to use exercise to maneuver our brains into an optimally functioning state.

For children, this is precisely the role played by recess; research shows that kids return from a session on the playground better able to focus their attention and to engage their executive function faculties. Yet at schools all over the country, recess has been reduced or even eliminated in order to generate more "seat time" spent on academic learning. The notion that time away from concentrated mental work is effectively time wasted is one of several wrongheaded notions we hold regarding breaks — wrongheaded, in this case, because the ability to attend to such work declines steadily over time, and is actually refreshed by a bout of bodily exertion. Parents, teachers, and administrators who want students to achieve academically should be advocating for an increase in physically active recess time.

Another misguided idea about breaks: they should be used to rest the body, so as to fortify us for the next round of mental labor. As we've seen, it's through *exerting* the body that our brains become ready for the kind of knowledge work so many of us do today. The best preparation for such (metaphorical) acts as wrestling with ideas or running through possibilities is to work up an (actual) sweat. Instead of languidly sipping a latte before tackling a difficult project, we should be taking an energetic walk around the block.

There's one more erroneous assumption about breaks to address: we imagine that we're replenishing the brain's depleted resources when we spend our breaks doing something that *feels* different from work — scrolling through Twitter, checking the news, looking at Facebook. The problem is that such activities engage the same brain regions and draw down the same mental capital we use to do our cognition-centric jobs. We resume our duties just as frazzled as before the pause, and maybe more so. Turning coffee breaks into what some public health experts call "movement breaks" allows us to return to our work a bit smarter than when we left it.

On his walks along the coastal California hills, Daniel Kahneman noticed that moving very fast "brings about a sharp deterioration in my ability to think coherently." This observation, too, is supported by research: scientists draw what they call an "inverted U-shaped curve" to describe the relationship between exercise intensity and cognitive function, with the greatest benefits for thinking detected in the moderate-intensity middle part of the hump. On the right downward slope of the curve, where high-intensity activity is charted, control over cognition does indeed start to slacken — but this is not always a bad thing. Very intense exercise, extended over a relatively long period, can induce a kind of altered state conducive to creative thought.

Such is the experience of Haruki Murakami, the celebrated Japanese novelist. Murakami is a committed runner, a veteran of more than two dozen marathons who logs as many as fifty miles a week. He has even authored a book about it, titled *What I Talk About When I Talk About Running*. "I'm often asked what I think about as I run," Murakami writes. "Usually the people who ask this have never run long distances themselves. I always ponder the question. What exactly do I think about when I'm running?" Not much, he concludes. That's kind of the point. "As I run, I don't think much of anything worth mentioning. I just run. I run in a void. Or maybe I should put it the other way: I run in order to *acquire* a void."

Scientists have a term for the "void" Murakami describes: "transient hypofrontality." *Hypo* means low or diminished, and *frontality* refers to the frontal region of the brain—the part that plans, analyzes, and critiques, and that usually maintains firm control over our thoughts and behavior. When all of our resources are devoted to managing the demands of intense physical activity, however, the influence of the prefrontal cortex is temporarily reduced. In this loose hypofrontal mode, ideas and impressions mingle more freely; unusual and unexpected thoughts arise. Scientists speculate that the phenomenon of transient hypofrontality may underlie all kinds of altered states, from dreaming to drug trips—but intense exercise may be the most reliable way to induce it. Low- and moderate-intensity exercise does not generate this disinhibiting effect. (Indeed, as we've seen, moderately intense physical activity actually *enhances* executive function.) Achieving transient hypofrontality generally requires exercising at one's "ventilatory threshold"—the point at which breathing becomes labored, corresponding to about 80 percent of the exerciser's maximum heart rate—for forty minutes or more.

It's a daunting summit to scale, but when it is reached, observes Kathryn Schulz, another writer-runner, it can "provoke a kind of Cartesian collapse": mind and body melding together in what she calls a "glorious collusion."

IN CONSIDERING THE IMPACT of motion on thought, the releasing or enhancing or disinhibiting effects of physical activity are only half the story. Also important are the many varied and nuanced ways that particular physical movements, which carry their own load of meaning and information, participate in our thinking processes. Over the past several decades, the field of embodied cognition has produced persuasive evidence that our thoughts—even, or especially, those of an abstract or symbolic nature—are powerfully shaped by the way we move our bodies. According to the conventional, brainbound understanding of cognition, we first have a thought, and then direct our bodies to move accordingly. This more recent corpus of research turns that causal arrow around so that it points in the opposite direction: we move our bodies, and our thoughts are influenced in turn. The exciting implication of such findings is that we can intentionally enhance our *mental* functioning through an application of *physical* activity — that we can, for example, improve our memory not through working our brains ever harder, but by looping in the meaning-bearing movements of our limbs.

When we're charged with learning and remembering new material, our tendency is to lean heavily on visual and auditory modes: reading it over, saying it aloud. This approach has its limits; in particular, research demonstrates that our memory for what we have *heard* is remarkably weak. Our memory for what we have *done,* however — for physical actions we have undertaken — is much more robust. Linking movement to the material to be recalled creates a richer and therefore more indelible "memory trace" in the brain. In addition, movements engage a process called *procedural memory* (memory of how to do something, such as how to ride a bike) that is distinct from *declarative memory* (memory of informational content, such as the text of a speech). When we connect movement with information, we activate both types of memory, and our recall is more accurate as a result — a phenomenon that researchers call the "enactment effect."

Fittingly, professional actors can shed light on the way physical enactment reinforces memory. Helga Noice, a professor emeritus of psychology at Elmhurst University in Illinois, and her husband, Tony Noice, a professor of theater at Elmhurst as well as a Chicago-area actor, have spent years studying actors' ability to memorize pages and pages of lines. They have determined that during performance, actors render written lines with 98 percent accuracy, on average; months after a play's run has ended, the Noices found, actors can still recall verbatim some 90 percent of the script. How do they do it? Actors' mental feats of memory, the Noices have concluded, are intimately connected to the movements they make with their bodies. In the course of their research, many actors commented that they never tried to learn their lines until a play had been "blocked" — that is, until all the physical movements to be made onstage had been planned out. "You've got to have these two tracks going simultaneously — 'This is what I say, and this is when and where I move' . . . One feeds the other," an actor observed in one of their interviews.

In a study they conducted in 2000, the Noices gathered together six actors from a repertory company who had earlier performed together in a production of *The Dining Room,* by the American playwright A. R. Gurney Jr. In this scene from the play, adult siblings Arthur and Sally debate what to do with the contents of their parents' house, which is being sold:

ARTHUR: You sure Mother doesn't want this stuff in Florida?
SALLY: She hardly has room for what she's got. She wants us to take turns. Without fighting.
ARTHUR: We'll just have to draw lots then.
SALLY: Unless one of us wants something, and one of us doesn't.
ARTHUR: We have to do it today.
SALLY: Do you think that's enough time to divide up a whole house?
ARTHUR: I have to get back, Sal. (*He looks in the sideboard.*) We'll draw lots and then go through the rooms taking turns. (*He brings out a silver spoon.*) Here. We'll use this salt spoon. (*He shifts it from hand to hand behind his back, then holds out two fists.*) Take your pick. You get the spoon, you get the dining room.
SALLY: You mean you want to start here?
ARTHUR: Got to start somewhere.

Even though the run of *The Dining Room* had ended five months earlier, and many of the actors had learned new roles since then, they still remembered the lines from Gurney's play that had been accompanied onstage by movement or gestures (as when Arthur holds out the spoon to Sally). Lines they had delivered while standing or sitting still, the Noices discovered, were much more likely to be forgotten.

In other studies, the Noices have shown that connecting words to movements improves recall among people who are *not* actors — college students, for example, and older people residing in assisted living facilities. Most activities advertised to forestall age-related memory loss — such as crossword puzzles, sudoku, and commercial brain-training programs like Lumosity — conform to our society's brainbound model of how thinking works: users sit still and use their heads. By contrast, Helga and Tony Noice have found that moving the body makes an essential contribution to strengthening memory — and other mental capacities as well.

In a series of studies conducted with people aged sixty-five to eighty-five, the Noices instructed participants in professional acting techniques and then led them in the rehearsal and performance of theatrical scenes. Before and after the four-week-long program, they tested participants on general cognitive capacities like word memory, verbal fluency, problem solving, and handling daily tasks, such as comparing nutrition labels, paying bills by check, and looking up phone numbers. Compared to people the same age who did not enroll in any program, or who took part in activities that did not involve movement (an art appreciation class, for example), individuals who had participated in the theatrical program became mentally sharper. Participants were apparently able to borrow strategies they learned in acting class — such as linking movement to material to be remembered — and apply them to the activities of everyday life.

Similar results have been found among younger people; again, movement seems to be key to remembering. In a study of undergraduates published in 2001, for example, the Noices reported that the effects of physical movement "previously found after lengthy real-world rehearsal and repeated performances by professional actors could be produced by giving a few minutes of instruction to non-actors with little or no performing experience." The difference such minimal instruction made in participants' ability to recall information, they noted, was "striking": students who incorporated movement into their learning strategy remembered 76 percent of the material, while those who engaged in "deliberate memorization" recalled only 37 percent.

The implications of the Noices' research are clear. First: information is better remembered when we're moving as we learn it. This is the case even when the movement is not a literal enactment of the meaning of the information to be recalled but simply a movement of the body, meaningfully related to the information and made at the same time the information is absorbed. Second: information that has become associated with a movement is better remembered when we can reproduce that same movement later, when we're calling it up from memory. This may be possible in some situations — for example, when giving a speech for which we have practiced accompanying gestures — but moving while learning is still beneficial even when those movements can't be replicated at the point of recall (during an exam, for instance).

Indeed, simply forming the *intention* to move in connection with a piece of information seems to tag that information with a mental marker of importance. Our natural egocentric bias leads us to preferentially attend to and remember that information that we have connected in some way to ourselves: my intention, my body, my movement. Concluded Helga and Tony Noice in one of their academic articles, "One might paraphrase Descartes and say, 'I move, therefore I remember.'"

Moving while learning can help us to remember information more accurately. It may also be the case that moving while learning can help us understand information differently: more deeply, "from the inside," as it were. A stray comment from an undergraduate working in her lab got psychologist Sian Beilock—then an assistant professor at Miami University in Oxford, Ohio—wondering about the body's role in the act of understanding. The student played on the university's hockey team, and he mentioned to his professor that when watching hockey on television, he seemed to understand the action as it unfolded in a way that was different from that of his friends who had not been out on the ice.

Beilock and her colleagues designed a study to test his impression. First, experimenters read action sequences drawn from hockey games ("The hockey player shot the puck") and from everyday life ("The child saw the balloon in the air") to two groups of study participants; one group was made up of experienced hockey players, while the other group had never played the sport. Then participants were shown pictures that either did or did not correspond to the action sequences they'd heard (for instance, an image depicting a child seeing a balloon in the air, or a child seeing a deflated balloon on the ground).

For each sentence-picture pair, all the participants were able to correctly identify whether the two matched or not—but when the action concerned a hockey game, the hockey players were much quicker than their non-playing counterparts to make the identification; they exhibited what Beilock calls "facilitated comprehension." Brain scans conducted on both groups showed that a particular neural region was activated more strongly in the hockey players' brains than in the non-players' brains when they listened to hockey-specific language: the left dorsal premotor cortex, responsible for executing well-practiced physical movements. This area isn't typically associated with language processing, but the players' personal history with the game gave them a body-based experience that they could connect to the words they heard. The surprising import of Beilock's study is that people who have *moved* in different ways go on to *think* in different ways — an insight that can be applied well beyond sports.

The research on using movement to enhance thinking identifies four types of helpful motion: congruent movements, novel movements, self-referential movements, and metaphorical movements. The first of these, congruent movements, express in physical form the content of a thought. With the motions of our bodies, we enact the meaning of a fact or concept. Congruent movements are an effective way to reinforce still tentative or emerging knowledge by introducing a corporeal component into the process of understanding and remembering. A familiar example is moving the body along a number line: children who are learning about math benefit from taking steps on an oversized number line placed on the floor as they count or as they carry out procedures like addition and subtraction. Moving their bodies up or down the number line is congruent with the mental operation of counting up or down; taking small steps is congruent with the mental operation of counting one unit at a time, while venturing a bigger leap is congruent with the mental operation of adding or subtracting a number of units at once. Students who practice connecting numbers with movements in this way later demonstrate more mathematics knowledge and skill.

Moving the body in ways congruent with thinking is beneficial in part because it helps students make the tricky transition from the concrete to the abstract. This is the challenge facing children when they first learn to read: they must forge connections between the solid stuff of the world and the abstract symbols we use to represent it. In everyday life, children typically encounter the word "ball" or "cup" when there's an actual ball or cup around, notes Arthur Glenberg, a professor of psychology at Arizona State University. But within the pages of a book, words must be understood in the absence of such real-world referents. Glenberg uses congruent movement to bridge this gap. His "Moved by Reading" intervention teaches children how to simulate (with concrete, physical action) the text they are reading (abstract symbols). Such simulation leads to large gains in learning; when children act out the words on the page, Glenberg has found, their reading comprehension can actually double.

In one such study, Glenberg asked first- and second-graders to read stories about life on a farm. The children were also provided with farm-related toys, such as a miniature barn, tractor, and cow. Half of the kids were directed simply to read the stories a second time. The other half were instructed to use the toys to enact what they were reading. After reading the sentence "The farmer drove the tractor to the barn," for example, the child would move the toy tractor over to the toy barn. Youngsters who acted out the sentences were better able to make inferences about the text, and they later remembered much more about the stories than their peers who merely reread them.

Other studies have shown that congruent movements of this type can help children with math as well. In another of Glenberg's experiments, elementary school students were asked to act out a zookeeper's distribution of food to his animals while figuring out how many fish went to each of the hippos and alligators. Glenberg reports that the students who moved in ways congruent with text of word problems were more accurate in their calculations, and more likely to reach the right answer, than children who completed the problem in their heads. It seems that enacting the "story" told within the math problem helps students identify the information important for its solution: enacting made them 35 percent less likely to be distracted by irrelevant numbers or other details included in the problem.

Technology—which so often seems designed to keep us sitting still in our chairs, eyes glued to our screens—could extend itself with movement by incorporating congruent motions into the way it operates. And indeed, research using touch screen devices shows that digital educational programs that encourage users to make hand movements that are congruent with the mental operation being taught support the successful learning of those operations. A program offering instruction in number line estimation, for example—a task that depends on understanding numerical magnitude as continuous and not discrete—achieves better results when the movement required to interact with the program is a continuous one (dragging a finger across the screen) as opposed to a discrete one (tapping the screen once).

Another kind of physical action capable of advancing our thinking is *novel* movements: movements that introduce us to an abstract concept via a bodily experience we haven't had before. Consider: When you step into your shower at home, how do you turn on the hot water? To answer that simple question, you simulated the familiar, well-practiced action in your head; maybe you even reached out and turned an imagined faucet handle. But how would you engage in thinking about an action you've never physically experienced? That's the dilemma facing physics students, who are expected to reason about phenomena like angular velocity and centripetal force without a felt sense of what they're like. Decades of research on physics education reveal the discouraging result: most students never achieve a firm grasp of the subject. Some studies have found that students' understanding of physics becomes *less* accurate after they have completed an introductory college physics course.

The conventional, and widely ineffective, approach to teaching physics is based on a brainbound model of cognition: individuals are expected, like computers, to solve problems by applying a set of abstract rules. Yet the fact is that—very *unlike* computers—humans solve problems most effectively by imagining themselves into a given scenario, a project that is made easier if the human in question has had a previous physical encounter on which to base her mental projections. Providing students with such physical encounters was the purpose behind a study designed by Sian Beilock, inspired by the work with hockey players we read about earlier.

In collaboration with Susan Fischer, an associate professor of physics at DePaul University in Chicago, Beilock designed a set of hands-on activities intended to introduce students to the forces they were studying in physics class—not as abstract concepts but as visceral experiences. One such activity, for example, employed a prop: two bicycle wheels mounted on a single axle, which could be held out in front of the body while the wheels were spun. When the axle was tilted from horizontal to vertical, the person holding it felt firsthand what physicists call torque—the resistive force that causes objects to rotate. Beilock and Fischer asked one group of undergraduates to hold the contraption in their hands and to experience what it felt like to tilt the axle; a second group of students simply watched while someone else demonstrated its use. Afterwards, members of both groups were tested on their understanding of the concept of torque.

Students who'd experienced torque with their own bodies, the experimenters found, achieved higher scores on the assessment. Their superior understanding was especially evident in their answers to the most challenging theoretical questions. What's more, brain scans showed that when they were asked to think about torque, the region of the brain that controls movement was activated only in those who'd had a direct physical encounter with the force. Even while lying immobile inside an fMRI machine — or while sitting still, taking an exam — these students were able to access a bodily experience of motion, access that gave them a deeper and more accurate understanding of the concept.

An implication that leaps out from this research: When demonstrations are incorporated into science class, students should not be relegated to the role of observer. Only those who physically participate will gain the deeper, from-the-inside understanding that comes from physical action. As education professor Dor Abrahamson puts it, "Learning is moving in new ways."

YET ANOTHER TYPE of motion with the capacity to improve the way we think is *self-referential movements:* movements in which we bring ourselves — in particular, our bodies — into the intellectual enterprise. Though it may seem "unscientific" to place oneself at the center of the action, scientists themselves frequently use their bodies as instruments of exploration, imagining themselves as the object of their investigation. In so doing, they cultivate a kind of "empathy with entities they are struggling to understand," notes Elinor Ochs, an anthropologist who has studied theoretical physicists at work in the laboratory. The world's most famous physicist, Albert Einstein, reportedly imagined himself riding on a beam of light while developing his theory of relativity. "No scientist thinks in equations," Einstein once claimed. Rather, he remarked, the elements of his own thought were "visual" and even "muscular" in nature.

Other scientists have described the imagined acts of embodiment that helped them make their discoveries. Geneticist Barbara McClintock, whose work on the chromosomes of corn plants earned her a Nobel Prize, recalled how it felt for her to examine the chromosomes through a microscope: "When I was really working with them I wasn't outside, I was down there. I was part of the system. I was right down there with them, and everything got big. I even was able to see the internal parts of the chromosomes — actually everything was there. It surprised me because I actually felt as if I were right down there and these were my friends." Virologist Jonas Salk, inventor of the polio vaccine, is another scientist who brought his body into his research. He once described how he went about his work in this way: "I would picture myself as a virus, or a cancer cell, for example, and try to sense what it would be like to be either. I would also imagine myself as the immune system, and I would try to reconstruct what I would do as an immune system engaged in combating a virus or cancer cell. When I had played through a series of such scenarios on a particular problem and had acquired new insights, I would design laboratory experiments accordingly."

While students are often encouraged to adopt a detached, objective perspective on science, research shows that they can benefit from engaging the "embodied imagination" — just as scientists do. Thinking and learning with our bodies takes advantage of humans' fundamentally egocentric mindset. We've evolved to understand events and ideas in terms of *how they relate to us,* not from some neutral or impartial perspective. Research has found that the act of self-reference — connecting new knowledge to our own identity or experience — functions as a kind of "integrative glue," imparting a stickiness that the same information lacks when it is encountered as separate and unrelated to the self. Adopting a first-person perspective doesn't mean we become limited by it; indeed, using the movements of our own bodies to explore a given phenomenon seems to promote the ability to alternate between viewing it from an internal perspective and from an external one, an oscillation that produces a deeper level of understanding.

Rachel Scherr, an assistant professor of physics at the University of Washington, has devised an educational role-playing program called "Energy Theater." One attribute of energy that students find difficult to grasp, Scherr notes, is that energy is always *conserved* — it doesn't get "used up," but instead is converted to a different form, as when the energy in the coiled spring of a pinball plunger is converted into the energy of the pinball's motion. Students may read about the conservation of energy in a textbook without truly grasping its implications; when, as part of Energy Theater, they *embody* energy, they begin to understand such implications in a visceral way. "Students who use movement to 'become' energy can fall back on the feeling of permanence and continuity conveyed by their own bodies," says Scherr. "*They* don't get 'used up,' and so they're better able to understand that energy doesn't, either." Scherr's research shows that students who have taken part in Energy Theater develop a more nuanced understanding of energy dynamics.

Another opportunity to recruit self-referential movement in the service of learning arises when students are endeavoring to understand the nature of complex, multistep, interactive processes — such as that bane of biology class, mitosis and meiosis. The many phases and processes entailed in the study of how cells divide and reproduce can easily overwhelm students' mental bandwidth, leading to superficial comprehension at best and utter confusion at worst. Having watched many of his undergraduates struggle to understand these centrally important concepts, Joseph Chinnici, an associate professor of biology at Virginia Commonwealth University, had an idea: Why not ask students to "become" human chromosomes — to understand cell division and reproduction from the inside, as it were, by acting out these processes with their own bodies?

After fine-tuning his approach over several years, Chinnici published an account of his method in the journal *The American Biology Teacher.* He begins by distributing baseball caps and t-shirts, each of which has been marked with a letter representing a gene: uppercase letters for dominant genes, lowercase letters denoting recessive genes. Once they have donned these items of clothing, students are guided through a kind of carefully choreographed waltz. In prophase, some of the "human chromosomes" pair up by linking arms. In metaphase, those chromosomes who have remained unpaired move to an area designated as the "spindle." On to anaphase, during which the paired-up students split apart and move to opposite poles of the spindle. Finally, they act out telophase, in which the spindle dissolves and the chromosomes unwind. Amid some awkward laughter and some momentarily furrowed brows, students find their way through this odd dance — seeing, and feeling, for themselves how the many moving parts interact.

Chinnici's study found that students who had engaged in role-playing mitosis and meiosis achieved a more accurate understanding of the concept — a result mirrored in other, similar studies. Researchers have examined the effects of having students embody the solar system's planets as they engage in direct and retrograde motion; of having students embody carbon molecules as they undergo the enzymatic reactions of the Krebs cycle; and of having students embody amino acids undergoing polymerization as they are synthesized into proteins. In each of these scenarios, students learned more and performed better when they were offered the opportunity to embody these entities rather than simply reading or hearing about them.

"Being it" — embodying a conceptual object — is a very different experience from "watching it," or viewing a conceptual object as "remote and separate from oneself," notes Carmen Petrick Smith of the University of Vermont, who has studied the effects of physically embodying mathematical concepts. Groups of students might form a triangle with their outstretched arms, for example, and then experiment with moving closer to and farther away from one another; in this way they come to understand that the *size* of a triangle can vary without changing the degree of the angles at its corners. Smith notes that such "body-based activities" have been shown to deepen students' understanding and strengthen their memory of mathematical concepts. Mathematics teachers have long incorporated manipulatives into their instruction — counting rods and cubes, for example. The research of Smith and others suggests that students learn even more when the "manipulatives" they employ are their own bodies.

ONE FINAL CATEGORY of thought-enhancing movements encompasses those that enact an analogy, whether explicit or implicit. The language we use is full of metaphors that borrow from our experience as embodied creatures; *metaphorical movements* reverse-engineer this process, putting the body through the motions as a way of prodding the mind into the state the metaphor describes. "Moving the body can alter the mind by unconsciously putting ideas in our head before we are able to consciously contemplate them on our own," as Sian Beilock has written. "Getting a person to move lowers his threshold for experiencing thoughts that share something in common with the movement."

To take one example: by moving our bodies, we activate a deeply ingrained and mostly unconscious metaphor connecting dynamic *motion* with dynamic *thinking.* Call to mind the words we use when we can't seem to muster an original idea — we're "stuck," "in a rut" — and those we reach for when we feel visited by the muse. Then we're "on a roll," our thoughts are "flowing." Research has demonstrated that people can be placed in a creative state of mind by physically acting out creativity-related figures of speech — like "thinking outside the box." Psychologist Evan Polman of the University of Wisconsin–Madison designed an experiment in which participants were asked to complete a creative thinking task. Some students carried out the assignment while sitting inside a five-foot-square cardboard box; others completed the task while sitting next to the box. The participants who did their thinking literally "outside the box" came up with a list of creative solutions that was, on average, 20 percent longer than the list produced by those who brainstormed inside the box.

Polman and his colleagues also tested the generative effect of enacting another metaphor: the use of the phrase "on one hand . . . on the other hand" to convey the consideration of multiple possibilities. This time, participants were asked to come up with novel uses for a new campus building complex; half of them were asked to hold one hand outstretched as they engaged in brainstorming, while the others were instructed to alternate holding out one hand and then the other. The study subjects who (unwittingly) acted out the metaphor "on the one hand . . . on the other hand" generated nearly 50 percent more potential uses for the building, and independent judges rated their ideas as more varied and more creative.

Such experiments suggest we can activate a particular cognitive process by embodying the metaphor that has come to be associated with it. Simply moving the body through space is itself a loose kind of metaphor for creativity—for new angles and unexpected vistas, for fluid thinking and dynamic change. The activation of this metaphor may help account for the finding that people are more creative during and after walking than when they are sitting still.

Daniel Schwartz, dean of the Stanford Graduate School of Education, often urges his doctoral students to walk with him as they brainstorm about their dissertations, rather than remaining seated in his office. In 2014 one of these students, Marily Oppezzo (now an instructor in medicine at the Stanford Prevention Research Center) decided to investigate empirically the effects of walking on creativity. In a series of experiments, Schwartz and Oppezzo administered several different tests of original thinking to groups of Stanford undergraduates, Stanford employees, and students from a nearby community college. Some students were asked to complete the tasks while taking a stroll through campus or while walking on a treadmill; others took the test while seated in a classroom.

For the first test, participants were asked to generate unexpected uses for ordinary objects, such as a brick or a paper clip. On average, students came up with four to six more uses for the items if they were walking rather than sitting down. Another test presented participants with an evocative image, such as "a light bulb blowing out," and asked them to come up with an analogous image (like, for example, "a nuclear reactor melting down"). Ninety-five percent of students who walked were able to do so, compared to only 50 percent of those who remained immobile. "Walking opens up the free flow of ideas," the authors conclude. Studies by other researchers have even suggested that following a meandering, free-form route—as opposed to a fixed and rigid one—may further enhance creative thought processes.

Although contemporary culture prescribes sitting still while thinking, a stroll through the history of literature and philosophy finds ample evidence of a counter-message. Remember Friedrich Nietzsche, from earlier in our journey. "Only thoughts which come from walking have any value," he maintained. Søren Kierkegaard felt similarly. "I have walked myself into my best thoughts," remarked the Danish philosopher. Walking is "gymnastics for the mind," observed the American writer Ralph Waldo Emerson. "I am unable to reflect when I am not walking; the moment I stop, I think no more, and as soon as I am again in motion, my head resumes its workings," averred the Swiss-born philosopher Jean-Jacques Rousseau. The French philosopher and essayist Michel de Montaigne lamented that his thoughts often came to him when he was on the move, at moments when "I have nothing to jot them down on"; this was wont to happen "especially on my horse, the seat of my widest musings."

These great thinkers were clearly on to something. We ought to be finding ways to integrate movement into all our daily activities, to tap into the mobile intelligence of "the mind at three miles per hour," as the contemporary writer Rebecca Solnit has called the mental state induced by walking. This could mean walking on a treadmill as we type at our computers, walking as we talk on the phone, walking as we conduct work meetings—even walking as we attend class.

Thinking while walking would seem to be a natural fit for the world of academe. A few years ago, philosophy professor Douglas Anderson of the University of North Texas got to wondering why he and his students stayed put in a lecture hall while the texts they studied so often extolled the merits of movement. He began teaching one of his courses, "Philosophy of Self-Cultivation," while on the move: professor and students walk about the campus as they discuss the week's assigned reading. Anderson says he has noticed a difference in his students as soon as they leave the room where the class initially assembles: their voices and expressions become more animated, they have more to say, their minds seem to work at a faster clip.

Included on Anderson's syllabus is, of course, "Walking," the essay by philosopher and naturalist Henry David Thoreau first delivered at the Concord Lyceum in 1851. "I think that I cannot preserve my health and spirits, unless I spend four hours a day at least — and it is commonly more than that — sauntering through the woods and over the hills and fields," he declared. That same year, Thoreau expanded on the theme in his journal. "How vain it is to sit down to write when you have not stood up to live!" he exclaimed. "Methinks that the moment my legs begin to move, my thoughts begin to flow."

3

.

Thinking with Gesture

G ABRIEL HERCULE bounded onto the stage of Startupboot-
camp's 2018 Demo Day wearing a slim gray suit, white shirt,
and red tie. Before he said a word, his confidence announced
itself in his smooth stride and fluid gestures.

"Two years ago I was hit by a van because the driver's eyes weren't on
the road," he began. Hercule widened his eyes and held out his hands,
palms up, as if to say, *Can you believe it?* "Fortunately, I made it out with
only minor scratches, but this left me with a very real sense that something needed to be done about automotive safety — especially in the commercial vehicle sector, where drivers face *huge* time pressures." He spread
his arms wide to emphasize the word "huge."

"One of the key lessons we learned" — the fingers of his right hand
formed a precise pincer gesture — "in the commercial automotive sector
is that the fleet manager is responsible for ensuring that *every delivery*" —
the words were punctuated by two jabs at the air — "is made at the right
place at the right time, a task that's much easier said than done."

Fleet managers, Hercule continued, are nervous: nervous about their
drivers getting lost, nervous about road accidents, nervous about pack-
ages arriving late. "This is why fleet managers wish they could be *in* the
vehicle with their drivers" — he made a downward tucking motion, emphasizing the word "in" — "to teach them better habits. And now they
can." Hercule paused for dramatic effect.

"I'd like to introduce the Atlas One, the very first head-up display that
sends driving information directly to the driver through a projected ho-

logram, with three key features to enhance driving behavior. All this, while keeping the driver's eyes on the road." Hercule's hands were in constant motion as he got to the heart of his pitch: they framed his gaze like a camera's viewfinder, then motioned toward himself as if information was flowing into his field of vision, then pointed from his eyes toward the imagined road before him.

In fact, what was before him was an audience of several hundred potential customers, partners, and investors, listening carefully to every word. What these attendees likely didn't realize is that they were being influenced at least as much by Hercule's gestures as by his speech. Researchers who study embodied cognition are drawing new attention to the fact that people formulate and convey their thoughts not only with words but also with the motions of the hands and the rest of the body. Gestures don't merely echo or amplify spoken language; they carry out cognitive and communicative functions that language can't touch. Where language is discrete and linear — one word following another — gesture is impressionistic and holistic, conveying an immediate sense of how things look and feel and move.

The special strengths of gesture are especially valuable in the effort to persuade or enlist others. Such movements visually place the gesturer at the center of the action, situating him at the locus of agency and control. When he talks, his words may describe or extol or explain — but when he gestures, he *acts* on the world (if only symbolically). At the same time, the gesturer's motions render an abstract idea in human-scale, embodied terms, an act of translation that makes it easier for onlookers to mentally simulate the gesturer's point of view for themselves. Perhaps most important, gesture generates the sense that an as yet immaterial enterprise is a palpable reality in the present moment. Using gesture in this way can confer an enormous advantage in the start-up world, one group of researchers notes, since "entrepreneurs are operating on the boundary of what is real and what is yet to happen." That's true for many of us, whether we're offering projections for the next quarter, presenting a proposal for a project, or explaining why a change we'd like to make would be well advised. Gesture brings an uncertain future into the observable present, imbues it with a realness that we can almost touch.

Jean Clarke, a professor of entrepreneurship and organization at Emlyon Business School in France, has spent years watching entrepreneurs like Gabriel Hercule make their case at demo days, incubators, and investment forums across Europe. In a study published in 2019, she and her colleagues reported that company founders who deployed "the skilled use of gesture" in their pitches were 12 percent more likely to attract funding for their new ventures. Such adept use of movement includes the presentation of "symbolic gestures" — movements that capture the overall meaning of the speaker's message — along with what are called "beat gestures": hand motions that serve to punctuate a particular point. When Hercule repeatedly pointed to his eyes and then to the view before him, he was making a symbolic gesture (*eyes on the road*); when he formed his fingers into a pincer, or jabbed at the air with his fist, he was emphasizing his contentions with beat gestures. Skilled gesturers don't leave this important element of their delivery to chance, Clarke notes; they practice the movements they intend to make just as they rehearse the words they plan to say.

Adding persuasive force to a presentation is just one role gesture can play in shaping the thinking we do, as both gesture makers and gesture observers. Research demonstrates that gesture can enhance our memory by reinforcing the spoken word with visual and motor cues. It can free up our mental resources by "offloading" information onto our hands. And it can help us understand and express abstract ideas — especially those, such as spatial or relational concepts, that are inadequately expressed by words alone. Moving our hands helps our heads to think more intelligently, and yet gesture is often scorned as hapless "hand waving," or disparaged as showy or gauche.

This is an attitude that Frederic Mishkin, an economist at Columbia Business School, knows well. Whether he's standing in front of a lecture hall or engaged in a casual conversation, Mishkin's hands are continually in motion, an emphatic complement to his speech. "I talk with my hands," he says. "I always have." Early in his career, however, one of his mentors became exasperated by his constant gesticulating. Seeking to break his protégé of the habit, he declared a rule for when Mishkin visited his office. "He made me sit on my hands as I talked with him," Mishkin recalls ruefully.

Such disdain for gesture is a cultural constraint at odds with the way humans naturally communicate. Indeed, linguists theorize that gesture was humankind's earliest language, flourishing long before the first word was spoken. Even now, gesture provides an alternate channel of communication every bit as significant as the verbal one. Gestures exert a powerful impact on how we understand and remember our interactions with others, but its influence operates largely below our awareness. We may choose our own words carefully, and listen closely to what others say, yet still fail to notice a substantial proportion of the communication that is actually occurring. A profusion of "extra-verbal meaning" is continually being offered and received.

At times, gesture works with language to more richly specify the speaker's meaning — to clarify or emphasize what is being said. On other occasions, gesture supplies meaning that is not found anywhere in the speaker's words. And at still other moments, gesture asserts meaning that contradicts or departs entirely from the speaker's verbal self-expression. Gesture conveys things we don't say; as we'll discover, gesture even conveys things we *can't* say — because we don't yet have the words.

All of us, then, are effectively bilingual: we speak one or more languages, but we are also fully fluent in gesture. Over the course of our species's evolutionary history, gesture was not superseded or replaced by spoken language — rather, it maintained its place as talk's ever present partner, one that is actually a step or two *ahead* of speech. Christian Heath, a professor of work and organization at King's College London, uses the close analysis of videotaped conversations to examine the dynamic interplay between physical movement and verbal expression. One doctor-patient dialogue recorded by Heath demonstrates in fine-grained detail how people often gesture first and speak second.

In Heath's video, the doctor on camera is saying of a particular class of medicines he is prescribing that "they help, sort of, you know, to dampen down inflammation" — but by the time he says "you know," he has already completed three downstrokes of his hand. For her part, the patient makes reference to the stress of her financial troubles and how she goes "round and round in circles" trying to catch up with her bills — but before she utters these words, her hands have already started moving in a circular pattern. In each exchange, gesture provides a preview of the concept that will be conveyed in words — and in both cases, the listener shows that he or she understands the sentiment (by nodding and murmuring) at the moment when he or she perceives the gesture, *before* the spoken part of the sentiment is uttered. From watching Heath's tapes, it's easy to conclude that most of our conversations are carried on with our hands, the words we speak a mere afterthought.

Research shows that we all engage in such "gestural foreshadowing," in which our hands anticipate what we're about to say. When we realize we've said something in error and we pause to go back to correct it, for example, we stop gesturing a couple of hundred milliseconds before we stop speaking. Such sequences suggest the startling notion that our hands "know" what we're going to say before our conscious minds do, and in fact this is often the case. Gesture can mentally prime a word so that the right term comes to our lips. When people are prevented from gesturing, they talk less fluently; their speech becomes halting because their hands are no longer able to supply them with the next word, and the next. Not being able to gesture has other deleterious effects: without gesture to help our mental processes along, we remember less useful information, we solve problems less well, and we are less able to explain our thinking. Far from tagging along as speech's clumsy companion, gesture represents the leading edge of our thought.

GESTURE WAS HUMANITY'S first language — and every human infant recapitulates this evolutionary history, becoming fluent in gesture in advance of acquiring even the rudiments of speech. Well before babies can talk, they are waving, beckoning, holding up their arms in a wordless signal: *Pick me up.* Pointing is one of children's first gestures, usually initiated around nine months of age; between ten and fourteen months, a more nuanced capacity for gesture begins to develop as fine-motor finger control improves. During this time, spoken language lags well behind what toddlers are able to express by moving their hands. Children can typically understand and act on a request to point to their nose, for example, a full six months before they are able to form the spoken word "nose." And, research suggests, they use their gestures to elicit from their caretakers the very words they need to hear as they learn about the world. A child will point to an unfamiliar object, for example, and an adult will often obligingly supply the name of that thing. When a parent "translates" her child's gesture into a word in this way, that word is particularly likely to enter the child's spoken vocabulary within a few months. Amazingly enough, as one researcher puts it, "Young children use their hands to tell their mothers what to say."

These early experiences with gesture lay the foundation for spoken language. Gesture constitutes a first attempt at the trick of making one thing (a movement of the body, the sound of a word) stand for another (a physical object, a social act). Connecting a wish — say, to get down from a high chair — with the spoken word "down" is a sophisticated mental move; performing a downward motion of the hand can function as an important intermediate step. Indeed, researchers have documented a link between a child's rate of gesturing at fourteen months and the size of that same child's vocabulary at four and a half years of age. Children learn to make these movements from the gesturing figures around them: adults. Studies show that children whose parents gesture a lot proceed to gesture frequently themselves, and eventually to acquire expansive spoken-word vocabularies.

Child development experts have long emphasized the importance of *talking* to children; an often cited 1995 study carried out by psychologists Betty Hart and Todd Risley estimated a 30-million "word gap" in the number of words heard spoken aloud by affluent and poor children by the time they start school. Since the publication of the Hart-Risley study, other research has confirmed that higher-income parents tend to talk more than lower-income parents, that they employ a greater diversity of word types, that they compose more complex and more varied sentences — and that these differences are predictive of child vocabulary. Now researchers are generating evidence that how parents *gesture* with their children matters as well — and that socioeconomic differences in how often parents use their hands when talking to children may be producing what we could call a "gesture gap."

High-income parents gesture more than low-income parents, research finds. And it's not just the *quantity* of gesture that differs but also the *quality:* more affluent parents provide a greater variety of types of gesture, representing more categories of meaning — physical objects, abstract concepts, social signals. Parents and children from poorer backgrounds, meanwhile, tend to use a narrower range of gestures when they interact with each other. Following the example set by their parents, high-income kids gesture more than their low-income counterparts. In one study, fourteen-month-old children from high-income, well-educated families used gesture to convey an average of twenty-four different meanings during a ninety-minute observation session, while children from lower-income families conveyed only thirteen meanings. Four years later, when it was time to start school, children from the richer families scored an average of 117 on a measure of vocabulary comprehension, compared to 93 for children from the poorer families.

Differences in the way parents gesture may thus be a little-recognized driver of unequal educational outcomes. Less exposure to gesture leads to smaller vocabularies; small differences in vocabulary size can then grow into large ones over time, with some pupils arriving at kindergarten with a mental word bank that is several times as large as that possessed by their less fortunate peers. And vocabulary size at the start of schooling is, in turn, a strong predictor of how well children perform academically in kindergarten and throughout the rest of their school years.

The good news: research shows that offering simple instructions to parents leads them to gesture more often; in turn, their children also gesture more. Any parent can adopt the strategies suggested by these intervention programs: Engage in frequent pointing with young children, and encourage the kids themselves to point. Incorporate this same gesture into the reading of picture books; point to particular words or illustrations, and ask children to point to what they see. Come up with simple gestures to pair with real-life referents — a clawing motion for cat, a wiggling index finger for a caterpillar — and be sure to say the word aloud as the gesture is demonstrated. Perhaps the most important fact to keep in mind, says Harvard education professor Meredith Rowe, is that a child's language development is malleable, and that parents play an important role in shaping that development. In a 2019 study, Rowe delivered this message to a group of socioeconomically diverse parents and caregivers, along with reminders to gesture more. At the conclusion of the intervention, she found that adults who'd received the programming were engaging in pointing thirteen times more often, on average, than those who had not undergone the gesture training; the children in their lives were pointing significantly more too.

As children grow older, gesture continues to act as an advance party, scouting out mental territory well ahead of where their words are. Rather astonishingly, researchers have found that children's "newest and most advanced ideas" about how to understand a concept or solve a problem often show up first in their gestures. Take, for example, a six-year-old girl faced with a classic "conservation" task. (Such tasks were first employed by the pioneering psychologist Jean Piaget to investigate the course of childhood cognitive development.) The girl is shown a tall, skinny glass full of water, the contents of which are then poured into a short, wide glass. Asked whether the amount of water remains the same, the girl answers no—but at the same time, her hands are making a cupping motion, indicating that she's beginning to understand that the wider shape of the second glass accounts for the way the same amount of water fills it at a lower level than the first glass.

This episode is drawn from a voluminous video archive compiled by Susan Goldin-Meadow, a professor of psychology at the University of Chicago. Goldin-Meadow has collected thousands of such video vignettes, recordings of individuals as they use words and gestures to explain how they solve problems. Across these many scenes, she has identified an intriguing pattern: When speech and gesture are both correct and congruent, it's a given that the speaker has mastered the material. When speech and gestures match but both are wrong, we can assume that the speaker is still far from "getting it." But when there's a *mismatch* between speech and gesture—when a person says one thing but does something else with her hands—then that individual can be said to be in a "transitional state," moving from the incorrect notion she's expressing in words to the correct one she is expressing in gesture.

In videos recorded by Goldin-Meadow of children carrying out conservation tasks, new understandings emerged first in gesture some 40 percent of the time. Such mismatches appear to be a common occurrence across development: when ten-year-olds solve math problems, one study reported, their gestures represent strategies different from those found in their speech about 30 percent of the time. Another study found that for fifteen-year-olds working on a problem-solving task, the rate of speech-gesture mismatch was 32 percent.

Furthermore, Goldin-Meadow has found, learners who produce such speech-gesture mismatches are especially receptive to instruction —ready to absorb and apply the correct knowledge, should a parent or teacher supply it. Even adults signal their readiness to learn through mismatches between what they're saying and how their hands are moving. In one experiment, for example, a group of college students was asked to learn about a set of "stereoisomers"—chemical compounds that feature the same number of atoms but that differ from one another in the way the atoms are spatially arranged. The extent to which the undergraduates made gesture-speech mismatches while learning "predicted their ability to profit from instruction," writes Goldin-Meadow, who was lead author of the study. "Namely, the more they expressed correct conceptual information in gesture that they did not verbalize in speech, the more they were subsequently able to learn." When our words and our movements diverge, it's our gestures that signal what happens next.

WHY WOULD OUR "most advanced ideas" appear in our gestures before surfacing in our speech? Researchers speculate that gesture helps give shape to an incipient notion still forming in our minds. At a moment when we cannot quite put words to a concept we're struggling to comprehend, we can still move our hands in a way that captures some aspect of our emerging understanding. We may then be able to use the experience of making and seeing our own gesture to help us locate the appropriate language. It's possible, too, that we feel freer to try out new ideas in movement before committing to them in speech; as Susan Goldin-Meadow puts it, "Gesture encourages experimentation."

The role played by gesture in consolidating our initially inchoate thoughts is revealed by the changes our hand motions undergo as we begin to master new material. At first we gesture profusely, and rather indiscriminately, as we attempt to wrap our heads around an unfamiliar idea. We gesture more when we are actively trying to apprehend or reason about a concept than when we are describing a concept we already understand. Gesturing also increases as a function of difficulty: the more challenging the problem, and the more options that exist for solving it, the more we gesture in response. Meanwhile, we engage in what one researcher calls "muddled talk" — speech that is a jumble of incompletely articulated notions. Our speech and our gesture are not yet coordinated or congruent; so cognitively demanding is the task of assimilating a new idea that we divide the work between our head and our hands, each going its own way for now.

The process may be messy, but it allows us to circle ever closer to the attainment of complex knowledge that would otherwise remain out of reach. While moving our hands around, we may find that our gestures summon insights of which we had previously been unaware; psychologist Barbara Tversky has likened gesturing to a "virtual diagram" we draw in the air, one we can use to stabilize and advance our emerging understanding. As our comprehension deepens, our language becomes more precise and our movements become more defined. Gestures are less frequent, and more coordinated in meaning and timing with the words we say. Our hand motions are now more oriented toward communicating with others and less about scaffolding our own thinking. Without gesture's initial assist, however, this happy state might never have been reached. Research shows that people who are asked to write on complex topics, instead of being allowed to talk and gesture about them, end up reasoning less astutely and drawing fewer inferences.

This same sequence unfolds not only for novices learning a subject for the first time, but also for experts venturing into uncharted territory. Gesture is especially useful in supporting the provisional understandings shared by team members who together are working their way toward new discoveries. This was evident in a study conducted by researchers at the University of California, San Diego. Amaya Becvar and two colleagues analyzed hours of videotape recorded at the lab meetings of a biochemistry research group on the UCSD campus. The scientists in this lab were studying the dynamics of blood clotting, with a particular focus on an enzyme called thrombin. Depending on which protein attaches to thrombin's "active site," blood clots are either formed or broken down. Determining how and why thrombin binds to one protein or another could guide the design of drugs to treat heart attacks and strokes — conditions that are caused by aberrant blood clots. The lab's scientists had a hunch that a key role was played by thrombomodulin, a protein that is a "binding partner" of thrombin.

At one of the lab's weekly meetings, a graduate student presented new research on thrombomodulin to two other graduate students and a professor, the group's research adviser. Responding to the new findings, the professor used her left hand to represent a thrombin molecule, curling her fingers into a claw and pointing to her palm. "That's the active site," she explained. "Our new theory is that thrombomodulin does something like this" — she curled each of her spread fingers more tightly. Or, she added, perhaps it has an effect "like this" — and she drew her fingers tightly together. Throughout the rest of the meeting, Becvar notes, the professor and the graduate students continually reproduced the "thrombin hand" gesture: pointing to it, talking at it, shifting the position of their fingers to represent various potential configurations. The scientists' creation of new knowledge was actively supported by the symbolic movements of their hands, Becvar concludes — a fact acknowledged by the title of her article, "Hands As Molecules."

Gestures are especially useful in the effort to understand concepts that words will always fail to capture fully — concepts that are visual and image-rich, that pertain to the relation between objects or ideas, or that concern entities beyond direct perception (those things tiny as an atom or vast as the solar system). Gestures are also particularly well suited to conveying *spatial* concepts. Professional geologists, for example, employ a range of specialized gestures to think and communicate about the way terrestrial layers bend, fold, and shift in space. When they refer to "subduction" — the sideways and downward movement of a plate of the earth's crust beneath another plate — they slide one hand across and under the other. When they invoke "angular unconformity" — rock strata that lie flat on top of other strata that are "beveled," or uneven — they hold one hand steady above the other hand which tilts side to side. Though these experts surely have the words to express their meaning, still they rely on their hands to do much of the work.

Students learning about geology for the first time can also benefit from using gesture. Kinnari Atit, an assistant professor of education at the University of California, Riverside, asked two groups of college students to explain how they would use Play-Doh to create three-dimensional models of geological features. The members of one group were permitted to use their hands to gesture; those in the second group were asked to employ only words. Before and after the exercise, both groups were tested on their ability to engage in what experts call "penetrative thinking." This is the capacity to visualize and reason about the interior of a three-dimensional object from what can be seen on its surface — a critical skill in geology, and one with which many students struggle. Study participants who gestured got significantly higher scores the second time they were tested on a measure of penetrative thinking, while participants who only explained verbally showed no improvement.

Such results suggest that the act of gesturing doesn't just help communicate spatial concepts to others; it also helps the gesturer herself understand the concepts more fully. Indeed, without gesture as an aid, students may fail to understand spatial ideas at all. "Strike and dip" is a basic geological concept that describes a rock layer's rotation from north (strike) and a rock layer's rotation from horizontal (dip). After reading a textbook-style introduction to the concept, "many college students made dramatic errors in recording an outcrop's strike and dip on a campus map," noted a team of researchers from Penn State. Better performance on this task, they found, was linked to greater use of gesture by the students.

For learners endeavoring to grasp spatial concepts, "language can actually get in the way," says Michele Cooke, a professor of geoscience at the University of Massachusetts Amherst. Cooke, who is hearing-impaired, has spent years leading outreach programs that engage deaf students in the study of geologic fault systems. She has noticed that these students are especially quick to master geological concepts and theories, a deftness she attributes to the skills of observation and spatial cognition they have developed as users of American Sign Language (ASL). People who are fluent in sign language, as Cooke is, have been found to have an enhanced ability to process visual and spatial information. Such superior performance is exhibited by hearing people who know sign language, as well as by the hearing impaired — suggesting that it is the repeated use of a structured system of meaning-bearing gestures that helps improve spatial thinking.

Cooke often employs a modified form of sign language with her (hearing) students at UMass. By using her hands, Cooke finds, she can accurately capture the three-dimensional nature of the phenomena she's explaining. She can efficiently direct her students' attention to particular features she wants to highlight. And she can divide the river of information she's presenting into two smaller streams, one verbal and one visual, thereby reducing her students' cognitive load — often heavy when novices are learning new ideas and vocabulary terms at the same time. Cooke asks the undergraduates enrolled in her courses to imitate her "ASL-based gestures" as they are introduced to new geological concepts, and encourages her students to use the hand motions when talking with one another in discussion groups. In Cooke's class, the single-minded focus on written and spoken language that characterizes so much university teaching and learning has been nudged aside, making room for the intuitive genius of gesture.

WE CAN DO the same, elevating gesture above its current status as a dismissed or disparaged adjunct to speech. A good place to start: making more gestures ourselves. Research shows that moving our hands advances our understanding of abstract or complex concepts, reduces our cognitive load, and improves our memory. Making gestures also helps us get our message across to others with more persuasive force. Studies demonstrate that hearing speech and seeing gesture at the same time evoke a stronger reaction from the brains of those listening and watching than does speech or movement on its own. Gesture works to amplify the impact of speech: the sight of a speaker making gestures effectively captures listeners' attention and directs it to the words being uttered. (The sight of someone making motions that are *not* gestures — moving while carrying out a functional act, such as stirring a cup of coffee with a spoon — does not have the same attention-grabbing effect.) One of the brain regions roused to attention by the sight of gesture is the auditory cortex, the part of the brain responsible for processing oral language. "Hand gestures appear to alert the auditory cortex that meaningful communication is occurring," says Spencer Kelly, a professor of psychology and neuroscience at Colgate University in New York.

When we make gestures as we explain a concept or tell a story, others better understand what we're saying; our hand movements clarify, specify, and elaborate on our speech in ways that aid our audience's comprehension. People are also more likely to remember what we've said when we deliver gestures along with our words. In one study, subjects who had watched a videotaped speech were 33 percent more likely to recall a point from the talk if it was accompanied by a gesture. This effect, detected immediately after the subjects viewed the recording, grew even more pronounced with the passage of time: thirty minutes after watching the speech, subjects were more than 50 percent more likely to remember the gesture-accompanied points.

It's just these benefits of observing gesture that should lead us to take a second step: seeking out educational resources, for ourselves and others, in which the instructor makes proficient use of physical movement. A number of studies have demonstrated that instructional videos that include gesture produce significantly more learning for the people who watch them: viewers direct their gaze more efficiently, pay more attention to essential information, and more readily transfer what they have learned to new situations. Videos that incorporate gesture seem to be especially helpful for those who begin with relatively little knowledge of the concept being covered; for all learners, the beneficial effect of gesture appears to be even stronger for video instruction than for live, in-person instruction.

Yet the most popular and widely viewed instructional videos available online largely fail to leverage the power of gesture, according to a team of psychologists from UCLA and California State University, Los Angeles. The researchers examined the top one hundred videos on YouTube devoted to explaining the concept of standard deviation, an important topic in the study of statistics. In 68 percent of these recordings, they report, the instructor's hands were not even visible. In the remaining videos, instructors mostly used their hands to point or to make emphatic "beat" gestures. They employed symbolic gestures — the type of gesture that is especially helpful in conveying abstract concepts — in fewer than 10 percent of the videos reviewed.

The takeaway: When selecting instructional videos for ourselves or for our children or our students, we should look for those in which the teacher's hands are visible and active. And if we ourselves are called upon to teach online — or even just to communicate via Zoom or another video-conferencing platform — we should make sure that others can see our moving hands. Research suggests that making these motions will improve our own performance: people who gesture as they teach on video, it's been found, speak more fluently and articulately, make fewer mistakes, and present information in a more logical and intelligible fashion.

Enacting gestures has another, more indirect benefit: when others (children, students, co-workers, employees) see us gesturing, they tend to make more hand motions themselves. But we need not wait for them to follow our example; we can explicitly encourage them to gesture. A simple request to "move your hands as you explain that" may be all it takes. For children in elementary school, for example, encouraging them to gesture as they work on math problems leads them to discover new problem-solving strategies — expressed first in their hand movements — and to learn more successfully the mathematical concept under study.

Another experiment, this one conducted with college students, found that those who were encouraged to gesture while solving spatial problems — such as rotating a mental object or visualizing the folding of a piece of paper — solved more of the problems correctly than students who were prohibited from gesturing, or even students who were allowed (but not encouraged) to move their hands. The gesturing students' improved capacity for spatial thinking persisted into the solving of a second round of spatial problems during which they were not permitted to gesture. The spatial skills supported by their initial gestures had become "internalized," the researchers suggest, and the internalized effects of gesturing on the students' thinking carried over to a new and different set of spatial problems. Even adults, when asked to gesture more, respond by increasing their rate of gesture production (and consequently speak more fluently); when teachers are told about the importance of gesturing to student learning, and encouraged to make more gestures during instruction, their students make greater learning gains as a result.

Encouraging others to gesture may have surprisingly powerful effects —like helping to close achievement gaps. The disparity in spatial-thinking skills between males and females is the largest known cognitive gender difference. A study led by psychologists at the University of Chicago found that five-year-old boys were already better than girls the same age at solving spatial-thinking problems that involved mentally fitting shapes together to make a whole. Upon closer analysis, however, this disparity was revealed to be not a *gender* difference so much as a difference in the propensity to *gesture:* the more children gestured while executing the task, the better their performance — and boys tended to gesture far more than girls. While 27 percent of boys gestured on all eight problems, for example, only 3 percent of girls did; 23 percent of girls did not gesture at all, compared to only 6 percent of boys.

The study's authors suggest that this discrepancy may emerge from differences in boys' and girls' experience: boys are more likely to play with spatially oriented toys and video games, they note, and may become more comfortable making spatial gestures as a result. Another study, this one conducted with four-year-olds, reported that children who were encouraged to gesture got better at rotating mental objects, another task that draws heavily on spatial-thinking skills. Girls in this experiment were especially likely to benefit from being prompted to gesture.

Another, more subtle approach to promoting gesture involves creating *occasions* for its use — setting up scenarios in which people are likely to move their hands. One such circumstance arises when people are asked to improvise: that is, to come up with an explanation or a narrative on the spot, in front of an audience. Improvisation is cognitively taxing, and in the face of its demands we tend to gesture more.

Wolff-Michael Roth is a cognitive scientist at the University of Victoria in Canada. His research on the role of gesture in the development of scientific literacy has led him to change the way he conducts his courses as a professor. Rather than presenting lectures in which he does most of the talking, Roth finds as many opportunities as he can to ask individual students to describe and explain the topics being covered in that day's class. Lacking a fully developed understanding, or even the relevant technical vocabulary, his students lean heavily on gesture to convey their budding knowledge — and this is just what their professor wants to see. "It is from the attempt of expressing themselves that understanding evolves, rather than the other way around," he maintains.

Roth is also a practitioner of another kind of occasion creation: he has observed, and research has confirmed, that people are more likely to gesture when they have something to gesture *at*. Providing what Roth calls "visual artifacts" — charts, diagrams, maps, models, photographs — induces speakers to gesture more, thus generating all the benefits for understanding that such hand motions confer. With his colleagues in the University of Victoria physics department, he has developed a set of visual depictions and concrete models that the physics professors now use to encourage gesture production in class. When standing next to such objects, students can simply point to parts or processes that they can't yet fully describe or explain, allowing them to engage in "more mature physics talk" than would otherwise be possible at this early stage of their studies. The use of gesture supplies a temporary scaffold that supports these undergraduates' still wobbly understanding of the subject as they fix their knowledge more firmly in place.

One more way to leverage the power of gesture: we can pay closer attention to the manner in which others move their hands. As we've discovered, people's newest and most advanced ideas often show up first in their gestures; moreover, individuals signal their readiness to learn when their gestures begin to diverge from their speech. In our single-minded focus on spoken language, however, we may miss the clues conveyed in this other mode. Research finds that even experienced teachers pick up on less than a third of the information contained in students' hand movements. But studies also demonstrate that we can train ourselves to attend more closely to gesture's corporeal code.

In a study carried out by Susan Goldin-Meadow and colleagues at the University of Chicago, a group of adults was recruited to watch video recordings of children solving conservation problems, like the water-pouring task we encountered earlier. They were then offered some basic information about gesture: that gestures often convey important information not found in speech, and that they could attend not only to what people say with their words but also to what they "say" with their hands. It was suggested that they could pay particular attention to the *shape* of a hand gesture, to the *motion* of a hand gesture, and to the *placement* of a hand gesture. After receiving these simple instructions, study subjects watched the videos once more. Before the brief gesture training, the observing adults identified only around 30 to 40 percent of instances when children displayed emerging knowledge in their gestures; after receiving the training, their hit rate shot up to about 70 percent.

With a little effort, it's possible to glean the information gesture holds, and once we do so, we have a host of new options. We can supply the insight the gesturer is reaching for — an insight for which, research suggests, she is already mentally primed; we can "translate" that individual's gesture into words ("It looks like you're suggesting that . . ."); and we can "second" her gesture by reproducing it ourselves, thereby reaffirming the promising strategy she has pointed to with her hands.

It's clear that *spontaneous* gestures can support intelligent thinking. There's also a place for what we might call *designed* gestures: that is, motions that are carefully formulated in advance to convey a particular notion. Geologist Michele Cooke's gestures, inspired by sign language, fall into this category; she very deliberately uses hand movements to help students understand spatial concepts that are difficult to communicate in words.

Designed gestures offer another benefit as well: they are especially effective at reinforcing our memory. That's because gesturing while speaking involves sinking multiple mental "hooks" into the material to be remembered — hooks that enable us to reel in that piece of information when it is needed later on. There is the auditory hook: we hear ourselves saying the words aloud. There is the visual hook: we see ourselves making the relevant gesture. And there is the "proprioceptive" hook; this comes from *feeling* our hands make the gesture. (Proprioception is the sense that allows us to know where our body parts are positioned in space.) Surprisingly, this proprioceptive cue may be the most powerful of the three: research shows that making gestures enhances our ability to think even when our gesturing hands are hidden from our view.

Kerry Ann Dickson, an associate professor of anatomy and cell biology at Victoria University in Australia, makes use of all three of these hooks when she teaches. Instead of memorizing dry lists of body parts and systems, her students practice pretending to cry (the gesture that corresponds to the lacrimal gland/tear production), placing their hands behind their ears (cochlea/hearing), and swaying their bodies (vestibular system/balance). They feign the act of chewing (mandibular muscles/mastication), as well as spitting (salivary glands/saliva production). They act as if they were inserting a contact lens, as if they were picking their nose, and as if they were engaging in "tongue-kissing" (motions that represent the mucous membranes of the eye, nose, and mouth, respectively). Dickson reports that students' test scores in anatomy are 42 percent higher when they are taught with gestures than when taught the terms on their own.

The acquisition of vocabulary is also central to learning a foreign language — and here, too, designed gestures can act as an aid to memory, says cognitive psychologist and linguist Manuela Macedonia. Today, Macedonia researches second-language acquisition as a senior scientist at Johannes Kepler University in Austria. But earlier in her career she was a language instructor, teaching Italian to German-speaking college students. Back then, Macedonia found herself increasingly frustrated with the conventional format of foreign-language courses: a lot of sitting, listening, and writing. That's not how anyone learns their native language, she notes. Young children encounter new words in a rich sensorimotor context: as they hear the word "apple," they see and touch the shiny red fruit; they may even bring it to their mouth, tasting its sweet flesh and smelling its crisp scent. All of these many hooks for memory are missing from the second-language classroom.

Macedonia sought to recover at least one: physical movement. She began pairing each vocabulary word with a corresponding gesture; after demonstrating the gesture for her students, she would ask them to perform the motion themselves while speaking the word aloud. Her students learned new words more easily this way, she found, and retained them more successfully over time. Macedonia eventually became a student again herself, writing her PhD thesis on the use of gestures to enhance verbal memory during foreign-language encoding. In the years since, she has continued to contribute to a growing body of evidence showing that enacting a gesture while learning a word helps cement that word in memory — perhaps by stimulating a more extensive network of areas in the brain.

In a study published in 2020, for example, Macedonia and a group of six coauthors compared study participants who had paired new foreign-language words with gestures to those who had paired the learning of new words with images of those words. The researchers found evidence that the motor cortex — the area of the brain that controls bodily movement — was activated in the gesturing group when they reencountered the vocabulary words they had learned; in the picture-viewing group, the motor cortex remained dormant. The "sensorimotor enrichment" generated by gesturing, Macedonia and her coauthors suggest, helps to make the associated word more memorable.

Macedonia has also begun experimenting with a form of extended technology that seems a natural fit with her previous work: an online language-learning platform that features a virtual agent, or "avatar," providing instruction in vocabulary. The avatar on the screen behaves just as Macedonia did as a teacher: it demonstrates a gesture, which the user imitates as he or she repeats the novel word. Evaluations of the platform show that users who follow the avatar in making a gesture achieve more lasting learning than those who simply hear the word. Gesturing students also learn more than those who observe the gesture but don't enact it themselves. Studies by other researchers have found that math students solve problems more quickly, and generalize their new knowledge more effectively, after they have been taught by an avatar that gestures than by one that does not. Online instructional resources — including commercial language-learning platforms like Duolingo and Rosetta Stone — might be made much more effective by adding an animated agent who induces users to engage in gesture.

In addition to reinforcing memory, designed gestures can lighten our mental load. Gesturing offloads our cognitive burden in much the

same way as making a list or drawing a diagram on a piece of paper—
except that we always have our hands with us, ready to assume some of
the weight. (Indeed, studies show that when people are given challenging problems to solve but are prohibited from using pencil and paper,
they gesture more to compensate.) A familiar example of such offloading is the way young children count on their fingers when working out
a math problem. Their fingers "hold" an intermediate sum so that their
minds are free to think about the mathematical operation they must ex-
ecute (addition, subtraction) to reach the final answer. More complex or
conceptual gestures serve a similar purpose for older children and adults.
Our hands offer our heads overflow capacity, such that we can manage a
greater volume of information overall, and can subject that information
to a greater number of acts of manipulation and transformation.

In pursuit of that additional capacity, some teachers purposely show
their students how to shift information onto their hands. Washington
State math teacher Brendan Jeffreys turned to gesture as a way of easing
the mental load carried by his students, many of whom come from low-
income households, speak English as a second language, or both. "Academic language—vocabulary terms like 'congruent' and 'equivalent' and
'quotient'—is not something my students hear in their homes, by and
large," says Jeffreys, who works for the Auburn School District in Auburn,
a small city south of Seattle. "I could see that my kids were stumbling over
those words even as they were trying to keep track of the numbers and
perform the mathematical operations." So Jeffreys devised a set of simple
hand gestures to accompany, or even temporarily replace, the unfamiliar
terms that taxed his students' ability to carry out mental math.

To signify that an angle is acute, Jeffreys taught them, "make Pac-Man
with your arms." To signify that it is obtuse, "spread out your arms like
you're going to hug someone." And to signify a right angle, "flex an arm
like you're showing off your muscle." For addition, bring two hands together; for division, make a karate chop; to find the area of a shape, "motion as if you're using your hand as a knife to butter bread."

Jeffreys's students enthusiastically adopted the gestures, and they now
employ them while talking in class, while doing homework, and even
while taking tests, he reports. As they become more adept at executing
the mathematical operations themselves, the students are often able to
reincorporate the academic language that had confused them initially—a

benefit of the mental spaciousness made possible by the load-lightening effects of gesture. Jeffreys's approach has proved so successful in helping his students learn math that he has been asked to extend it to all twenty-two schools in his district. He is now at work on developing a set of gestures to support students in reading and writing: motions that indicate terms like "character," "setting," "summary," and "main point."

As Brendan Jeffreys's students have discovered, our hands are impressively flexible tools. They can represent so many things: an entrepreneur's vision for his product; an infant's step toward spoken language; a teacher's clue that a student is ready to learn. Hands can be a prompt, a window, a way station—but what they ought never have to be is still.

PART II

THINKING
WITH OUR
SURROUNDINGS

4

■

Thinking with Natural Spaces

A S THE SUMMER OF 1945 came to an end, the artist Jackson Pollock was approaching a breaking point. New York City, where he lived in a downtown apartment, felt increasingly frenetic and chaotic. His ever-present struggles with drinking and depression seemed to be worsening. His wife, the painter Lee Krasner, worried about his mental health.

In August, Pollock and Krasner went to visit friends on the East End of Long Island, then a quiet home to farmers and fishermen, as well as a few artists and writers. Pollock felt both soothed and stimulated by the place: the light, the green, the cool breezes blowing in from Long Island Sound. After returning from the trip, Pollock sat for three days, thinking, on a couch in the Eighth Street apartment. When he rose to his feet he had a plan: he and Lee would move to the East End.

In short order, Pollock and Krasner relocated to a ramshackle farmhouse near the sleepy hamlet of Springs, Long Island. Pollock spent hours on the house's back porch, gazing out at the trees and at the marshland stretching down to Bonac Creek. The move to Springs would inaugurate a years-long period of relative peace for the volatile painter. "It was a healing place," says Audrey Flack, a fellow artist who spent time with the couple on Long Island. "And they were in great need of being healed."

Nature changed Pollock's thinking—gently tempering his raging intensity—and it also changed his art. In New York, Pollock worked at an easel, painting intricate, involved designs. In Springs, where he worked in a converted barn full of light and views of nature, he began spreading his canvases on the floor and pouring or flinging paint from above. Art critics view this period of Pollock's life as the high point of his career, the years when he produced "drip painting" masterpieces like *Shimmering Substance* (1946) and *Autumn Rhythm* (1950). The fulcrum for this turn of fortune was the time Pollock spent musing in his New York apartment; the artist said it was then that he realized he would always be homeless when inside. Out of doors, he found his home.

ARTISTS LIKE Jackson Pollock are not the only people whose mental activity is shaped by their surroundings; all of us think differently depending on where we are. The field of cognitive science commonly compares the human brain to a computer, but the influence of place reveals a major limitation of this analogy: while a laptop works the same way whether it's being used at the office or while we're sitting in a park, the brain is deeply affected by the setting in which it operates. And nature provides particularly rich and fertile surroundings with which to think. That's because our brains and bodies evolved to thrive in the outdoors; our ancient forebears practiced a lifestyle that would look, to us, like "a camping trip that lasts a lifetime," as a pair of ecologists has put it.

Over hundreds of thousands of years of dwelling outside, the human organism became precisely calibrated to the characteristics of its verdant environment, so that even today, our senses and our cognition are able to easily and efficiently process the particular features present in natural settings. Our minds are tuned to the frequencies of the organic world. No such evolutionary adjustment has prepared us for the much more recent emergence of the world in which we now spend almost all our time: the built environment, with its sharp lines and unforgiving textures and relentless motion. We've set up camp amid the high-rises and highways of our modern milieu, but our minds are not at ease in this habitat. The mismatch between the stimuli we evolved to process and the sights and sounds that regularly confront our senses has the effect of depleting our limited mental resources. We are left frazzled, fatigued, and prone to distraction, simply as a function of the hours we spend in a setting for which we are biologically ill-equipped.

Just how much of our lives unfolds inside buildings and vehicles is revealed by scientists' time use studies: only about 7 percent of our time is spent outdoors. That's much, much less than for our nature-dwelling ancestors, of course, but it's meager even when compared to that of Americans twenty years ago. More than 60 percent of American adults report spending five hours or less outside in nature each week. Children, too, engage in outdoor recreation far less frequently than earlier generations; only 26 percent of mothers report that their kids play outside every day. Such trends are likely to continue: more than half of Earth's humans now live in cities, and by 2050 that figure is predicted to reach almost 70 percent.

Yet despite these massive shifts in culture, our *biology* remains identical to that of our progenitors. Even now, our brains and bodies respond to nature in ways that reveal the deep imprint of our evolution in the outdoors. In fact, we can draw a direct line from the kind of landscapes we enjoy today to the settings in which our species evolved. One place to begin drawing that line is with the crowds of people who throng to a half-mile-wide strip of land in the middle of Manhattan Island. Forty-two million people visit New York's Central Park each year — ambling across its wide Sheep Meadow, browsing its fragrant gardens, circling its shimmering reservoir. Why are so many drawn to these 840 acres? Its creator knew the reason. "Natural scenery," wrote landscape architect Frederick Law Olmsted, "employs the mind without fatigue and yet enlivens it; and thus, through the influence of the mind over the body, gives the effect of refreshing rest and reinvigoration to the whole system."

Residents and tourists alike love Central Park's rolling hills, copses of trees, and dappled bodies of water. But though the park's features look natural, they were in fact almost entirely man-made. When Olmsted started his work in 1858, the piece of land with which he had to contend was an unpromising expanse of swamps and rocky outcroppings, available for parkland only because real estate developers could not build on it. Over the course of the next fifteen years, more than three thousand laborers moved some 10 million cartloads of rocks and dirt, and planted an estimated 5 million trees and shrubs, in service to Olmsted's vision. That vision was drawn from the visits Olmsted made to legendary estates in England, such as Birkenhead Park and Trentham Gardens, which had themselves been contrived by landscape architects in the late eighteenth and early nineteenth centuries.

The preferences around which Olmsted was shaping Central Park were older still — ancient, in fact, reaching back to humanity's early days on the African savanna. The particular environment in which our species evolved left us with a set of predilections that persist to this day. For it's not just *any* kind of nature that appeals to us. Much of what is natural is unpleasant or even threatening: predators, storms, deserts, swamps. In order to survive, humans evolved strong, and shared, preferences for certain kinds of natural spaces — spaces that look safe and resource-rich. We like wide grassy expanses, dotted by loose clumps of trees with spreading branches, and including a nearby source of water. We like the capacity to see long distances in many directions from a protected perch, aspects that geographer Jay Appleton memorably named "prospect" and "refuge." And we like a bit of mystery — a beckoning promise of more to be revealed around the bend.

The world's greatest landscape designers intuited these preferences and incorporated them into their work. Starting in the mid-eighteenth century, Lancelot Brown — better known by his nickname, "Capability" — transformed more than 250 English estates, moving hills and planting trees in order to achieve his vision of an idealized countryside. Brown was succeeded by another British landscape architect, Humphry Repton, who drew "before" and "after" sketches for prospective clients; the "after" pictures promised shady clusters of trees giving way to views of open meadows and glimmering ponds.

Repton's and Brown's designs inspired Frederick Law Olmsted, along with many others, but they were no mere fashion. The preferences they elevate transcend time, culture, and nationality. They are shared by people all over the world — from Australia to Argentina, from Nigeria to Korea — including people who must go to great lengths to emulate the archetype. Landowners in the bone-dry southwest United States irrigate their properties to evoke the lush, grassy savanna. Gardeners in Japan prune their trees so that the boughs resemble the spreading branches of the trees of East Africa. Such choices reflect the brain's very particular evolved history — the "ghosts of environments past," in the phrase of biologist Gordon Orians.

What we imagine to be aesthetic preferences are really survival instincts honed over millennia, instincts that helped us find promising places to forage and to rest. When, today, we turn to nature when we're stressed or burned out — when we take a walk through the woods or gaze out at the ocean's rolling waves — we are engaging in what one researcher calls "environmental self-regulation," a process of psychological renewal that our brains cannot accomplish on their own.

IT'S NOT SIMPLY that we *prefer* such settings. They actually help us to think better—in part by relieving our stress and reestablishing our mental equilibrium. Drivers who travel along tree-lined roads, for example, recover more quickly from stressful experiences, and handle emerging stresses with more calm, than do people who drive along roads crowded with billboards, buildings, and parking lots. Laboratory studies of people who are given a challenging math test or are subjected to sharp questioning by a panel of judges report that subsequent contact with nature calms their nervous system, returning them to a state of psychological balance in the wake of such trying experiences. The more stressed individuals are, the more benefit they derive from exposure to nature.

The sights and sounds of nature help us rebound from stress; they can also help us out of a mental rut. "Rumination" is psychologists' term for the way we may fruitlessly visit and revisit the same negative thoughts. On our own, we can find it difficult to pull ourselves out of this cycle—but exposure to nature can extend our ability to adopt more productive thought patterns. Gregory Bratman, an assistant professor at the University of Washington, asked study participants to undergo a brain scan and to complete a measure of ruminative thinking before taking a ninety-minute walk outside. Half the participants strolled through a quiet, leafy natural area; the other half walked alongside a busy roadway. Upon returning to the lab, all the participants took the ruminative thinking measure and had their brains scanned for a second time.

People who'd spent the previous hour and a half in nature had become less preoccupied by the negative aspects of their lives; in addition, an area of the brain associated with rumination, the subgenual prefrontal cortex, was less active than before the nature walk. The people who had walked alongside a busy roadway gained no such relief. Rumination is especially common in those who are depressed, and research has shown that a walk in nature lifts the mood of people diagnosed with depression. It also improves their memory. The obsessive cycling through negative thoughts that many depressed people experience consumes a significant portion of their mental resources, adversely affecting their ability to recall important information—a deficit that time in nature helps ameliorate.

Yet another way that nature helps us think better is by enhancing our ability to maintain our focus on the task in front of us. People who have recently spent time amid outdoor greenery catch more errors on a proofreading assignment, for example, and provide quicker and more accurate answers on a fast-paced cognitive test, than do people who have just finished a walk in an urban setting. Working memory—our ability to hold in mind information relevant to the problem we're currently solving—also benefits from time spent in a natural setting. In a study led by Marc Berman, a psychologist at the University of Chicago, participants who walked through an arboretum for just under an hour scored 20 percent higher on a test of working memory compared to those who spent the same amount of time navigating busy city streets.

Time spent in nature can even relieve the symptoms of attention deficit/hyperactivity disorder (ADHD). A pair of researchers at the University of Illinois, Andrea Faber Taylor and Ming Kuo, were intrigued by reports from parents that their ADHD-affected children seemed to function better after exposure to nature. Putting this possibility to an empirical test, they had children aged seven to twelve take a supervised walk in a park, in a residential neighborhood, or in a busy area of downtown Chicago. Following the walks, the youngsters who'd spent time in the park were better able to focus than the children in the other two groups—so much so, in fact, that on a test of their ability to concentrate, they scored like typical kids without ADHD. Indeed, Taylor and Kuo point out, a twenty-minute walk in a park improved children's concentration and impulse control as much as a dose of an ADHD drug like Ritalin. "'Doses of nature' might serve as a safe, inexpensive, widely accessible new tool in the tool kit for managing ADHD symptoms," the researchers concluded.

All these salutary effects on our mental function can be understood as a process of *restoration:* time spent outdoors gives us back what the built environment so relentlessly drains away. More than a century ago, psychologist William James drew a distinction that bears closely on our understanding of nature's restorative powers. There are two kinds of attention, wrote James in his 1890 book *The Principles of Psychology:* "voluntary" and "passive." Voluntary attention takes effort; we must continually direct and redirect our focus as we encounter an onslaught of stimuli or concentrate hard on a task. Navigating an urban environment —with its hard surfaces, sudden movements, and loud, sharp noises—

requires voluntary attention. Passive attention, by contrast, is effortless: diffuse and unfocused, it floats from object to object, topic to topic. This is the kind of attention evoked by nature, with its murmuring sounds and fluid motions; psychologists working in the tradition of James call this state of mind "soft fascination."

The respite from insistent cognitive demands that nature provides gives our supply of mental resources an opportunity to renew and regenerate. As we've seen, these resources are finite and are soon exhausted — not only by the clamor of urban living, but also by the stringent requirements of academic and professional work. Just as our brains did not evolve to react with equanimity to speeding cars and wailing sirens, neither did they evolve to read, or to perform advanced math, or to carry out any of the highly abstract and complex tasks we ask of ourselves every day. Though we manage to meet these demands, our near-universal struggles with attention and focus (not to mention motivation and engagement) suggest that we ought to pay more heed to the *supply* side of our attentional economy — that is, not simply drawing down our mental resources but also ensuring their regular replenishment.

We can do so by simply going outside. We don't need to wait for perfect weather, or to find our way to some unspoiled wilderness; any form of nature, under any conditions, will do. There is, however, an optimal attitudinal stance we can adopt: what researchers call "open monitoring," or a curious, accepting, nonjudgmental response to all we encounter. Dor Abrahamson, a professor of education at the University of California, Berkeley, suggests borrowing the technique of "soft gazing" from the traditional Chinese practice of meditative movement known as tai chi. When soft gazing was introduced at a workshop he organized on bringing mindfulness to education, Abrahamson reports that one participant immediately noticed how it departed from his usual mode. "There is hard looking-at," the participant pointed out, "and there is the soft receiving of images." The orientation we adopt when in nature is ideally one of such informal mindfulness. (For those intending to establish a more formal practice of meditation, research shows that doing so *in nature* makes the habit easier to implement and maintain.)

When we seek psychological restoration outdoors, it's best to leave our devices behind; research has found that using a smartphone while outside "substantially counteracts the attention enhancement effects" of being in nature. One exception might be the use of an app like ReTUNE — an example of extended technology developed by University of Chicago psychologist Marc Berman and doctoral student Kathryn Schertz. (The name is an acronym that stands for Restoring Through Urban Nature Experience.) ReTUNE is like a conventional GPS system, but programmed with a different set of values: instead of providing its users with the speediest route, it offers them the path with the greatest number of trees, the largest proportion of flowers, the highest frequency of birdsong.

Say a visitor to Chicago wants to walk from the university's campus, in the city's Hyde Park neighborhood, to Promontory Point, a peninsula that juts out into Lake Michigan. If she consulted Waze, the popular mapping app owned by Google, its recommended route would take a brisk twenty-eight minutes: three blocks north on South Blackstone Avenue, then a right on East 56th Street and a left onto South Shore Drive. It's the most direct way to make the journey, but the experience would be one of glass and brick, concrete sidewalks and asphalt roads, honking cars and hurried commuters.

The same coordinates, entered into ReTUNE, would produce a markedly different suggested route: a stroll that begins in the Midway Plaisance, an urban oasis of green; then a ramble through Jackson Park, with its lush gardens and sparkling lagoons; emerging, finally, onto a path that runs alongside the shores of Lake Michigan, winding its way to Promontory Point. The trip would take a comparatively less efficient thirty-four minutes to complete, but our visitor would likely arrive at her destination clearer-headed and more relaxed.

There are values embedded in the technology we use, and as we scroll and tap we often unthinkingly adopt these priorities as our own. By extending technology with the thought-enhancing properties of nature, ReTUNE encourages us to question the elevation of efficiency above all other values, including our own mental well-being. ReTUNE makes its own commitment clear: it's not the speed of the journey, it's what you see along the way.

THE RETUNE APP assigns each potential route a "Restoration Score," based on the presence of auditory and visual features that connote "naturalness." Its co-inventor, Marc Berman, is one of a number of researchers working to answer the question: What makes nature natural? This investigation may seem befuddlingly circular — but if science could determine, at a granular level, precisely which features of nature work their effects on our bodies and brains, such information could be used to inform the design of buildings and landscapes that actively enhance their users' mood, cognition, and health. Through close analysis of natural and man-made scenes, Berman and others have begun to compile a kind of taxonomy of naturalness.

Natural landscapes, they have found, feature less variation in hue than urban settings (that is, their colors range from green to yellow to brown to more green) and more "color saturation," or pure, undiluted color. Natural environs also present the eye with fewer straight lines, and more curving shapes, than do urban ones. Finally, unlike man-made designs, in which edges tend to be spaced apart (think of a row of windows on the face of an office building), in natural vistas, edges tend to cluster densely together (picture the overlapping edges of a tree's many leaves). Applying these characteristics as a filter, Berman and his team have designed a computer model that can predict with 81 percent accuracy whether a given image depicts a scene that actual human beings have rated as highly natural. Other researchers have identified additional features that distinguish natural tableaux from their artificial counterparts: natural settings incorporate dynamic and diffuse light; gentle, often rhythmic movement; and muted sounds that repeat with variations, such as ocean waves or birdsong.

It's not the case that nature is simpler or more elementary than man-made environments. Indeed, natural scenes tend to contain *more* visual information than do built ones — and this abundance of visual stimulation is a condition we humans crave. Roughly a third of the neurons in the brain's cortex are dedicated to visual processing; it takes considerable visual novelty to satisfy our eyes' voracious appetite. But balanced against this desire to *explore* is a desire to *understand;* we seek a sense of order as well as an impression of variety. Nature meets both these needs, while artificial settings often err on one side or the other. Built environs may be monotonous and under-stimulating: picture the unvarying glass and metal façades of many modern buildings, and the uniform rows of beige cubicles that fill many offices. Or they may be overwhelming and over-stimulating, a barrage of light and sound and motion: call to mind New York's Times Square or Tokyo's Shibuya Crossing.

Nature is complex, it's true, but its complexity is of a kind that our brains are readily able to process. When we're surrounded by nature, we experience a high degree of "perceptual fluency," notes Yannick Joye, a senior researcher at the ISM University of Management and Economics in Lithuania. Basking in such ease gives our brains a rest, Joye explains, and also makes us feel good; we react with positive emotion when information from our environment can be absorbed with little effort.

The perceptual fluency we experience in regard to nature emerges from the way the various elements of a natural vista interact with one another. Natural scenes are more *coherent,* lacking the jarring disjunctions common in man-made settings (a rococo building next to a garish billboard beside a severe modern sculpture). Natural scenes also offer more *redundant* information. The shape and color of a leaf or a hillside repeat again and again, a fact that facilitates the brain's habit of making predictions. When in nature, we have a good sense from what we just looked at of what we will take in with the glance that follows—unlike in an urban setting, where we never know what we'll come upon next. "Natural environments are characterized by a deep degree of perceptual predictability and redundancy, whereas urban scenes tend to consist of perceptually divergent objects," Joye notes. "These divergences compete for our visual attention and make urban scenes substantially less easy to grasp and process."

Fractals are one form of redundancy that has attracted particular attention from scientists. A fractal pattern is one in which the same motif is repeated at differing scales. Picture the frond of a fern, for example: each segment, from the largest at the base of the plant to the tiniest at its tip, is essentially the same shape. Such "self-similar" organization is found not only in plants but also in clouds and flames, sand dunes and mountain ranges, ocean waves and rock formations, the contours of coastlines and the gaps in tree canopies. All these phenomena are structured as forms built of smaller forms built of still smaller forms, an order underlying nature's apparently casual disarray.

Fractal patterns are much more common in nature than in man-made environments. Moreover, nature's fractals are of a distinctive kind. Mathematicians rank fractal patterns according to their complexity on a scale from 0 to 3; fractals found in nature tend to fall in a middle range, with a value of between 1.3 and 1.5. Research shows that, when presented with computer-generated fractal patterns, people prefer mid-range fractals to those that are more or less complex. Studies have also demonstrated that looking at these patterns has a soothing effect on the human nervous system; measures of skin conductance reveal a dip in physiological arousal when subjects are shown mid-range fractals. Likewise, people whose brain activity is being recorded with EEG equipment enter a state that researchers call "wakefully relaxed" — simultaneously alert and at ease — when viewing fractals like those found in nature.

There is even evidence that our ability to think clearly and solve problems is enhanced by encounters with these nature-like fractals. Study participants asked to carry out information-search, map-reading, and location-judgment tasks within computerized "fractal landscapes" did so most efficiently and effectively when the complexity of the fractals fell within the middle range. Another experiment, led by Yannick Joye, required subjects to complete challenging puzzles during and after exposure to fractal patterns of varying complexity. They solved the puzzles most quickly, easily, and accurately when the fractals they saw had a structure similar to those found in nature. It seems that our brains are optimized to process the fractal characteristics of natural scenes; hundreds of thousands of years of evolution have "tuned" our perceptual faculties to the way visual information is structured in natural environments. We may not take conscious note of fractal patterns, but at a level deeper than awareness, these patterns reverberate.

Research on the effects of fractals on the human psyche was pioneered by Richard Taylor, a professor of physics, psychology, and art at the University of Oregon. As his title suggests, Taylor has a dizzyingly broad range of interests. A number of years ago, he was studying the fractal patterns present in the flow of electricity when the forms he saw reminded him of something: Jackson Pollock's paintings. Pursuing this unexpected connection, Taylor analyzed a group of Pollock's later works and determined that they too exhibited a fractal pattern, with a value in the mid-range of 1.3 to 1.5. Astonishingly, Pollock's drip paintings — the ones he made after moving from New York City to Springs, Long Island — turn out to bear the visual signature of nature. Like Capability Brown and Humphry Repton, the landscape architects who unconsciously re-created the characteristic features of the East African savanna, Pollock seems to have tapped into humanity's ancient affinity for the natural world.

Take, for example, a drip painting Pollock completed in 1948, known only as *Number 14*. Rendered in stark black, white, and gray, the artwork bears no immediate resemblance to the verdant vegetation that surrounded the home he shared with Lee Krasner in Springs. A longer look at the painting, however, draws the viewer deeper into its swoops and whorls, its intricate nest of thinning and thickening lines. Says Richard Taylor admiringly, "If someone asked, 'Can I have nature put onto a piece of canvas?', the best example there has ever been of that is 1948's *Number 14*."

TIME SPENT IN NATURE relieves stress, restores mental equilibrium, and enhances the ability to focus and sustain attention. We spend most of our time indoors, however — so might we find ways to make inside more like outside? Environmental psychologist Roger Ulrich has asked this question, and answered it in the affirmative, regarding a particular kind of built setting: hospitals and other health care facilities. Our ancient biological wiring allows us to "exploit nature like a drug," says Ulrich, now a professor of architecture at Chalmers University in Sweden, and he means this almost literally. Several decades ago, he demonstrated that exposure to nature relieved pain and promoted healing in patients recovering from surgery.

Ulrich's study, carried out at a hospital in suburban Philadelphia, found that patients who occupied rooms with a view of trees required fewer painkillers, experienced fewer complications, and had shorter hospital stays than patients whose rooms looked out on a brick wall. Nurses also recorded many fewer negative notes about patients' state of mind — "upset and crying," "needs much encouragement" — for those individuals who had a view of greenery. (Ulrich recounts that the inspiration for the study emerged from his own long-ago experience: "As a teenager, I had some serious illnesses that forced me to spend time at home in bed," he says. "My window was my compass of stability. Every day, I watched the trees in the wind. There was something endlessly calming about it.")

Further research by Ulrich and other scientists has confirmed that exposure to natural elements helps diminish patients' pain and hasten their recovery. This work helped inspire a revolution in the design of health care facilities, leading to the renovation and construction of buildings that provide patients and staff with natural light and views of greenery. It also led to a raft of research that has sought to identify more precisely the effects of the out-of-doors on our bodies. This work has established that nature is indeed a highly reliable and effective "drug," bioengineered by evolution over thousands of years. It appears to affect us all in the same fashion: within twenty to sixty seconds of exposure to nature, our heart rate slows, our blood pressure drops, our breathing becomes more regular, and our brain activity becomes more relaxed. Even our eye movements change: We gaze longer at natural scenes than at built ones, shifting our focus less frequently. We blink less often when viewing nature than when looking at urban settings, indicating that nature imposes a less burdensome cognitive load. We remember details of natural scenes more accurately, and brain scans show that a larger portion of the visual cortex is activated when we look at nature, along with a larger number of the brain's pleasure receptors.

Of course, it's not only hospital patients who can benefit from regular doses of this "drug." Our homes, schools, and workplaces could all become more cognitively congenial spaces if they were to incorporate elements of what is known as *biophilic design*. In a book published in 1984, the Harvard University biologist E. O. Wilson advanced what he called the "biophilia hypothesis": the notion that humans have an "innate tendency to focus on life and lifelike processes," an "urge to affiliate with other forms of life." This urge is powerful, Wilson argued, and our thinking (as well as our health and well-being) suffers when it is suppressed, as it must be when we spend most of our time surrounded by inorganic forms and materials. Fortunately, he adds, an alternative path has already been mapped out for us: nature itself provides a comprehensive guide to the conditions in which our minds and bodies work best.

For example: We know that our brains thrive on the coherent organization and redundant information embedded in plant life—so why not bring the green stuff inside? That's what the managers of Second Home, a network of co-working spaces in Europe, have done. More than a thousand plants fill the Second Home headquarters in London; when the space opened in 2014, E. O. Wilson was brought in to speak to the Second Home staff. "Everything we do at Second Home is inspired by nature and biophilia," says Rohan Silva, one of Second Home's cofounders. Silva adds that the Second Home office in Lisbon, Portugal, was modeled on a greenhouse and is home to even more vegetation—more than two thousand plants, representing one hundred different varieties, including tillandsias, philodendrons, and monsteras. Research indicates that the presence of indoor plants does in fact improve employees' attention and memory and increases their productivity levels; for students, too, a "green wall" in the classroom, sprouting living plants, has been shown to enhance their ability to focus.

Greenery is only one part of nature's bounty, of course, and practitioners of biophilic design have begun to incorporate other organic elements into new construction — of schools, office buildings, factories, and even skyscrapers. The fifty-five-story Bank of America Tower, completed in 2009, sits at the corner of Bryant Park in midtown Manhattan; when people arriving at the entrance reach for a door handle, what they feel is not steel or plastic but wood. "We wanted the first thing visitors touched to have a grain, an imprint of nature," says Bill Browning, an environmental strategist who helped design the space. Inside the lobby, the natural theme continues: the ceilings are covered in bamboo, and the walls are constructed of stone in which the fossils of tiny shells and sea creatures are visible. Even the shape of the tower was inspired by nature; it emulates the fractal form of a quartz crystal.

Biophilic design is an emerging discipline, but a handful of studies have begun to suggest that working and learning in buildings inspired by nature can grant some of the same benefits for cognition as actually being outdoors. In a study published in 2018, for example, a group of researchers from Harvard's T. H. Chan School of Public Health asked participants to spend time in an indoor environment featuring biophilic elements (potted plants, a bamboo-wood floor, windows with a view of greenery and a river) as well as an indoor environment lacking such elements (a windowless, carpeted, fluorescent-lit space). Participants were outfitted with wearable sensors that monitored their blood pressure and their skin conductance; tests of mood and cognitive function were administered following each visit. After just five minutes in the biophilic surroundings, the subjects' positive emotions increased, their blood pressure and skin conductance dropped, and their short-term memory improved by 14 percent over the level recorded following a visit to the non-biophilic environment.

It may not come as a surprise that the people in this study felt better in the space with windows: Who doesn't prefer a glimpse of sun and sky to the sickly glow of fluorescent panels? Yet for decades, planners and builders believed that thinking happens best under even, unchanging light, shielded from the undesirable distractions and eye-straining glare ostensibly introduced by windows. These beliefs still shape the spaces in which we learn and work today — spaces that were often deliberately engineered to eliminate the natural variations in sunlight that we experience when outside. Such subtle shifts in illumination, we now know, help keep us alert and help regulate our biological clocks. Indeed, research shows that people who experience natural light during the day sleep better, feel more energetic, and are more physically active; one study found that employees exposed to daylight through windows in their offices sleep an average of forty-six minutes more per night than workers who labor in windowless spaces. The technology giant Google has determined that employees who have desks positioned near windows report being more creative and productive than do workers located farther away from sources of natural light. The company has issued guidelines for how much daylight "Googlers" should be getting at work, and has asked some employees to wear light sensors around their necks to determine whether their workplace makes the grade.

In addition to brainbound notions about the ideal conditions for thinking, the buildings many of us occupy today were shaped by the energy crisis of the 1970s, when windows were commonly blocked or removed in the name of energy conservation. Today some educational and business leaders are recognizing the folly of depriving students and workers of natural light. At the H. B. Plant High School in Tampa, Florida, for example, the large windows that were closed off in the 1970s were recently restored, letting daylight shine in for the first time in some four decades. When new schools and office buildings are constructed, some architects are now thinking in terms of balancing occupants' need for natural light with the obligation to reduce costs and conserve energy. At PS 62 in Staten Island, New York — also known as the Kathleen Grimm School of Leadership and Sustainability — the school building generates as much energy as it consumes, thanks to solar panels, a wind turbine, and heating and cooling systems powered by underground geothermal wells. Completed in 2015, the structure achieves "daylight autonomy" as much as 90 percent of the time: that is, its classrooms and hallways are illuminated almost entirely by the sun. In addition to saving funds and protecting the environment, says principal Lisa Sarnicola, the thoughtful design of the school building enlivens the students' education. "It changes the whole mood of the building," she says. "It makes the children happy."

There is even evidence that views of nature are associated with improved academic performance. John Spengler, a professor at the Chan School of Public Health at Harvard, employed an ingenious method of gauging the "greenness" of school grounds: with satellite images, taken by a NASA spacecraft from four hundred miles above the earth. Spengler and his colleagues analyzed aerial photographs of public schools in Massachusetts, determining the amount of vegetation present on their grounds, and compared these measurements with scores on the MCAS, the statewide assessment of academic skill, earned by students in grades three through ten. Controlling for race, gender, family income, and English as a second language, among other factors, the results "showed that a higher surrounding greenness contributes to a better English and mathematics academic performance in students of all grades," the researchers reported.

Another, more direct test of the effect of natural surroundings was carried out by William Sullivan, a professor of landscape architecture at the University of Illinois, and his colleague Dongying Li. High school students were randomly assigned to a room with a view of greenery, a room with a view of a building or parking lot, or a windowless room. The researchers taxed the mental resources of the participants with a series of challenging activities, including completing a proofreading exercise, giving a speech, and solving a set of mental math problems. Then Sullivan and Li administered a test of attention to each participant, gave them a ten-minute break, and administered the attention test a second time — all within each participant's assigned room. The high school students who were able to gaze out the window at greenery during their break scored 13 percent higher on the second administration of the attention test; the students who looked out on built surroundings, or at windowless walls, did not improve at all.

The benefits of a room with a natural view hold for office workers as well as students. A study carried out for the California Energy Commission examined the effects of window access on the performance of two groups of people employed by the Sacramento Municipal Utility District. Workers in the district's call center were found to process calls 6 to 12 percent faster when they had the best possible view (a nearby window looking onto a green vista) versus having no view. And workers in the district's main office were found to score 10 to 25 percent higher on tests of mental function and memory recall when they had the best possible view versus having no view.

Even a brief glance out the window can make a difference in our mental capacity. Researchers from the University of Melbourne in Australia found that a forty-second "micro-break" spent looking out at a roof covered with flowering meadow plants led study participants to perform better on a cognitive test than did an equally short break spent looking at a bare concrete expanse. Participants who gazed at the green roof were more alert, made fewer errors, and were more in control of their attention. We can seek out such "microrestorative opportunities" throughout the day, replenishing our mental resources with each glance out the window.

MICRORESTORATIVE EXPERIENCES have their place in busy workdays and school schedules. But longer stretches spent in nature, when we can arrange them, may change how we think in deeper and more subtle ways, altering how we experience time and how we think about the future. "Oh, these vast calm measureless mountain days," exclaimed naturalist and author John Muir about the time he spent hiking the Sierras in the American West. In May 1903 Muir guided Theodore Roosevelt, then president of the United States, on a three-day camping trip through California's Yosemite Valley. "The first night we camped in a grove of giant sequoias," Roosevelt later recalled. "It was clear weather, and we lay in the open, the enormous cinnamon-colored trunks rising about us like the columns of a vaster and more beautiful cathedral than was ever conceived by any human architect."

As the two men toured the glories of Yosemite, Muir warned the president that the country's bountiful natural beauty was in danger of being despoiled if it was not preserved by official decree. Muir's urgent message was received. Over the next few years, Roosevelt more than tripled the square miles reserved for natural forests, doubled the number of national parks, and established seventeen national monuments, including the Grand Canyon. Speaking to an audience in Sacramento soon after he parted from Muir, Roosevelt explained why the sequoias under which he had reclined should be protected from lumbering: "I hope for the preservation of the groves of giant trees simply because it would be a shame to our civilization to let them disappear," he remarked. "They are monuments in themselves." These and other "natural resources," he continued, should be "handed on unimpaired to your posterity. We are not building this country of ours for a day. It is to last through the ages."

The capacity to delay immediate gratification in favor of longer-term interests — what Roosevelt called "posterity" — is one that is strengthened by spending time in nature, research has found. Meredith Berry, a psychologist at the University of Florida, showed images of nature (mountains, forests) and of urban settings (buildings, roads) to study participants, who were then asked questions that measured their tendency to engage in "future discounting"—that is, to prefer a smaller payoff right now over a larger reward later. People who'd been exposed to pictures of nature were more likely to postpone gratification, responding with less impulsivity and more self-control than people who had looked at pictures of cities.

Arianne van der Wal, a psychologist at Leiden University in the Netherlands, asked participants in her study to walk through a green, leafy area of Amsterdam, or through a busy, built-up part of the city. Afterwards, she reported, the people who'd just experienced nature were 10 to 16 percent more likely to restrain their impulse to satisfy their immediate desires. Similar results have been found with children: kids aged eight to eleven were more capable of delaying gratification after watching a video of nature than after viewing a video about cities. When we see or experience an urban setting, we are primed to feel competitive, to believe we need to grab what's available. Nature, by contrast, inspires a feeling of abundance, a reassuring sense of permanence.

The orientation toward the future that is evoked by nature may also be related to the way it changes our sense of time. In another of Berry's studies, participants were asked to estimate how many seconds or minutes had passed in a given period; people who had just viewed scenes of nature, she found, perceived time as passing more slowly. Likewise, research has demonstrated that people who take a walk in a natural setting overestimate how long the walk lasted, while people who have walked in an urban setting accurately estimate the amount of time that elapsed. Our perception of time is malleable, subject to the influence of situational cues;

by reducing our arousal and increasing our attentional capacity, exposure to nature grants us a more expansive sense of time, and a more generous attitude toward the future.

Nature may also lead us to think more *creatively*. Children's play is more imaginative when they are outdoors than when they are inside, research has shown; natural play spaces are less structured and more varied, and the props children may come across (leaves, pebbles, pinecones) have no purpose predetermined by teachers or parents. For adults as well, spending time in nature can promote innovative thinking. Scientists theorize that the "soft fascination" evoked by natural scenes engages what's known as the brain's "default mode network." When this network is activated, we enter a loose associative state in which we're not focused on any one particular task but are receptive to unexpected connections and insights. In nature, few decisions and choices are demanded of us, granting our minds the freedom to follow our thoughts wherever they lead. At the same time, nature is pleasantly diverting, in a fashion that lifts our mood without occupying all our mental powers; such positive emotion in turn leads us to think more expansively and open-mindedly. In the space that is thus made available, currently active thoughts can mingle with the deep stores of memories, emotions, and ideas already present in the brain, generating inspired collisions.

Nature's contributions to creative thought can be felt on a lunchtime stroll through the park. But there may be something especially fruitful about an extended foray into wilder terrain. David Strayer calls it the "three day effect." Strayer, a psychologist at the University of Utah, had long observed that his most original ideas came to him on his overnight trips into Utah's rugged outback. For a study published in 2012, he administered a measure of creative thinking to groups of hikers in Alaska, Colorado, Maine, and Washington state; some completed the measure before setting off on their expedition, while others did so after they had spent three days in the wilderness. The results confirmed his own experience: the backpackers' responses were 50 percent more creative following extended exposure to nature. Strayer, who also studies the distracting effects of digital media, believes that an enforced separation from phones and computers played no small part in the hikers' increased ability to think creatively. Our electronics are deliberately engineered to grab our attention and not let it go; our devices work against the diffuse mental processes that generate creativity, and escaping into nature is one of the only ways we can leave them behind.

There's another way in which the digital respite we get in nature could enhance our creativity. The time we spend scrutinizing our small screens leads us to *think* small, even as it enlarges and aggrandizes our sense of self. Nature's vastness — the unfathomable scale of the ocean, of the mountains, of the night sky — has the opposite effect. It makes us feel tiny, even as it opens wide our sense of the possible. It does all this through an emotion that we confront most commonly in nature: awe. Dacher Keltner, a professor of psychology at the University of California, Berkeley, has led much of the recent research on awe; he calls it an emotion "in the upper reaches of pleasure and on the boundary of fear."

One of the pleasurably fearsome things about awe is the radically new perspective it introduces. Our everyday experience does not prepare us to assimilate the gaping hugeness of the Grand Canyon or the crashing grandeur of Niagara Falls. We have no response at the ready; our usual frames of reference don't fit, and we must work to accommodate the new information that is streaming in from the environment. Consider the physical behavior that accompanies awe: we stop, we pause, we stare with eyes wide and features slack, as if to let in more of the scene that has so astonished us. The experience of awe, Keltner and other researchers have found, prompts a predictable series of psychological changes. We become less reliant on preconceived notions and stereotypes. We become more curious and open-minded. And we become more willing to revise and update our mental "schemas": the templates we use to understand ourselves and the world. The experience of awe has been called "a reset button" for the human brain. But we can't generate a feeling of awe, and its associated processes, all on our own; we have to venture out into the world, and find something bigger than ourselves, in order to experience this kind of internal change.

Scientists who study awe have also found that it alters the way we regard other people. Brain scans of people who are experiencing awe find that the region of the brain that contributes to our sense of occupying and orienting ourselves in space becomes less active. This diminished activity would seem to underlie the feeling we have when awestruck that the boundaries between ourselves and others have become more permeable, that we are part of a larger, connected whole. In behavioral terms, people act more prosocially and more altruistically following an experience of awe. They share and cooperate more in games after having watched a video of sublime scenes of nature; after having gazed up at a grove of old-growth trees, an experiment found, people were more likely to bend down and help pick up pens that a stranger (actually a confederate of the researcher) had dropped on the ground.

The "functional" account of awe—biologists' and psychologists' attempt at explaining *why* we feel this emotion—proposes that it spurs humans to put aside their individual interests in the service of a collective project. Members of the species who were inclined to feel awe, the story goes, were better able to band together to accomplish essential tasks. By extending ordinary thinking with awe at nature's immensity, humankind may have ensured its own survival—a reminder to us, perhaps, to look away from our small screens long enough to confront the dangers that threaten our kind and our planet today.

THE OCCASIONS FOR awe available to our forebears are the same ones that move us now: mountains, oceans, trees, sky. But there is one all-encompassing encounter with the natural world that, to date, has been experienced by only a handful of our contemporaries: the view of Earth from outer space, as glimpsed by astronauts. So emotionally overwhelming is this sight—and so consistent are the psychological consequences for the few who have seen it—that scientists have given it a name: "the overview effect." "If somebody'd said before the flight, 'Are you going to get carried away looking at the earth from the moon?' I would have said, 'No, no way,'" recalled Alan Shepard, an American astronaut who walked on the moon in 1971. "But yet when I first looked back at the earth, standing on the moon, I cried."

This emotional reaction to the sight of Earth—to its beauty, to its fragility—is one persistent motif noted by David Yaden, a research fellow at Johns Hopkins University who has studied astronauts' accounts of their flights. A second recurring theme is the dissolution of the boundaries and barriers that divide those of us here on the ground. From up in outer space, recounted astronaut Rusty Schweickart, "you identify with Houston and then you identify with Los Angeles and Phoenix and New Orleans . . . and that whole process of what it is you identify with begins to shift when you go around the Earth . . . You look down and see the surface of that globe you've lived on all this time, and you know all those people down there and they are like you, they *are* you . . . You recognize that you're a piece of this total life." Astronaut Edgar Mitchell recalled an "explosion of awareness" upon seeing Earth from space, an experience that gave him an "overwhelming sense of oneness and connectedness."

Paradoxically, the intense feelings of connection felt by space travelers are often pierced by equally powerful feelings of dislocation and alienation. When they're not marveling at the sight of Earth on the horizon, astronauts must contend with what many describe as a stultifying environment, tightly confined, filled with highly technical instruments but with none of the features that ease or delight the human mind. The result may be boredom, ennui, anxiety, even aggression toward other crew members. How to preserve the psychological well-being of astronauts has emerged as a compelling question as the prospect of long-term space travel draws closer. One possible answer has already shown some promise: have them grow green things.

When NASA astronaut Michael Foale arrived at the joint American-Russian space station Mir in 1997, one of the tasks he was assigned was tending to the station's greenhouse. He carried out a variety of experiments intended to investigate how plants grow in space; if astronauts are to spend months and even years onboard, they will need to have a supply of fresh food. And indeed, Foale did succeed in growing, and tasting, what the astronauts called "space broccoli." That required figuring out how to use light to guide the broccoli seedlings in the right direction; without gravity, they didn't know to grow "up." Without bees to pollinate the mizuna (a salad green), Foale himself had to do the job, gently transferring pollen from one plant to another with a toothpick.

Foale took such care with his plants that the team on the ground at NASA nicknamed him "Farmer Foale." But the space garden was of more than utilitarian value. "I loved the greenhouse experiment," said Foale at a press conference after returning from Mir. Maintaining a greenhouse on a long-range space flight, such as a trip to Mars, "would be essential," Foale added, "because there would be so little to do in that long period in between planets, and growing and tending plants is certainly a very soothing thing to do." Of the plants under his care, he said, "I enjoyed looking at them every morning for about ten, fifteen minutes. It was a moment of quiet time."

Psychologists who study the psyches of astronauts — many of whom already spend months at a time on space stations orbiting Earth — use a telling word for the malaise that many experience: such individuals, they say, are "homesick." It's possible to perceive an unsettling similarity between these space travelers, starved for the Earth they can see only through a window, and ourselves — sealed in our capsules of home and car and office, separated from the verdant vegetation, fresh air, and ever-shifting sunlight that is our natural habitat. When we're outdoors, "we feel 'at home' because we remain, in part, the children of our ancestral past," the architect Harry Francis Mallgrave has written. That past lives on in the way we think when we open the door and go outside.

5

■

Thinking with Built Spaces

J ONAS SALK WAS STUCK. For years the young medical researcher had been working sixteen hours a day, seven days a week, in a small basement laboratory in Pittsburgh, trying to develop a vaccine for polio. In the spring of 1954, exhausted and out of ideas, Salk realized he had to take leave of the lab to refresh his mind. He found the solitude and tranquility he sought at the Basilica of Saint Francis of Assisi, a thirteenth-century monastery in central Italy.

For several weeks Salk read and thought and walked amid the monastery's whitewashed colonnades and arches, its quiet courtyards, its chapel filled with light pouring in from tall windows. During this time he experienced an intellectual breakthrough — one he attributed to the buildings themselves. "The spirituality of the architecture there was so inspiring that I was able to do intuitive thinking far beyond any I had done in the past," Salk would later write. "Under the influence of that historic place I intuitively designed the research that I felt would result in a vaccine for polio. I returned to my laboratory in Pittsburgh to validate my concepts and found that they were correct."

But this is not the end of the story. Less than ten years after his visit to Assisi, Salk had the opportunity to build an intellectual community from the ground up — to create a space where scientists like him would do their thinking. Together with his architect, Louis Kahn, he set out to design the ideal place for reflection and discovery. They modeled the new institute on a monastery. In his discussions with Kahn, Salk specifically invoked the basilica at Assisi; the architect knew it well, having visited it and sketched it years earlier. Kahn imbued his design with references to the monastery both subtle and overt.

The resulting complex of buildings—the Salk Institute in La Jolla, California, completed in 1965—has been recognized as a masterpiece of modern architecture. Its structures are monumental, even austere. But they are also carefully designed to fit the needs of the people who work inside them. Natural light is everywhere, even on the floors located belowground, thanks to sunken courtyards and light wells. The laboratories feature wide swaths of unobstructed space; through an ingenious construction technique, Kahn managed to create entire floors free of interior columns. Mechanical elements like ducts and piping are located on alternating floors, so that repairs and updates can be made without disturbing the scientists as they work. Set apart from the labs are the scientists' studies, each of which offers a view of the Pacific Ocean. The researchers who work there, including a number of Nobel Prize winners, report that it is an optimal place to think. Salk professed himself pleased with the result: "I would say the building is as close to perfection as anything possible."

For centuries, architects like Louis Kahn have thought about how to create spaces that evoke particular states of mind. There is an even longer tradition of folk architecture embedded in the structures ordinary people have created for themselves, tweaking and tuning inherited forms. Now the emerging field of "neuroarchitecture" has begun to examine empirically how the brain responds to buildings and their interiors, and to theorize about how these reactions might be shaped by our evolutionary history and by the biological facts of our bodies.

Given all this thought and care, it's puzzling and dispiriting that many of us learn and work and live in spaces that don't help us think effectively. One reason for this: our culture does not hold the built environment in especially high regard; many individuals and institutions seem of the view that it doesn't much matter, that we should be able to engage in productive mental labor no matter what the setting. Second, thoughtful design takes time and money, and there is always the pressure to build quickly and cut costs. A third reason, it must be said, is architects and designers themselves, and the tyranny of their bold ideas. As we'll see in this chapter, their edgy experiments and avant-garde adventures have often created difficulties for the people who must live inside them every day.

Even Louis Kahn could fall into this trap. A few years before he designed the Salk Institute, he received a commission to build the Richards Medical Research Laboratories at the University of Pennsylvania. Kahn's design was hailed as a triumph by architectural critics; it even earned him an exhibit at the Museum of Modern Art. But from the perspective of the people who worked in the building, it was a disaster: cramped, dark, confusing to navigate — hardly the sort of place to evoke a eureka moment of discovery. Kahn corrected course with his next project, placing the needs of the buildings' occupants at the heart of the plan for the Salk Institute.

During this busy time in Kahn's professional life, an equally robust intellectual project was unfolding halfway across the country, in Oskaloosa, Kansas. Psychologist Roger Barker had set out to discover why people behave as they do by recording their activities in minute detail as they went about their everyday lives. With his colleague Herbert Wright, he set up the Midwest Psychological Field Station in Oskaloosa (population 750), and began following a group of children from the instant they woke up in the morning to the moment they were put to bed at night.

From their exhaustive observations, a distinct pattern emerged. As one scholar notes, "Barker and his colleagues found that there was a great deal of order, consistency, and predictability in the children's behavior." But this order was not a product of the children's personalities, nor their intelligence, nor any other internal quality. Rather, the factor that overwhelmingly determined the way the children acted was the *place* in which they were observed. As Barker himself reported, "The characteristics of the behavior of a child often changed dramatically when he moved from one region to another, e.g. from classroom, to hall, to playground, from drugstore to street, from baseball game to shower room."

Barker's "Midwest Study," which ultimately extended over a quarter-century, generated reams of evidence that the spaces in which we spend our time powerfully shape the way we think and act. It is *not* the case that we can muster the ability to perform optimally no matter the setting — a truth that architects have long acknowledged, even as our larger society has dismissed it. Christopher Alexander, author of the classic book *A Pattern Language* and an architect who celebrates the hard-earned wisdom embedded in folk architecture, laments "the arrogance of the belief that the individual is self-sufficient, and not dependent in any essential way on his surroundings." To the contrary, Alexander writes, "a person is so far formed by his surroundings, that his state of harmony depends entirely on his harmony with his surroundings." He adds: "Some kinds of physical and social circumstances help a person come to life. Others make it very difficult."

Today, we too often learn and work in spaces that are far from being in harmony with our human nature, that in fact make intelligent and effective thinking "very difficult." Yet the built environment — when we know how to arrange it — can produce just the opposite effect: it can sharpen our focus; it can sustain our motivation; it can enhance our creativity and enrich our experience of daily life. A tour through recent research in psychology and neuroscience, and through the varied kinds of places that humans have long created, can show us how to turn space into an extension of our minds.

APART FROM OFFERING shelter from the elements, the most critical function of a built interior is simply to give us a quiet place to think. Such protected space is necessary because thinking — at least of the kind the modern world expects of us — doesn't come naturally to the human animal. Throughout the long history of our species, we did our cogitating out in the open, in the moment, often on the run, relying on instinct and memory far more than considered reflection or careful analysis. It was only when we found ourselves compelled to concentrate in a sustained way on abstract concepts that we needed to sequester ourselves in order to think. To attend for hours at a time to words, numbers, and other symbolic content is a tall order for our brains. Maintaining this intensely narrow focus is a highly unnatural activity, and our minds require external structure in order to pull it off.

Historically, society's demand for increasingly abstract thought combined with the growing density of human habitation to create a need for such structure: that is, for walls. Walls became necessary as a way of relieving the mental strain that comes along with closely packed populations of unfamiliar others. For most of human history, after all, people lived with their family members in one-room dwellings. Everyone they knew lived not far from their front door, and it was useful to keep track of others' comings and goings. Even medieval kings and queens lived in a single large hall, filled with their chosen attendants and counselors. But with the emergence of cities and their crowds of unacquainted people, urban residents began to seek out spaces in which to read, think, and write — alone.

"The wall was designed to protect us from the cognitive load of having to keep track of the activities of strangers," observes Colin Ellard, an environmental psychologist and neuroscientist at the University of Waterloo in Canada. "This became increasingly important as we moved from tiny agrarian settlements to larger villages and, eventually, to cities — where it was too difficult to keep track of who was doing what to whom." The privacy afforded by walls represented a truly revolutionary extension of the mind, maintains John Locke, professor of linguistics at Lehman College of the City University of New York. "Our distant ancestors could see each other at all times, which kept them safe but also imposed a huge cognitive cost," he notes. "When residential walls were erected, they eliminated the need to look around every few seconds to see what others were doing." The result, he says, was that "a human vigil, one beginning with ancestors that we share with apes, was reduced to manageable proportions, freeing up many hours of undistracted time per day."

An early example of such walls can be found amid the hubbub of today's Manhattan, tucked away inside the Metropolitan Museum of Art. There, among the Grecian urns and the colonial-era silver, is a tiny gem of a room, re-created as it was in fifteenth-century Perugia: the *studiolo* of Federico da Montefeltro, Duke of Urbino. Federico, whose title called on him to be variously a royal, a politician, and a warrior, lived in the town of Gubbio in what is now central Italy. The walls of the study allowed the duke, a lover of literature, architecture, and mathematics, to retreat from the company of the townspeople he ruled into quiet study and contemplation. And because his *studiolo* was constructed in Renaissance Italy, these were no ordinary walls. Craftsmen from Siena, Florence, and Naples created elaborate trompe l'oeil murals made entirely of inlaid wood, a technique called *intarsia*.

In slivers of rosewood, oak, and beech, these designs depict in precise detail (and in linear perspective, a then newly invented technique) simulated cabinets full of precious objects — each one a symbol of what the duke most admired and to which he most aspired. A lute and a harp showed he was a man of culture; a mace and a pair of spurs represented his skill in battle; a bound volume of Virgil's *Aeneid* was a sign of his erudition. Incorporated into every corner of the space were mottos and motifs that represented the duke's personal, familial, and regional identity.

Federico's version of a private study was extraordinary, but in the centuries that followed, as people continued to flock from the countryside to the cities, such "thinking rooms" served a need that became more and more common. The fad for the *studiolo* spread north through Europe, as people of means added such rooms to their homes. Like Federico's study, these spaces often featured displays of meaningful or sacred objects: collections of books, scientific and musical instruments, religious relics. And they carved out a space of undisturbed quiet — space that made deep, fresh thinking possible. For one of the sixteenth century's most original thinkers, Michel de Montaigne, the study became a central metaphor for the freedom of thought he prized. (Montaigne, it should be noted, was not only a man of letters but also mayor of the city of Bordeaux.) Amid the hustle and bustle of social and commercial life, "we must reserve a back room, wholly our own and entirely free, wherein to settle our true liberty, our principal solitude and retreat," wrote Montaigne in his essay "Of Solitude." Montaigne's term for this room — *arrière-boutique,* or literally "behind the shop" — speaks to the close relationship between busy engagement and quiet withdrawal. In this room, Montaigne added, "we must for the most part entertain ourselves with ourselves."

For many of us, that space has now largely disappeared; starting in the middle of the twentieth century, the walls that had gone up hundreds of years earlier began to come down. In every sort of building — homes, schools, offices — walls that had once been welcomed as guardians of discrete or dedicated spaces were now disparaged as interruptions of a desired "openness." An unstructured space of potential came to feel preferable to a closed-off space of definition. This development was especially apparent in the places where we work. By the beginning of our own century, some 70 to 80 percent of American office workers labored in open-plan rooms.

Why did the wall-less workspace triumph over the private office? For one thing, it's cheaper. Open-plan office space can cost as much as 50 percent less per employee than more traditional office layouts because of its smaller footprint and lower interior construction costs. But there was also a theoretical rationale behind the enthusiasm for open offices

—a bold idea, if you will. The idea is this: take down the walls, throw everybody together in one big room, and communication will flow, with increased collaboration and creativity sure to follow.

The notion of promoting collisions in a shared space comes with an appealing historical and intellectual pedigree. The coffeehouse, as author Steven Johnson has told us in his influential writings on "where ideas come from," is the arena where the modern world was born. These buzzy gathering places, Johnson writes, "fertilized countless Enlightenment-era innovations; everything from the science of electricity, to the insurance industry, to democracy itself." New ideas, he argues, arise out of "the collisions that happen when different fields of expertise converge in some shared physical or intellectual space."

For example: during the years he lived in London, from 1764 to 1775, Benjamin Franklin spent many hours at a coffeehouse near St. Paul's Cathedral. In his book *The Invention of Air*, Johnson recounts how Franklin mingled there with a group of "freethinkers"—scientists, mathematicians, philosophers—who stimulated and inspired one another through their wide-ranging conversations. "There should be a plaque to commemorate that coffeehouse," Johnson has said. "It was really a tremendously generative space." Today's leaders and managers have seized on this notion: get people to "collide" with one another, the thinking goes, and magic will happen. How better to promote collisions than to remove the physical barriers that would keep them from happening?

And, in fact, it *is* the case that people who work near one another are more likely to communicate and collaborate. This finding was first demonstrated more than forty years ago by Thomas Allen, a professor at MIT who drew what has come to be known as the "Allen curve." The curve describes a consistent relationship between physical distance and frequency of communication: the rate of people's interactions declines exponentially with the distance between the spaces where they work. This would mean, for example, that people sitting six feet apart are four times more likely to talk regularly than people seated sixty-five feet apart. Allen found that fifty meters (about 165 feet) was the cutoff point for regular information exchange; beyond that distance, routine communication effectively ceased. People who are located close to one another are more likely to encounter one another, and it's these encounters that spark informal exchanges, interdisciplinary ideas, and fruitful collaborations.

Allen further observed that shared spaces through which every member of an organization passes at least once a day are especially useful encounter promoters. He offered as an example MIT's "Infinite Corridor," an 825-foot-long hallway that runs through several buildings, effectively extending from one side of the campus to the other. (That's longer than two football fields, although MIT students are less apt to be interested in football than in . . . other things. Undergraduates at the university celebrate the moment each school year when the sun lines up in just the right spot to beam its rays directly down the long hallway. They call it MIThenge.) More recent research has confirmed Allen's original findings; in the age of texting, email, and Slack, the Allen curve still applies. Online communication, it seems, is no substitute for the offhand conversation, the casual exchange carried out in person.

BUT: FRUITFUL PROXIMITY is one thing; continual distraction born of unenclosed, unprotected spaces is another. Appealing though it may be to modern sensibilities, the coffeehouse was always a terrible model for a place in which complex, cognitively demanding work is to be carried out. That's because the conditions created by the open-plan workspace are in direct conflict with the unalterable realities of human biology. The brain evolved to continually monitor its immediate environment — to be, in effect, distractible, lest nearby sounds or movements signal a danger to be avoided or an opportunity to be seized. And organizational environments are full of the kind of stimuli that distract us most.

First: humans are especially attuned to the presence of *novelty*, to whatever appears new and different. The pull of the novel on our attention is an efficient evolved strategy; it would be a waste of our time and energy to keep noticing the many things around us that don't change from day to day. But our selective attraction to the fresh and new becomes a problem when we operate in environments that are hubs of constant activity and change. Psychologist Fabrice Parmentier, who researches the effects of acoustic distraction, reports that unexpected sounds "ineluctably break through attentional filters," and end up distracting those who hear them — "regardless of the sound's informational value." Our attention, he notes, is subject to "involuntary capture" by any sudden or surprising noise.

Second: we are especially attuned to the sound of *speech*, especially when the words are distinct enough to make out. Any ambient noise can grab our attention, but intelligible speech is particularly distracting, because its semantic meaning is processed by our brains whether or not we want to be listening. What's more, the speech we can't help overhearing is processed by the same brain regions we employ to carry out the kind of knowledge work done at the office, like analyzing data or writing a report. Involuntarily auditing speech, and trying to complete tasks involving words or other symbols, means drawing on the same limited resource, with the result that we have "less brain" to devote to each. In a 2014 study conducted by researchers from the University of Gävle in Sweden, participants were asked to write short essays under five different acoustic conditions. Background noise in the five conditions ranged from 0.08 to 0.71 on a measure called the Speech Transmission Index — that is, from completely unintelligible speech, to somewhat intelligible speech, to crystal-clear speech. The participants' writing fluency, the investigators reported, dropped "drastically" at Speech Transmission Index values above 0.23 — levels that, they note, "would not be at all uncommon" in an open-plan office.

Third: we are especially attuned to the nuances of *social interactions,* alert to what people say to one another, and to what we think they *will* say. In an effort to master our interpersonal world, we're constantly making predictions about what will unfold in the social exchanges that go on around us. Our habit of projecting forward makes it especially hard for us to tune out one-sided conversations, such as those we hear when others near us are talking on the phone. Research by Lauren Emberson, an assistant professor of psychology at Princeton University, has found that we are more distracted, and our cognitive performance more impaired, when we overhear what she calls a "halfalogue" than when we catch both sides of an in-person dialogue. When we hear only half of a conversation, it's more difficult to predict when the speaker will pause or resume talking, and what that person will say to their unheard-by-us conversational partner. In a study Emberson published in 2010, participants asked to complete verbal and motor tasks started to make mistakes just moments after they began hearing a halfalogue.

What about wearing headphones? That takes the problem and puts it directly into our ears. Like overheard speech, music with lyrics competes for mental resources with activities that also involve language, such as reading and writing. Music has been found to impair performance on tasks that are difficult or complex, as well as those requiring creativity. And it's not just the words. Music, with its insistent beat and catchy riffs, is engineered to grab and hold our attention. Studies suggest that music with high intensity, fast tempos, and frequent variation is more distracting than more low-key, laid-back music. (Students who try to study while listening to high-intensity music like hip hop, one researcher found, are subject to what he memorably called the "attention drainage effect.") Music disrupts cognition for young people just as much as it does for adults. Perhaps most regrettably, people's intellectual performance while listening to music they prefer is "significantly poorer" than when listening to music they dislike.

What's true for our sense of hearing holds for vision as well. We can't choose not to see what enters our field of vision; it is nearly impossible to prevent our gaze from darting toward a visual stimulus that is new or in motion. And just as our ears prick up at the sound of people's speech, our eyes are drawn toward the human visage. The brain automatically prioritizes the processing of faces, even when we're trying hard to focus on a page or a screen. Our attention is pulled especially powerfully to the *gaze* of other people; we are uncannily sensitive to the feeling of being observed. Once we spot others' eyes on us, the processing of eye contact takes precedence over whatever else our brains were working on. An awareness of being looked at even increases our physiological arousal, as revealed by a spike in skin conductance. (Recall that when our nervous system becomes aroused, we start to perspire in a barely perceptible way. This slight sheen of sweat momentarily turns our skin into a better conductor of electricity.) When we glimpse a face in which the eyes are closed or directed elsewhere, we register no change in skin conductance—but that same measure shows a jolt whenever we see eyes looking our way.

All this visual monitoring and processing uses up considerable mental resources, leaving that much less brainpower for our work. We know this because of how much better we think when we *close* our eyes. Eye closure "helps people to disengage from environmental stimulation and thereby enhances the efficiency of cognitive processing," one team of researchers reports. Temporarily relieved of such stimulation, people experience less cognitive load, are better able to engage in visualization, and can more readily retrieve elusive information when faced with one of those frustrating "tip-of-the-tongue" moments. They're also much better at recalling details, both visual and auditory. One study reported a 23 percent increase in correct answers when participants closed their eyes as they answered questions about a film they had just watched.

Of course, we can hardly go about our work or our learning with our eyes closed. We have to rely on elements of physical space to save us from our own propensity for distraction — to impose the "sensory reduction" that supports optimal attention, memory, and cognition. "Good fences make good neighbors," wrote poet Robert Frost; likewise, good walls make good collaborators.

WALLS, AND the protected spaces they create, shield us from distraction. But they do more: they also provide us with *privacy*, a state that bears a surprising relationship to creativity.

The open-plan coffeehouse model now enjoying such popularity is performance oriented, almost exhibitionist in nature; think of Benjamin Franklin and his fellow debaters holding forth at the coffeehouse near St. Paul's. But putting oneself on display consumes mental resources, leaving less brainpower for the work itself. (The act of self-presentation may be particularly draining for members of certain groups. One recent study, conducted in a British government agency that switched from enclosed offices to an open-plan workspace, found that the heightened imperative to engage in self-presentation in such settings fell most heavily on women, for whom appearance is considered especially important.)

When people are relieved of the cognitive load imposed by their environment, they immediately become more creative, neuroscientist Moshe Bar has found. Bar, who directs the Gonda Multidisciplinary Brain Research Center at Bar-Ilan University in Israel, reports that when he taxed subjects' mental resources as they completed a test of creative thinking, they came up with more "statistically common" (that is, conventional and commonplace) associations. In his study, Bar found that "a high mental load consistently diminished the originality and creativity" of his subjects' responses. His explanation: when our minds are otherwise occupied, we resort to mental shortcuts — convenient stereotypes, familiar assumptions, well-worn grooves. These are the thoughts that come most readily to mind, that take the least mental energy to generate. It requires abundant cognitive resources to inhibit these stale, reflexive responses and to reach beyond them for ideas that are fresher and more original.

Privacy supports creativity in another way: it offers us the freedom to experiment unobserved. When our work is a performance put on for the benefit of others, we're less likely to try new approaches that might fail or look messy. Ethan Bernstein, an associate professor at Harvard Business School, has investigated the relationship between privacy and innovation at a mobile phone factory in China. In a study published in 2012, he found that granting the workers greater privacy — concealing their activities behind a curtain — led them to become more innovative and more productive. They came up with faster and more effective ways of doing their work when the process of experimentation was shielded from view.

A similar dynamic prevails in white-collar work, says Bernstein, where the surveillance that employees experience may be digital in nature. Professionals are less likely to play around with new ideas or approaches when they know that an all-seeing electronic eye is tracking their every keystroke. It's not simply that staffers fear their bosses will think they're goofing off or breaking the rules. Being subject to oversight at all times is a disempowering experience, and feeling powerless discourages exploration and creativity. Conversely, a number of studies have found that a sense of privacy leads to feelings of empowerment, which in turn lead to greater creativity.

Lastly, the benefits of privacy extend to our communications with colleagues. Effective collaboration often requires a degree of discretion — a retreat from public scrutiny that's hard to find in an open office. Research has found that employees have fewer and more superficial work-related conversations in wall-less spaces, often because they are leery of discussing delicate or confidential matters in the open. Ben Waber, whose firm, Humanyze, tracks employees' activities via wearable sensors, found that after companies move from a workplace with enclosed offices to an open-plan workspace, interactions among employees actually *decrease* during most times of the day. "This may be due to employees putting on headphones or the difficulty of engaging in conversation when dozens of people are within earshot," Waber notes. Other studies have found that as workspaces become more open, trust and cooperativeness among coworkers declines. The open-plan office seems to discourage exactly the kind of behavior it was intended to promote.

Such negative effects are especially pronounced in offices where employees have no assigned workspace at all, an increasingly common arrangement. The drawbacks of "hot-desking" or "hoteling," as it's called, point to another way in which space can be used to extend our minds (but too often is not). When we operate within a space over which we feel ownership — a space that feels like it's *ours* — a host of psychological and even physiological changes ensues. These effects were first observed in studies of a phenomenon known as the "home advantage": the consistent finding that athletes tend to win more and bigger victories when they are playing in their own fields, courts, and stadiums. On their home turf, teams play more aggressively, and their members (both male and female) exhibit higher levels of testosterone, a hormone associated with the expression of social dominance.

But the home advantage is not limited to sports. Researchers have identified a more general effect as well: when people occupy spaces that they consider their own, they experience themselves as more confident and capable. They are more efficient and productive. They are more focused and less distractible. And they advance their own interests more forcefully and effectively. A study by psychologists Graham Brown and Markus Baer, for example, found that people who engage in negotiation within the bounds of their own space claim between 60 and 160 percent more value than the "visiting" party.

Benjamin Meagher, an assistant professor of psychology at Kenyon College in Ohio, has advanced an intriguing theory that may explain these outcomes. The way we act, the way we think, and even the way we perceive the world around us differ when we're in a space that's familiar to us — one that we have shaped through our own choices and imbued with our own memories of learning and working there in the past. When we're on our home turf, Meagher has found, our mental and perceptual processes operate more efficiently, with less need for effortful self-control. The mind works better because it doesn't do all the work on its own; it gets an assist from the structure embedded in its environment, structure that marshals useful information, supports effective habits and routines, and restrains unproductive impulses. In a familiar space over which we feel ownership, he suggests, "our cognition is distributed across the entire setting." The place itself helps us think.

With ownership comes control, and a sense of control over their space — how it looks and how it functions — leads people to perform more productively. This was demonstrated in dramatic fashion in an experiment conducted by psychologists Craig Knight and Alex Haslam. In the study, volunteers were given a set of tasks to perform under one of four conditions: they worked in a lean office (spare, uncluttered); in an enriched office (decorated with posters and potted plants); in an empowered office (participants could arrange the room as they liked); and in a disempowered office (the room was rearranged in front of them, without their cooperation or consent).

In the lean office, found Knight and Haslam, participants invested a low level of effort in their assigned work; they were listless and lackadaisical. In the disempowered office, subjects' productivity was similarly mediocre; in addition, they were very, very unhappy. "I wanted to hit you," one participant confessed to the experimenter in a follow-up interview, describing how he felt as "his" office was rearranged to the researcher's liking. In the enriched office, participants worked harder and were more productive; in the empowered office, people performed best of all. They got 30 percent more done there than in the lean office, and about 15 percent more than in the enriched office. The size of such effects is large enough to make any employer sit up and take notice: three people working in empowered offices accomplished almost as much as four people in lean offices.

Perhaps the most important form of control over one's space is authority over who comes in and out—a point missed by those who believe that our workspaces should resemble a bustling coffeehouse. The informal exchanges facilitated by proximity are indeed generative. But the value of such interactions can be extracted only if it is also possible, when necessary, to avoid interacting at all. Consider again in this light the denizens of the St. Paul's coffeehouse, who surely had private studies to which to retreat at home, or the professors who traverse MIT's Infinite Corridor —on their way to their own quiet, book-lined offices.

It's certainly the case that the nature of today's work demands frequent consultation and cooperation with others. What we may not realize is that good work also requires periods of abstention from such exchanges—a phenomenon that organizational psychologists call "intermittent collaboration." Research on intermittent collaboration is based on the understanding that complex problem solving proceeds in two stages, the first of which entails gathering the facts we need to clarify the nature of the problem and begin constructing a solution. In this stage, communication and collaboration are essential. But there is a second phase, equally vital: the process of generating and developing solutions, and figuring out which of these solutions is best. During this phase, studies find, excessive collaboration is actually *detrimental*.

The reason can be found in our nature as a group-dwelling species. We are exquisitely sensitive to social pressure, easily drawn into consensus and conformity. When we're constantly in touch with others, we all end up gravitating toward the same pretty-good-but-not-great answers. Research finds that people who keep lines of communication perpetually open consistently generate middling solutions — nothing terrible, but nothing exceptional either. Meanwhile, people who isolate themselves during the solution-generation phase tend to come up with a few truly extraordinary solutions — along with a lot of losers. The best of all worlds is enjoyed by those who engage in cycles of sociable interaction and quiet focus. Just as we need walls to protect us from our propensity to be distracted, so we require walls to shield us from our susceptibility to social pressure.

WHAT ARRANGEMENT of space could support this way of thinking and working? A surprisingly apt model can be found in the one adopted by Jonas Salk and Louis Kahn: the monastery. In the popular imagination, monks are solitary, hermit-like creatures — but historically they have lived within a communal setting that balanced time spent alone in study and contemplation with time spent with others in robust social interaction. Richard Irvine, an anthropologist at the University of Cambridge, explored this equilibrium in an ethnographic study he conducted at Downside Abbey, a Benedictine monastery in Somerset, England, where the way of life has remained little changed for centuries.

In describing the abbey's architecture, Irvine observes that the buildings reflect their inhabitants' daily cycles of intense engagement and hushed withdrawal, accommodating communal spaces like the library, the refectory, the workshop, and the courtyard, as well as the monks' solitary cells. The monastery even has its own version of the Infinite Corridor — in the form of the cloister, a long passageway that "serves as the key element of connection in the architecture of the monastery," Irvine notes. "The cloister is a space of movement which facilitates frequent interpersonal encounter, as the monks pass through it regularly to reach the common areas of Abbey Church (where they attend six communal services a day) [and the] refectory (where they join together in silence to eat three meals a day)."

Though organizational psychologists have only recently recognized its value, the monks of Downside Abbey have been practicing intermittent collaboration for more than four hundred years. "While the monastery is the site of frequent interpersonal encounter, the importance of solitude is also structured into the timetable through the commitment to twice-daily private prayer, as well as the *summum silentium* (complete silence, sometimes referred to as 'the great silence') at the end of the day," Irvine explains. This silence "restricts interaction and gives the monk opportunity to be alone." (He notes that members of the community have available to them another way of ensuring intermittent interaction: they can pull up the hood of their vestment, "hence covering the ears and closing off part of the peripheral vision, thus [making them] less inclined to distraction from other people.")

The ancient arrangement of space in a monastery bears some resemblance to today's "activity-based workspaces," which nod to the human need for both social interaction and undisturbed solitude by providing dedicated areas for each: a café-style meeting place, a soundproof study carrel with a door. Too often, however, such offices still fail to provide their occupants with that most effective mental extension: a private space, persistent and therefore familiar, over which they have a sense of ownership and control. Such a space can generate benefits even beyond those we've already covered because they are so well positioned to meet two additional pressing human needs: the need to claim one's own identity and the need to belong to a larger group. For this extension of mind by physical space, there can be no better model than the *studiolo*.

Recall the richly decorated walls of the study belonging to Federico da Montefeltro, Duke of Urbino. They surrounded their occupant with visual reminders of who he was: an aesthete, a warrior, a scholar. There was the lute and the harp, the mace and the pair of spurs, the bound volume of Virgil. On display as well were symbols representing his membership in meaningful groups. Worked into the intricately inlaid paneling was the Montefeltro *imprese* — a kind of family crest — featuring an image of an ostrich holding an arrow in its beak; inscribed underneath the bird was a swaggering motto, rendered in German and first proclaimed by Federico's grandfather: *"Ich kann ein großes Eisen essen,"* or "I can eat a big iron." The walls also featured the emblem of the Order of the Garter, England's highest chivalric order, awarded to Federico by King Edward IV.

Self-referential images and messages are not mere decorations — whether they're built into the paneling of a duke's splendidly outfitted retreat or tacked to the walls of an office worker's cubicle. Research shows that in the presence of cues of *identity* and cues of *affiliation,* people perform better: they're more motivated and more productive. The first of these are the tangible signs and signals we employ to support our self-conception: we're the kind of person who likes cats, or rock climbing, or "Far Side" cartoons. We use our space to advertise our hobbies, to show off our awards and honors, to express an unexpected creative streak or a quirky sense of humor. Such displays may sometimes be aimed at informing other people of who we are (or who we'd like to be), but often they are intended for a more intimate audience: ourselves. For a study published in the *Academy of Management Journal,* researchers examined the workspaces of people holding a variety of jobs — from engineer to event planner, from creative director to real estate agent. The investigators found that about one-third of the personal tokens these professionals had incorporated into their workspaces were positioned so as to be visible only to their owners. Of the objects whose stated purpose for their owners was "reminding them of their goals and values," 70 percent were placed out of sight of others.

Why would we need such reminders? While our sense of self may feel stable and solid, it is in fact quite fluid, dependent on external structure for its shape. We can sense this truth when we travel to a foreign country. Everything around us is strange and unfamiliar, producing in us a dizzy sense of discombobulation. This disorientation can be pleasurable, if exhausting, while on a holiday in a faraway land — but in our day-to-day lives, we need to cultivate a steady sense of identity in order to function effectively. The material things we arrange around us help us maintain that sturdy self-conception. As the psychologist Mihaly Csikszentmihalyi has written, we keep certain objects in view because "they tell us things about ourselves that we need to hear in order to keep our selves from falling apart."

Moreover, we need to have close at hand those prompts that highlight particular *facets* of our identity. Each of us has not one identity but many

—worker, student, spouse, parent, friend—and different environmental cues evoke different identities. Daphna Oyserman, a psychologist at the University of Southern California, notes that signals from the environment function to bring one of these many personas to the fore, with real effects on our thought and behavior. "Which identity is salient in the moment influences both what one pays attention to and what one chooses to do," she writes. One particularly striking example: research has found that cues that remind Asian American girls of their ethnicity improve their performance on math tests, while cues that remind them of their gender undermine their performance. For all of us, the objects on which our eyes come to rest each day reinforce what we're doing in that place, in that role.

Through the ups and downs of our lives at school and at work, the reassuring stability of meaningful material objects can also help us manage our moods and emotions. When we engage in such "environmental self-regulation," we rely on cues *outside* ourselves to maintain the kind of equilibrium *inside* ourselves that facilitates the pursuit of our goals. In a study of mid-level professionals, Gregory Laurence, a professor of management at the University of Michigan–Flint, found that incorporating personal items into their workspaces helped them relieve the "emotional exhaustion" brought on by a stressful job. Especially for employees whose office settings did not afford much privacy, being able to personalize their work area—with photographs, posters, comic strips, mugs—helped them "carve out their own space, inscribe it with personal meaning, and thus create a kind of sanctuary at work," write Laurence and his coauthors.

Laurence is not the only researcher to detect an almost spiritual cast to the way people regard their personal space. In an ethnographic study of research and development professionals in the US headquarters of the Japanese technology company Hitachi, authors Ryoko Imai and Masahide Ban note that the employees they studied did "intensive reading, writing, and most importantly, thinking in the comfort and solitude of [their] own cubicles . . . Often stocked with personal comfort items, familiar references, [and] favorite tools, a dedicated private space served as a sacred space to rejuvenate and regroup." It's possible to see shades of the monastery and the *studiolo* in these modern-day workstations and cubicles—evidence of the persistent human need to imbue our spaces with meaning and significance.

Yet many organizations discourage or even ban the display of personal items in employees' workspaces. Such "clutter" may be frowned upon as an irrelevant distraction from the work at hand, or as an obstacle to achieving the clean, spare aesthetic associated with admired leaders like the late Apple Computer founder Steve Jobs. For psychologists Craig Knight and Alexander Haslam, such dictates bring a different figure to mind: Frederick Winslow Taylor, the early-twentieth-century engineer who introduced "scientific management" to American corporations. Taylor specifically prohibited workers from bringing their personal effects into the factories he redesigned for maximum speed and minimum waste. Stripped of their individuality, he insisted, employees would function as perfectly efficient cogs in the industrial machine.

Knight and Haslam, who conducted the study of "lean" versus "enriched" and "empowered" versus "disempowered" working environments described earlier, believe that an embrace of neo-Taylorism is a mistake, especially in an era when employees are expected to behave not as generic cogs but as critical and creative thinkers. As their own research has demonstrated, people are *less* productive in a lean, featureless office. "Making people feel they are in the wrong place, that it isn't their place, is a very powerful way of undermining performance," says Haslam. Although managers and administrators may fear that the presence of personal tokens will reduce organizational cohesion and loyalty, research evidence suggests that just the opposite is the case: employees feel *more* committed to their company when they are able to see themselves reflected in it. A sense of ownership extends from the individual to the organization, and it flows through physical space.

THE WAY SPACE is arranged can recognize our individuality, with positive effects on our motivation and performance. It can also affirm — or deny — our sense of *belonging* within an organization. Members of groups that have historically been excluded, marginalized, or negatively stereotyped are especially attuned to the signals of belonging or non-belonging they encounter in the environments they enter. Such signals are pervasive and powerful, though they are rarely the focus of discussions about bias. "When we think of prejudice, most of us think of it as a problem of *people*," says Mary Murphy, a professor of psychological and brain sciences at Indiana University Bloomington. But, she points out, inequalities of experience and outcome are also generated by features of the "organizational setting," of which the built environment is a significant part. Murphy and her colleagues have advanced a theory of "prejudiced places," which they define as places that "unequally tax the emotions, physiology, cognitive function, and performance of some groups more than others."

When we regard prejudice as a property of people alone — as an entity that exists inside individuals' heads — we fail to see the full picture of how bias operates within institutions, and we miss out on opportunities to push back against it. According to what Murphy calls the "prejudice-in-places model," systemic inequality will not be eradicated by identifying and rooting out a few bad actors who hold racist or sexist beliefs. Indeed, prejudice can be perpetuated by places even when the people within them are endeavoring to act in egalitarian ways. Precisely because physical places exert such a profound effect on the behavior of the people who work and learn within them, altering aspects of these spaces might offer the most effective route to reducing bias. Efforts to change people's beliefs more directly may provoke resistance or resentment — and, in fact, research has repeatedly found unimpressive results for conventional diversity workshops and training sessions.

"Prejudiced places" might well have described the environments Sapna Cheryan encountered in the summer of 2001. Recently graduated from college, Cheryan was interviewing for internships at tech firms located in California's Bay Area. At one of the companies she visited, she found herself put off by a workspace that looked like a computer geek's basement hangout: "the action figures, the Nerf guns, the soda cans stacked up to make a model of the Golden Gate Bridge," as Cheryan recalls it now. The firm's adherence to a very particular aesthetic seemed designed to promote an exclusive conception of its ideal employee — one that felt unwelcoming or even alienating to her, a young woman and person of color. Cheryan encountered a very different setting when interviewing at the software company Adobe; there, the workspace was bright and inviting. Cheryan accepted Adobe's offer and stayed on at the firm for five years. Her next move was to a psychology PhD program at Stanford University, where she enrolled with the aim of investigating how cues in the physical environment affect the way people think.

In an experiment she conducted while still a graduate student,

Cheryan commandeered space in Stanford's Gates Computer Science Building, creating what she called a "stereotypical" classroom and a "non-stereotypical" classroom. The stereotypical classroom was filled with soda cans, books of science-fiction fantasy, and *Star Trek* and *Star Wars* posters. The non-stereotypical classroom featured accoutrements like nature posters, literary novels, and bottles of water. After spending time in each room, undergraduates were surveyed about how interested they were in computer science and how well they thought they would perform in that discipline.

Following a few minutes in the stereotypical room, male students reported a high level of interest in pursuing computer science; female students indicated much less interest than their male counterparts. But after they spent time in the non-stereotypical room, women's interest in computer science increased markedly—actually exceeding that of the men. Subsequent research by Cheryan has found that women exposed to a non-stereotypical classroom are more likely to predict that they will perform well in computer science courses, whereas men tend to predict that they will succeed regardless of which room they encounter. That's important, says Cheryan, because "we know from past work in psychology that how well you expect to do in a certain environment can determine how you actually perform."

Cheryan called the phenomenon documented in her study "ambient belonging," defined as individuals' sense of fit with a physical environment, "along with a sense of fit with the people who are imagined to occupy that environment." Ambient belonging, she proposed, "can be ascertained rapidly, even from a cursory glance at a few objects." In the research she has since produced, Cheryan has explored how ambient belonging can be enlarged and expanded—how a wider array of individuals can be induced to feel that crucial sense of "fit" in the environments in which they find themselves. The key, she says, is not to *eliminate* stereotypes but to *diversify* them—to convey the message that people from many different backgrounds can thrive in a given setting. Just such an effort was undertaken at the University of Washington, where Cheryan is now a professor. UW revamped its computer science lab, applying a fresh coat of paint, hanging new artwork, and arranging the seating to encourage more social interaction. Five years later, the proportion of undergraduate computer science degrees earned by women at UW reached 32 percent, higher than at any other flagship public university in the country.

Cheryan and others are now exploring how to create a sense of ambient belonging in online "spaces," an example of extending technology with what we know to be true about physical spaces in the offline world. As is the case in "real life," research finds that members of historically stigmatized groups are especially attuned to cues of exclusion that appear on digital platforms, such as online courses. And just as in non-digital realms, these signals of non-belonging can negatively affect their levels of interest, and their expectations of success, in subjects like computer science, as well as influencing how actively they engage these topics.

René Kizilcec, an assistant professor of information science at Cornell University, investigated the effect of adding a "gender-inclusive" image and statement to an advertisement for an online computer science course that ran on Facebook. The image featured a group of eight women of diverse ages and ethnicities; the statement read: "The history of computer programming is a history of WOMEN. You can join this epic journey." Female viewers clicked through to find out more at a rate that was 26 percent higher than for a similar ad that did not include such cues. When gender-inclusive signals were embedded in the enrollment page for the course, women were 18 percent more likely to sign up.

In another study led by Kizilcec, the inclusion of a diversity statement on the Web pages of online courses in science, technology, engineering, and mathematics increased the enrollment of students of lower socioeconomic status — another group that is often underrepresented in STEM fields. "This is an equal opportunity course that offers you a supportive and inclusive space to learn," the statement read. "Everyone, no matter their age, gender, or nationality, can be successful in this course. People like you are joining from all over the world and we value this diversity." Such "psychologically inclusive design," as Kizilcec calls it, aims to lower barriers to participation by "strategically placing content and design cues in the environment" — an effort that is as important in the online world as it is in the material one.

BRINGING EMPIRICAL RESEARCH to bear on our experience of the built environment is a relatively new phenomenon. For centuries, architects and builders have worked on the basis of tradition and intuition. Architect Louis Kahn, for example, consciously drew on the forms and styles of the past; he relied as well on his tactile sense of the material world, and on what the site or the structure demanded. In one of his characteristically gnomic remarks, Kahn described an imagined exchange between himself and the stuff of which his buildings were made. "You say to brick, 'What do you want, brick?'" he began. "Brick says to you, 'I like an arch.' If you say to brick, 'Arches are expensive, and I can use a concrete lintel over an opening. What do you think of that, brick?' Brick says, 'I like an arch.'"

Jonas Salk observed of his architect that Kahn had "the vision of an artist, the understanding of a philosopher," and "the knowledge of a metaphysician." Sometime in the future, architects may need to add to that arsenal "the expertise of a neuroscientist." The burgeoning discipline of neuroarchitecture has begun to explore the way our brains respond to the built settings we encounter—finding, for example, that high-ceilinged places incline us toward more expansive, abstract thoughts. That's not hard to believe when one pictures the domed ceiling of the reading room at the Bibliothèque Nationale in Paris, rounded like a bouquet of balloons, or the ethereal, cloud-studded ceiling of the Rose Main Reading Room at the New York Public Library in Manhattan. Symmetrical shapes, meanwhile, impress upon us a sense of power and robustness. Call to mind the Taj Mahal, built of white marble in the Indian city of Agra by the seventeenth-century Mughal emperor Shah Jahan. It is perfectly symmetrical, from the pair of minarets arrayed on either side, down to the tiles inlaid on the floor. (It's noteworthy that the Salk Institute is also symmetrical, and indeed has been compared to the Taj Mahal.)

As intensely social creatures, we respond positively to architectural designs that resemble faces—like the beloved Villa Rotonda, an Italian Renaissance villa designed by architect Andrea Palladio. Curved shapes, it has been found, induce in us a sense of ease and comfort; babies as young as one week old prefer to look at curved objects rather than sharp ones. The attraction goes so deep that it is found across generations, across cultures—even across *species*. Consider the undulating sculpture *Sheep Piece*, created in 1971–72 by the British artist Henry Moore. A study that scanned the brains of rhesus monkeys found that the areas of their brains associated with the experience of reward and pleasure lit up when they looked at it. The human animal is similarly delighted by the sight of gentle curves, like those that characterize the building Frank Gehry designed to house the Guggenheim Museum in Bilbao, Spain. When the eminent architect Philip Johnson first saw it, he was so moved that he began to weep.

Though its insights are suggestive, neuroarchitecture is far from a mature discipline. For now, when building and designing we still need to rely on our stock of conventional forms—and from these we must choose carefully, knowing how profoundly setting can shape the way we think and act. We've seen how fully the coffeehouse model has come to dominate our notion of how workplaces should be arranged, for example. It is quintessentially modern in its rejection of bounded spaces and closed social circles, its embrace of transparency and openness—but it bears both the advantages and the drawbacks of this stance. The appeal of such places, so suited to our current sensibilities, can blind us to the virtues of other models—premodern forms, such as the monastery and the *studiolo,* that served their inhabitants well before the coffeehouse was even invented.

But perhaps more concerning than spaces that affect our thinking in problematic ways are spaces that decline to shape us at all. Richard Coyne, a professor of architectural computing at the University of Edinburgh in Scotland, has lamented the "cognitive deficiency of non-places," those spaces—all too common in the modern world—that are empty of cues or associations. Recall the major finding from Roger Barker's "Midwest Study": physical places influence our thinking and behavior far more than personality or other factors. This is possible because places offer our minds so much to work with—a "rich layering of custom, history, and meaning," as Coyne writes. He continues: "A sign saying 'wait here' would be superfluous in the vestibule of the cathedral or temple, as the appropriate behavior or action is already inscribed in the architecture and ritual practices of the place. Neither would we require a text saying 'Think of God,' or 'Consider your finitude' in such places. In fact it could be said that we are already caught up in such thought by virtue of being in the sacred place." But such rich signification is missing from non-places. What meaning or message is inscribed on a featureless chain store, or a generic hotel lobby, or the bleak urban "plaza" that surrounds many a skyscraper? What thoughts are inspired, what emotions are stirred by a row of beige cubicles, or a classroom housed inside a windowless trailer? We are set adrift in such spaces, alienated and purposeless. This is not simply a question of aesthetics; it is a question of what we think, how we act, who we *are*.

Not surprisingly, Louis Kahn was one who understood the transformative effect places could have on the human psyche. Kahn, deeply versed as he was in the history of architecture, once reflected on the felt impact of the soaring design of public baths in ancient Rome. "If you look at the Baths of Caracalla," said Kahn, "we all know that we can bathe just as well under an eight-foot ceiling as we can under a 150-foot ceiling." But, he went on, "there's something about a 150-foot ceiling that makes a man a different kind of man."

6

■

Thinking with the Space of Ideas

B EN PRIDMORE IS FAMOUS for his astonishingly accurate memory. A three-time winner of the World Memory Championship, Pridmore has pulled off feats such as reciting without error almost a hundred historic dates after just five minutes' study, correctly recalling the order of more than 1,400 randomly shuffled playing cards, and committing to memory the digits of pi to thousands of decimal places. He was prominently featured in the best-selling *Moonwalking with Einstein*, a 2011 book by journalist Joshua Foer detailing the triumphs of memory prodigies.

Yet Pridmore, a resident of the town of Redditch in the United Kingdom, can't be counted on to remember his "lucky hat"—a black fedora that brought him good fortune at memory competitions, until he left it behind on a train. Pridmore sometimes forgets to bring his briefcase or important papers to his job as an accountant, and he admits that he is hopeless at remembering friends' birthdays. "I am famously bad at being able to remember people's names and faces," he concedes. He became a celebrated memory champion only through the application of what is known as the "method of loci": a mental strategy that draws on the powerful connection to *place* that all humans share.

The method of loci is a venerable technique, invented by the ancient Greeks and used by educators and orators over many centuries. It works by associating each item to be remembered with a particular spot found in a familiar place, such as one's childhood home or current neighborhood. For Ben Pridmore, this place is Queen Elizabeth's Grammar School, which he attended as a child growing up in Horncastle, England. Preparing to recall the sequence of, say, a randomly shuffled deck of cards, he imagines placing each card, in order, in a succession of physical locations he would pass by if strolling through his old school: through the front door, down the corridor, past the sixth-form common room, into the classroom where math was taught. The method of loci — also known as the "memory palace" strategy, or in Pridmore's phrasing, the "journey technique" — is remarkably effective. On their own, bits of data like the number or suit shown on a playing card are quickly forgotten. But when linked to a physical place we know well, that same information can be durably integrated into memory.

Pridmore is not the only memory champion to make use of the method of loci. Indeed, research conducted with other memory contest winners has concluded that the strategy of tying new information to preexisting memories of physical space is the key to the extraordinary performance of many of these "memory athletes." One such study was conducted by Eleanor Maguire, a professor of cognitive neuroscience at University College London. "Using neuropsychological measures, as well as structural and functional brain imaging, we found that superior memory was not driven by exceptional intellectual ability or structural brain differences," Maguire and her coauthors reported. "Rather, we found that superior memorizers used a spatial learning strategy, engaging brain regions such as the hippocampus that are critical for memory and for spatial memory in particular." The difference between "superior memorizers" and ordinary people, Maguire determined, lay in the *parts* of the brain that became active when the two groups engaged in the act of recall; in the memory champions' brains, regions associated with spatial memory and navigation were highly engaged, while in ordinary people these areas were much less active.

What sets memory champs apart, then, is their conscious cultivation of an ability every one of us comes by naturally — the capacity to find our way around and to remember where we've been. Research has found that all of us seem to use the brain's built-in navigational system to construct mental maps, not just of physical places but of the more abstract landscape of concepts and data — the space of ideas. This repurposing of our sense of physical place to navigate through purely mental structures is reflected in the language we use every day: we say the future lies "up ahead," while the past is "behind" us; we endeavor to stay "on top of things" and not to get "out of our depth"; we "reach" for a lofty goal or "stoop" low to commit a disreputable act. These are not merely figures of speech but revealing evidence of how we habitually understand and interact with the world around us. Notes Barbara Tversky, a professor of psychology and education at Teachers College in New York: "We are far better and more experienced at spatial thinking than at abstract thinking. Abstract thought can be difficult in and of itself, but fortunately it can often be mapped onto spatial thought in one way or another. That way, spatial thinking can substitute for and scaffold abstract thought."

Scientists have long known that the hippocampus is centrally involved in our ability to navigate through physical space. More recently, researchers have shown that this region is engaged in organizing our thoughts and memories more generally: it maps abstract spaces as well as concrete ones. In a study published in 2016, neuroscientist Branka Milivojevic, of the Donders Institute for Brain, Cognition and Behaviour in the Netherlands, scanned the brains of a group of volunteers as they watched the 1998 movie *Sliding Doors*. In this romantic comedy-drama, the main character, Helen — played by actress Gwyneth Paltrow — meets two different fates. In one storyline, she makes it onto a train and returns home in time to find her boyfriend in bed with another woman. In a second, parallel storyline, she misses the train and remains oblivious to her boyfriend's infidelity. As the study participants watched the film, Milivojevic and her collaborators observed activity in their hippocampi identical to that of people who are mentally tracing a path through a physical space. Milivojevic proposes that the viewers of *Sliding Doors* were effectively navigating through the events of the movie, finding their way along its branching plotline and constructing a map of the cinematic territory as they went. We process our firsthand experiences in the same manner, she submits.

Some researchers have even suggested that the way our sense of space helps organize mental content can explain the puzzling phenomenon of "infantile amnesia" — the fact that we can't recall much about our earliest years. Because very young children are not able to move through space under their own locomotion, the theory goes, they may lack a mental scaffold on which to hang their memories. Children's impressions of their own experiences may become well enough structured to be memorable only once kids are able to move about of their own volition. As adults, our memories continue to be tagged with a sense of the physical place where the original experience occurred. When re-listening to a podcast or audiobook, for example, we may find that we spontaneously recall the place where we first heard the words. The automatic place log maintained by our brains has been preserved by evolution because of its clear survival value: it was vitally important for our forebears to remember where they had found supplies of food or safe shelter, as well as where they had encountered predators and other dangers. The elemental importance of *where* such things were located means the mental tags attached to our place memories are often charged with emotion, positive or negative — making information about place even more memorable.

All of us possess this powerful place-based memory system simply by virtue of being human — but some, like Ben Pridmore and his memory champion peers, make far better use of it. The rest of us can learn to do what they do, as demonstrated in a study led by neuroscientist Martin Dresler of Radboud University in the Netherlands. Dresler and his collaborators (who included Boris Nikolai Konrad, himself a decorated memory athlete) tested two dozen of the world's top memory competitors, comparing their performance on a word-memorization task to that of a group of ordinary citizens. Not surprisingly, the memory champions came out ahead, correctly recalling an average of seventy-one words from the list of seventy-two words they were given, with a number of them turning in perfect scores. Regular people recalled an average of only twenty-nine words. After six weeks of training using the method of loci, however, these previously undistinguished experimental participants turned in an impressively improved performance, more than doubling their average score.

Associating information to be learned with a sense of physical space can help people remember in real-world situations too — whether it's a high school student memorizing verb conjugations, a medical resident learning a litany of diseases and their symptoms, or a best man practicing a speech for a rehearsal dinner. Undergraduates who take the course on civil liberties taught by Charles Wilson, a professor of political science at the University of North Georgia, have to learn a multitude of new facts and ideas. Wilson helps them do so by showing them how to link individual pieces of information to particular locations within a space they know well: the campus cafeteria, familiarly called the "Chow Hall."

For students struggling to recall the provisions set forth in the Bill of Rights, for example, Wilson encourages them to imagine stepping up to the Chow Hall soup tureens: the soup course, which comes first in a meal, thus becomes associated with the First Amendment. Then on to the station where bread is sliced; here Wilson recommends that students picture a pile of the severed limbs of grizzly bears, a scenario sure to prompt a memory of the Second Amendment and its assurance of the right to "bear arms."

That image is a bit grisly, but Wilson and his students have found that imagery that is garish or outlandish more readily triggers recall. Obeying that principle, Wilson's students learn to associate the landmark court case of *McDonald v. City of Chicago* with an image of the fast food clown Ronald McDonald, wearing a Chicago Bulls jersey, filling his plate at the salad bar. And so on through the remaining eight amendments in the Bill of Rights, each one paired with an imagined sampling of the Chow Hall's culinary offerings. Wilson's students enjoy the exercise, and they say it helps them enormously in remembering material for the course. Many students have told him that they've started using the method of loci to study for their other classes as well.

The human brain is not well equipped to remember a mass of abstract information. But it is perfectly tuned to recalling details associated with places it knows — and by drawing on this natural mastery of physical space, we can (as Martin Dresler showed) more than double our effective memory capacity. Extending our minds via physical space can do more than improve our recall, however. Our powers of spatial cognition can help us to think and reason effectively, to achieve insight and solve problems, and to come up with creative ideas. Such powers are especially generative when permitted to operate not on imagined space, as in the method of loci, but on the real thing: tangible, three-dimensional space, of the kind our minds and bodies are so accustomed to navigating.

Our culture tends to valorize doing things in one's head; we are awed by mathematicians who can engage in elaborate mental calculations, chess grandmasters who can plot a long progression of moves in their mind's eye — and, yes, memory champions, who can remember so much without reference to external prompts. But true human genius lies in the way we are able to take facts and concepts *out* of our heads, using physical space to spread out that material, to structure it, and to see it anew. The places we make for ideas can take many forms: a bank of computer screens, the pages of a field notebook, the surface of a workshop table — or even, as one celebrated author demonstrates, an expanse of office wall.

"BRILLIANT," "masterful," and "monumental" are words regularly used to describe the work of historian Robert Caro. He was awarded a Pulitzer Prize for *The Power Broker,* his sweeping biography of the urban planner Robert Moses. Assigned in many college courses, the book has sold more than 400,000 copies and has never been out of print since its publication in 1974. For the past four decades, Caro has been writing about the mid-century political maestro Lyndon B. Johnson — four volumes and counting, including *Master of the Senate* (another Pulitzer winner) and *Means of Ascent.* In total, he has written more than four thousand well-honed, fact-rich pages of prose over the course of an acclaimed career.

But at first Caro struggled even to wrap his head around his subjects. While researching and reporting *The Power Broker,* he was overwhelmed by the volume of information he had collected. "It was so big, so immense," he has said. "I couldn't figure out what to do with the material." Caro's books are too colossal to be held entirely in mind — even their author's mind. Nor is the space of a typewritten page (Caro does not use a computer) nearly big enough to contain the full sweep of his storytelling. In order to complete these massive projects, Caro has to extend his thinking into physical space. One entire wall of his office on Manhattan's Upper West Side is taken up by a cork board four feet high and ten feet wide; the board is covered with a detailed outline of Caro's current work in progress, plotting its trajectory from beginning to end. (So thorough is Caro that he must know the *last* sentence of a book before he starts composing its first lines.)

As he writes, the stretch of wall becomes another dimension in which to think. "I can't start writing a book until I've thought it through and can see it whole in my mind," he told a visitor to his office. "So before I start writing, I boil the book down to three paragraphs, or two or one — that's when it comes into view. That process might take weeks. And then I turn those paragraphs into an outline of the whole book. That's what you see up here on my wall now." In another interview, Caro explained how the outline wall helps him stay in the zone. "I don't want to stop while I'm writing, so I have to know where everything is," he explained. "It's hard for me to keep in the mood of the chapter I'm writing if I have to keep searching for files."

Out of necessity, Caro found his way to a mode of thinking and working that would not have been possible had he tried to keep his voluminous material entirely in his head. "When thought overwhelms the mind, the mind uses the world," psychologist Barbara Tversky has observed. Once we recognize this possibility, we can deliberately shape the material worlds in which we learn and work to facilitate mental extension — to enhance "the cognitive congeniality of a space," in the words of David Kirsh, a professor at the University of California, San Diego.

To understand how this works, let's take a closer look at what Caro's wall is doing for his mind. On the most basic level, the author is using physical space to *offload* facts and ideas. He need not keep mentally aloft these pieces of information or the complex structure in which they are embedded; his posted outline holds them at the ready, granting him more mental resources to think *about* that same material. Keeping a thought in mind — while also doing things to and with that thought — is a cognitively taxing activity. We put part of this mental burden down when we delegate the representation of the information to physical space, something like jotting down a phone number instead of having to continually refresh its mental representation by repeating it under our breath.

Caro's wall turns the mental "map" of his book into a stable external artifact. This is the second way in which the cork board in Caro's office extends his ability to think: looking it over, he can now see — far more clearly and concretely than if the map had remained inside his head — how his ideas relate to one another, how the many paths taken by his narrative twist and turn, diverge and converge. Although Caro tailored his longtime method to suit his particular style of working, the strategy he came up with is similar to one that has received substantial empirical support from psychology: an approach known as *concept mapping*. A concept map is a visual representation of facts and ideas, and of the relationships among them. It can take the form of a detailed outline, as in Robert Caro's case, but it is often more graphic and schematic in form.

Research has revealed that the *act* of creating a concept map, on its own, generates a number of cognitive benefits. It forces us to reflect on what we know, and to organize it into a coherent structure. As we construct the concept map, the process may reveal gaps in our understanding of which we were previously unaware. And, having gone through the process of concept mapping, we remember the material better—because we have thought deeply about its meaning. Once the concept map is completed, the knowledge that usually resides inside the head is made visible. By inspecting the map, we're better able to see the big picture, and to resist becoming distracted by individual details. We can also more readily perceive how the different parts of a complex whole are related to one another.

Joseph Novak, now a professor emeritus of biology and science education at Cornell University, was investigating the way children learn science when he pioneered the concept-mapping method in the 1970s. Although the technique originated in education, Novak notes that it is increasingly being applied in the world of work—where, he says, "the knowledge structure necessary to understand and resolve problems is often an order of magnitude more complex" than that which is required in academic settings. Concept maps can vary enormously in size and complexity, from a simple diagram to an elaborate plan featuring hundreds of interacting elements.

Robert Caro's map, for example, is *big*: big enough to stand in front of, to walk along, to lean into and stand back from. The sheer expansiveness of his outline allows Caro to bring to bear on his project not only his purely cognitive faculties of reasoning and analysis but also his more visceral powers of navigation and wayfinding. Researchers are now producing evidence that these ancient evolved capacities can help us to think more intelligently about abstract concepts—an insight that showed up first in, of all places, a futuristic action film.

THE SCENE FROM the 2002 movie *Minority Report* is famous because, well, it's just so cool: Chief of Precrime John Anderton, played by Tom Cruise, stands in front of a bank of gigantic computer screens. He is reviewing evidence of a crime yet to be committed, but this is no staid intellectual exercise; the way he interacts with the information splayed before him is active, almost tactile. He reaches out with his hands to grab and move images as if they were physical objects; he turns his head to catch a scene unfolding in his peripheral vision; he takes a step forward to inspect a picture more closely. Cruise, as Anderton, physically navigates through the investigative file as he would through a three-dimensional landscape.

The movie, based on a short story by Philip K. Dick and set in the year 2054, featured technology that was not yet available in the real world — yet John Anderton's use of the interface comes off as completely plausible, even (to him) unexceptional. David Kirby, a professor of science, technology, and society at California Polytechnic State University, maintains that this is the key to moviegoers' suspension of disbelief. "The most successful cinematic technologies are taken for granted by the characters" in a film, he writes, "and thus, communicate to the audience that these are not extraordinary but rather everyday technologies."

The director of *Minority Report*, Steven Spielberg, had an important factor working in his favor when he staged this scene. The technology employed by his lead character relied on a human capacity that could hardly be more "everyday" or "taken for granted": the ability to move ourselves through space. For added verisimilitude, Spielberg invited computer scientists from the Massachusetts Institute of Technology to collaborate on the film's production, encouraging them "to take on that design work as if it were an R&D effort," says John Underkoffler, one of the researchers from MIT. And in a sense, it was: following the release of the movie, Underkoffler says, he was approached by "countless" investors and CEOs who wanted to know "Is that real? Can we pay you to build it if it's not real?"

Since then, scientists have succeeded at building something quite similar to the technology that Tom Cruise engaged to such dazzling effect. (John Underkoffler is now himself the CEO of Oblong Industries, developer of a *Minority Report*–like user interface he calls a Spatial Operating Environment.) What's more, researchers have begun to study the cognitive effects of this technology, and they find that it makes real a promise of science fiction: it helps people to think more intelligently.

The particular tool that has become the subject of empirical investigation is the "large high-resolution display" — an oversized computer screen to which users can bring some of the same navigational capacities they would apply to a real-world landscape. Picture a bank of computer screens three and a half feet wide and nine feet long, presenting to the eye some 31.5 million pixels (the average computer monitor has fewer than 800,000 pixels). Robert Ball, an associate professor of computer science at Weber State University in Utah, has run numerous studies comparing people's performance when interacting with a display like this to their performance when consulting a conventionally proportioned screen.

The improvements generated by the use of the super-sized display are striking. Ball and his collaborators have reported that large high-resolution displays increase by more than tenfold the average speed at which basic visualization tasks are completed. On more challenging tasks, such as pattern finding, study participants improved their performance by 200 to 300 percent when using large displays. Working with the smaller screen, users resorted to less efficient and more simplistic strategies, producing fewer and more limited solutions to the problems posed by experimenters. When using a large display, they engaged in higher-order thinking, arrived at a greater number of discoveries and achieved broader, more integrative insights. Such gains are not a matter of individual differences or preferences, Ball emphasizes; *everyone* who engages with the larger display finds that their thinking is enhanced.

Why would this be? Large high-resolution displays allow users to deploy their "physical embodied resources," says Ball, adding, "With small displays, much of the body's built-in functionality is wasted." These corporeal resources are many and rich. They include *peripheral vision,* or the ability to see objects and movements outside the area of the eye's direct focus. Research by Ball and others shows that the capacity to access information through our peripheral vision enables us to gather more knowledge and insight at one time, providing us with a richer sense of context. The power to see "out of the corners of our eyes" also allows us to be more efficient at finding the information we need, and helps us to keep more of that information in mind as we think about the challenge before us. Smaller displays, meanwhile, encourage a narrower visual focus, and consequently more limited thinking. As Ball puts it, the availability of more screen pixels permits us to use more of our own "brain pixels" to understand and solve problems.

Our built-in "embodied resources" also include our *spatial memory*: our robust capacity, exploited by the method of loci, to remember where things are. This ability is often "wasted," as Ball would have it, by conventional computer technology: on small displays, information is contained within windows that are, of necessity, stacked on top of one another or moved around on the screen, interfering with our ability to relate to that information in terms of where it is located. By contrast, large displays, or multiple displays, offer enough space to lay out all the data in an arrangement that persists over time, allowing us to leverage our spatial memory as we navigate through that information.

Researchers from the University of Virginia and from Carnegie Mellon University reported that study participants were able to recall 56 percent more information when it was presented to them on multiple monitors rather than on a single screen. The multiple monitor setup induced the participants to orient their own bodies toward the information they sought — rotating their torsos, turning their heads — thereby generating memory-enhancing mental tags as to the information's spatial location. Significantly, the researchers noted, these cues were generated "without active effort." Automatically noting place information is simply something we humans do, enriching our memories without depleting precious mental resources.

Other embodied resources engaged by large displays include *proprioception,* or our sense of how and where the body is moving at a given moment, and our experience of *optical flow,* or the continuous stream of information our eyes receive as we move about in real-life environments. Both these busy sources of input fall silent when we sit motionless before our small screens, depriving us of rich dimensions of data that could otherwise be bolstering our recall and deepening our insight.

Indeed, the use of a compact display actively *drains* our mental capacity. The screen's small size means that the map we construct of our conceptual terrain has to be held inside our head rather than fully laid out on the screen itself. We must devote some portion of our limited cognitive bandwidth to maintaining that map in mind; what's more, the mental version of our map may not stay true to the data, becoming inaccurate or distorted over time. Finally, a small screen requires us to engage in *virtual* navigation through information — scrolling, zooming, clicking — rather than the more intuitive *physical* navigation our bodies carry out so effortlessly. Robert Ball reports that as display size increases, virtual navigation activity decreases — and so does the time required to carry out a task. Large displays, he has found, require as much as 90 percent less "window management" than small monitors.

Of course, few of us are about to install a thirty-square-foot screen in our home or office (although large interactive displays are becoming an ever more common sight in industry, academia, and the corporate world). But Ball notes that much less dramatic changes to the places where we work and learn can allow us to garner the benefits of physically navigating the space of ideas. The key, he says, is to turn away from choosing technology that is itself ever faster and more powerful, toward tools that make better use of our own human capacities — capacities that conventional technology often fails to leverage. Rather than investing in a lightning-quick processor, he suggests, we should spend our money on a larger monitor — or on multiple monitors, to be set up next to one another and used at the same time. The computer user who makes this choice, he writes, "will most likely be more productive because she invested in the human component of her computer system. She has more information displayed at one time on her monitor, which, in turn, enables her to take advantage of the human side of the equation."

THE "TECHNOLOGY" THAT allows us to explore the space of ideas need not be digital. Sometimes the most generative tools are the simplest: a pencil, a notebook, an observing gaze. For a young Charles Darwin, such modest equipment provided the key to developing a theory that would change the world. In 1831 Darwin was twenty-two years old, recently graduated from Christ's College, Cambridge, uncertain whether to pursue a conventional career as a doctor or parson — or to follow his burgeoning interest in natural history. In August of that year, he received a letter from his former tutor at Cambridge, asking if he would be interested in serving as a naturalist on a two-year expedition aboard the HMS *Beagle*. Darwin accepted, and in December he began his seaborne apprenticeship to Captain Robert FitzRoy.

The young man carefully observed and emulated the actions of the experienced captain. Darwin had never kept a journal before coming aboard the *Beagle*, for example, but he began to do so under the influence of FitzRoy, whose naval training had taught him to keep a precise record of every happening aboard the ship and every detail of its ocean-going environment. Each day, Darwin and FitzRoy ate lunch together; following the meal, FitzRoy settled down to writing, bringing both the formal ship's log and his personal journal up to date. Darwin followed suit, keeping current his own set of papers: his field notebooks, in which he recorded his immediate observations, often in the form of drawings and sketches; his scientific journal, which combined observations from his field notebooks with more integrative and theoretical musings; and his personal diary. Even when Darwin disembarked from the ship for a time, traveling by land through South America, he endeavored to maintain the nautical custom of noting down every incident, every striking sight he encountered.

The historian of science and Harvard University professor Janet Browne has remarked upon the significance of this activity of Darwin's: "In keeping such copious records, he learned to write easily about nature and about himself. Like FitzRoy, he taught himself to look closely at his surroundings, to make notes and measurements, and to run through a mental checklist of features that ought to be recorded, never relying entirely on memory and always writing reports soon after the event." She adds, "Although this was an ordinary practice in naval affairs, it was for Darwin a basic lesson in arranging his thoughts clearly and an excellent preparation for composing logical scientific arguments that stood him in good stead for many years afterwards." But Darwin's careful note keeping did not simply help him learn to "arrange his thoughts clearly" and "compose logical scientific arguments." The projection of the internal workings of his mind onto the physical space of his journal created a conceptual map he was able to follow all the way to his theory of evolution. Some twenty-five years before the publication of his epochal book *The Origin of Species,* the entries in the journal Darwin kept throughout his expedition moved his thinking forward, step by tentative step.

On October 10, 1833, for example, Darwin discovered a fossilized horse tooth on the banks of the Rio Paraná in northeastern Argentina. Alongside the tooth he found the fossilized bones of a *Megatherium,* a giant ground sloth. Writing in his journal, Darwin puzzled over the fact that although these fossilized remains were apparently of the same age, horses still populated the earth in great numbers, while the *Megatherium* was long extinct. Eighteen months later, on April 1, 1835, Darwin came upon a "fossil forest" high in the Andes — what he described in a letter to his tutor in Cambridge as "a small wood of petrified trees." Again he pondered the implications of this discovery in his journal, noting that one possible explanation for the existence of the forest was a long-ago "subsidence," or a sinking of the land under the sea, where the trees would have become calcified by marine sediment. Darwin knew that such dramatic topographical shifts — down and then up again, to the mountainous elevation at which he encountered the fossil forest — were not endorsed by the thinking of the day, which assumed geological stability since the time of Earth's creation. "I must confess however that I myself cannot quite banish the idea of subsidence, enormous as the extent of movement required assuredly is," he confided in his journal.

Such precise and yet open-minded observations helped keep Darwin moving on a steady path toward a conclusion that now appears inevitable, but which hardly seemed so at the time. In 1849, at forty years of age — his voyage on the HMS *Beagle* behind him but the publication of *The Origin of Species* still to come — Darwin advised those who would follow in his footsteps to "acquire the habit of writing very copious notes, not all for publication, but as a guide for himself." The naturalist must take "precautions to attain accuracy," he continued, "for the imagination is apt to run riot when dealing with masses of vast dimensions and with time during almost infinity."

When thought overwhelms the mind, the mind uses the world — and researchers have reported some intriguing findings about why this use of the (physical, spatial) world is so beneficial for our thinking. As with the creation of concept maps, the *process* of taking notes in the field — whether that field is a sales floor, a conference room, or a high school chemistry lab — itself confers a cognitive bonus. When we simply watch or listen, we take it all in, imposing few distinctions on the stimuli streaming past our eyes and ears. As soon as we begin making notes, however, we are forced to discriminate, judge, and select. This more engaged mental activity leads us to process what we're observing more deeply. It can also lead us to have *new* thoughts; our jottings build for us a series of ascending steps from which we can survey new vistas.

Erick Greene, a professor of ecology and evolutionary biology at the University of Montana, has relied on his field notebooks throughout his long career. His stacks of spiral-bound notebooks contain descriptions of macaws and parrots flying at dusk to roost in palm swamps in Peru; of the "wahoo" alarm calls of olive baboons in the Okavango Delta of Botswana, warning one another of approaching lions; of teenage male sperm whales flipping up their tails as they begin their hour-long dives to catch giant squid in a deepwater trench off New Zealand. But his notes are not merely a record of what he has observed and experienced; they are, he says, "the main source of ideas that take my research in new directions."

Seeking to give his students a sense of this process, Greene assigned a field notebook exercise in the upper-level ecology class he teaches to UM undergraduates. He asked them to "pick one thing" and observe it carefully over the entire semester; the object of study could be a single tree, a bird feeder, a beaver dam, the student's own garden. This was not, he stressed to his class, a rote exercise in recording but a highly generative activity, the starting point of scientific discovery. "One of the main things I wanted to get across is that one of the hardest parts of science is coming up with new questions," says Greene. "Where do fresh new ideas come from? Careful observations of nature are a great place to start." In addition to making observations of their chosen site over time, students were required to come up with at least ten research questions inspired by what they saw.

As Greene's students discovered, the very act of noticing and selecting points of interest to put down on paper initiates a more profound level of mental processing. Things really get interesting, however, when we pause and look back at what we've written. Representations in the mind and representations on the page may seem roughly equivalent, when in fact they differ significantly in terms of what psychologists call their "affordances" — that is, what we're able to *do* with them. External representations, for example, are more *definite* than internal ones. Picture a tiger, suggests philosopher Daniel Dennett in a classic thought experiment; imagine in detail its eyes, its nose, its paws, its tail. Following a few moments of conjuring, we may feel we've summoned up a fairly complete image. Now, says Dennett, answer this question: How many stripes does the tiger have? Suddenly the mental picture that had seemed so solid becomes maddeningly slippery. If we had *drawn* the tiger on paper, of course, counting its stripes would be a straightforward task.

Here, then, is one of the unique affordances of an external representation: we can apply one or more of our physical senses to it. As the tiger example shows, "seeing" an image in our mind's eye is not the same as seeing it on the page. Daniel Reisberg, a professor emeritus of psychology at Reed College in Oregon, calls this shift in perspective the "detachment gain": the cognitive benefit we receive from putting a bit of distance between ourselves and the content of our minds. When we do so, we can see more clearly what that content is made of—how many stripes are on the tiger, so to speak. This measure of space also allows us to activate our powers of *recognition*. We leverage these powers whenever we write down two or more ways to spell a word, seeking the one that "looks right." The curious thing about this common practice is that we *do* tend to know immediately which spelling appears correct—indicating that this is knowledge we already possess but can't access until it is externalized.

A similar phenomenon has been reported by researchers investigating science learning. In a study published in 2016, experimenters asked eighth-grade students to illustrate with drawings the operation of a mechanical system (a bicycle pump) and a chemical system (the bonding of atoms to form molecules). Generating visual explanations of how these systems work led the students to achieve a deeper level of understanding; without any additional instruction, participants were able to use their own drawings as a "check for completeness and coherence as well as a platform for inference," the researchers note. Turning a mental representation into shapes and lines on a page supported the students' growing understanding, helping them to elucidate more fully what they already knew about these scientific systems. At the same time, the explicitness— the *definiteness*—of the drawings they made revealed with ruthless rigor what the students did not yet know or understand, leading them to fill in the gaps thus exposed.

So external representations are more definite than internal ones— and yet, in another sense, they are also more usefully *ambiguous*. When a representation remains inside our heads, there's no mystery about what it signifies; it's *our thought*, and so "there can be neither doubt nor ambiguity about what is intended," notes Daniel Reisberg. Once we've placed it on the page, however, we can riff on it, play with it, take it in new directions; it can almost seem as if we ourselves didn't make it. And indeed, researchers who have observed artists, architects, and designers as they create report that they often "discover" elements in their own work that they did not "put there," at least not intentionally.

Gabriela Goldschmidt, professor emeritus of architecture at the Technion-Israel Institute of Technology, explains how this works: "One reads off the sketch more information than was invested in its making. This becomes possible because when we put down on paper dots, lines, and other marks, new combinations and relationships among these elements are created that we could not have anticipated or planned for. We discover them in the sketch as it is being made." Architects, artists, and designers often speak of a "conversation" carried on between eye and hand; Goldschmidt makes the two-way nature of this conversation clear when she refers to "the backtalk of self-generated sketches."

Research by Goldschmidt and others shows that those who are skilled at drawing excel in managing this lively dialogue. Such studies find, for example, that expert architects are far more adept at identifying promising possibilities within their existing sketches than are relative novices. In one in-depth analysis of an experienced architect's methods, researchers determined that fully 80 percent of his *new* ideas came from reinterpreting his *old* drawings. Expert architects are also less likely than beginners to get stuck perseverating on a single unproductive concept; they are proficient at recombining disparate elements found in their sketches into new and auspicious forms.

From these observations of expert draftsmen, some promising prescriptions can be drawn. When setting out to generate new ideas, we should begin with only the most general plan or goal; early on in the process, vagueness and ambiguity are more generative than explicitness or definition. Think of the task not in linear terms — tracing a direct line from point A to point B — but rather as a cycle: think, draw, look, rethink, redraw. Likewise, don't envision the mind telling the pencil what to do; instead, allow a conversation to develop between eye and hand, the action of one informing the other. Finally, we ought to postpone judgment as long as possible, such that this give-and-take between perception and action can proceed without becoming inhibited by preconceived notions or by critical self-doubt.

Across the board, in every field, experts are distinguished by their skillful use of externalization; as cognitive scientist David Kirsh has written of video game virtuosos, "Better players use the world better." Skilled artists, scientists, designers, and architects don't limit themselves to the two-dimensional space of the page. They regularly reach for three-dimensional models, which offer additional advantages: users can manipulate the various elements of the model, view the model from multiple perspectives, and orient their own bodies to the model, bringing the full complement of their "embodied resources" to bear on thinking about the task and the challenges it presents.

David Kirsh has made close observations of the way architects use physical mock-ups of the buildings they are designing; when they interact with the models they have constructed, he maintains, "they are literally *thinking* with these objects." Interactions carried out in three dimensions, he says, "enable forms of thought that would be hard if not impossible to reach otherwise." Kirsh calls this the "cognitive extra" that comes from moving concrete objects through physical space—a mental dividend that made the difference for one scientist struggling with a seemingly insoluble problem.

A DREARY DAY in February 1953 found James Watson in low spirits. He and his collaborator Francis Crick—both young scientists at the Cavendish Laboratory in Cambridge, England—had been working for months to determine the structure of DNA, the molecule that contains the genetic code for living things. That morning a colleague had urged him "not to waste any more time with my harebrained scheme," as Watson later recounted in his autobiography. Hoping to demonstrate that his proposed arrangement of the four chemical bases that make up DNA—adenine, guanine, cytosine, and thymine—was true to life, he had asked the machinists at the Cavendish workshop to solder models of the bases out of tin. The models were taking too long to be finished, however, and Watson felt as if he had "run into a stone wall." Finally, "in desperation," he took on the job himself, and spent the afternoon fashioning models out of stiff cardboard.

Watson continues the story: "When I got to our still empty office the following morning, I quickly cleared away the papers from my desktop so that I would have a large, flat surface on which to form pairs of bases held together by hydrogen bonds." At first he tried fitting his cardboard bases together in a fashion dictated by his latest thinking about how the elements of DNA might be arranged—but, Watson related, "I saw all too well that they led nowhere." So he "began shifting the bases in and out of various other pairing possibilities."

Then realization dawned: "Suddenly I became aware that an adenine-thymine pair held together by two hydrogen bonds was identical in shape to a guanine-cytosine pair held together by at least two hydrogen bonds." Moving around the pieces of his cardboard model, Watson began to envision the chemical bases embedded in a double helix structure. "All the hydrogen bonds seemed to form naturally," he noted; "no fudging was required to make the two types of base pairs identical in shape." His "morale skyrocketed," he recalled, as the pieces quite literally came together before his eyes. Just then his partner, Crick, made his appearance, and Watson wasted no time in announcing the breakthrough: "Upon his arrival Francis did not get more than halfway through the door before I let loose that the answer to everything was in our hands."

The final step of Watson and Crick's long journey of discovery demonstrates the value of what psychologists call *interactivity*: the physical manipulation of tactile objects as an aid to solving abstract problems. The fact that Watson had to make his models himself is telling. Outside the architect's studio—or the kindergarten classroom—interactivity is not widely employed; our assumption that the brain operates like a computer has led us to believe that we need only input the necessary information in order to generate the correct solution. But human minds don't work that way, observes Frédéric Vallée-Tourangeau, a professor of psychology at Kingston University in the UK. The computer analogy "implies that simulating a situation in your head while you think is equivalent to living through that situation while you think," he writes. "Our research strongly challenges this assumption. We show instead that people's thoughts, choices, and insights can be transformed by physical interaction with things. In other words, thinking with your brain alone—like a computer does—is not equivalent to thinking with your brain, your eyes, and your hands."

A series of studies conducted by Vallée-Tourangeau and his colleagues all follow a similar pattern. Experimenters pose a problem; one group of problem solvers is permitted to interact physically with the properties of the problem; a second group must think through the problem in their heads. Interactivity "inevitably benefits performance," he reports. This holds true for a wide variety of problem types—from basic arithmetic, to complex reasoning, to planning for future events, to solving creative "insight" problems. People who are permitted to manipulate concrete tokens representing elements of the problem to be solved bear less cognitive load and enjoy increased working memory. They learn more, and are better able to transfer their learning to new situations. They are less likely to engage in "symbol pushing," or moving numbers and words around in the absence of understanding. They are more motivated and engaged, and experience less anxiety. They even arrive at correct answers more quickly. (As the title of one of Vallée-Tourangeau's studies puts it, "Moves in the World Are Faster Than Moves in the Head.")

Given the demonstrated benefits of interactivity, why do so many of us continue to solve problems with our heads alone? Blame our entrenched cultural bias in favor of brainbound thinking, which holds that the only activity that matters is purely mental in kind. Manipulating real-world objects in order to solve an intellectual problem is regarded as childish or uncouth; real geniuses do it in their heads.

This persistent oversight has occasionally been the cause of some irritated impatience among those who do recognize the value of externalization and interactivity. There's a classic story, for example, concerning the theoretical physicist Richard Feynman, who was as well known for authoring popular books such as *Surely You're Joking, Mr. Feynman!* as for winning the Nobel Prize (awarded to him and two colleagues in 1965). In a post-Nobel interview with the historian Charles Weiner, Weiner referred in passing to a batch of Feynman's original notes and sketches, observing that the materials represented "a record of the day-to-day work" done by the physicist. Instead of simply assenting to Weiner's remark, Feynman reacted with unexpected sharpness.

"I actually did the work on the paper," he said.

"Well," Weiner replied, "the work was done in your head, but the record of it is still here."

Feynman wasn't having it.

"No, it's not a *record*, not really. It's *working*. You have to work on paper and this is the paper. Okay?"

Feynman wasn't (just) being crotchety. He was defending a view of the act of creation that would be codified four decades later in Andy Clark's theory of the extended mind. Writing about this very episode, Clark argues that, indeed, "Feynman was actually *thinking* on the paper. The loop through pen and paper is part of the physical machinery responsible for the shape of the flow of thoughts and ideas that we take, nonetheless, to be distinctively those of Richard Feynman." We often ignore or dismiss these loops, preferring to focus on what goes on in the brain — but this incomplete perspective leads us to misunderstand our own minds. Writes Clark, "It is because we are so prone to think that the mental action is all, or nearly all, on the inside, that we have developed sciences and images of the mind that are, in a fundamental sense, inadequate." We will "begin to see ourselves aright," he suggests, only when we recognize the role of material things in our thinking — when we correct the errors and omissions of the brainbound perspective, and "put brain, body, and world together again."

PART III

THINKING
WITH OUR
RELATIONSHIPS

7

.

Thinking with Experts

GERMANY HAS LONG BEEN Europe's economic powerhouse; among the nation's many strengths, observers often single out its distinctive system of apprenticeships. Every year, about half a million young Germans move directly from high school into well-designed apprentice programs operated inside companies, where they learn technical skills such as welding, machining, and electrical engineering. This deeply entrenched system has for decades enabled Germany's manufacturing sector to thrive. But as in other Western countries, the dominance of industry in Germany is giving way to a more information-centric economy, creating demand for skills like computer programming. This change has brought new challenges, and students and instructors have struggled to adapt.

At the University of Potsdam — a school of some twenty thousand students, located outside Berlin — a key step for undergraduates who want a career in tech is a course on theoretical computer science. Yet year after year, the rate at which students failed the course was stunning: as high as 60 percent. The problem seemed related to the course's highly abstract content; sitting passively in lectures, students simply weren't grasping the meaning of concepts like "parsing algorithms," "closure properties," and "linear-bounded automata." Then a group of computer science professors hit upon a solution, one that harked back to Germany's historical strength. Led by Potsdam professor Christoph Kreitz, the faculty members reimagined the class as an apprenticeship, albeit of a special sort. The course was reorganized around making the internal thought processes of computer scientists "visible" to students — as visible as a carpenter fitting a joint or a tailor cutting a bolt of cloth.

This is what's known as a *cognitive apprenticeship,* a term coined by Allan Collins, now a professor emeritus of education at Northwestern University. In a 1991 article written with John Seely Brown and Ann Holum, Collins noted a crucial difference between traditional apprenticeships and modern schooling: in the former, "learners can see the processes of work," while in the latter, "the processes of thinking are often invisible to both the students and the teacher." Collins and his coauthors identified four features of apprenticeship that could be adapted to the demands of knowledge work: *modeling,* or demonstrating the task while explaining it aloud; *scaffolding,* or structuring an opportunity for the learner to try the task herself; *fading,* or gradually withdrawing guidance as the learner becomes more proficient; and *coaching,* or helping the learner through difficulties along the way.

Christoph Kreitz and his colleagues incorporated these features of traditional apprenticeships into their course redesign, reducing the amount of time students spent in lectures and increasing the length and frequency of small-group sessions led by tutors. In these sessions, students didn't listen to a description of computer science concepts, or engage in a discussion about the work performed by computer scientists; they actually *did* the work themselves, under the tutors' close supervision. The results of these changes were dramatic: the proportion of students failing the course shrank from above 60 percent to less than 10 percent.

The kind of shift Kreitz and his colleagues made at Potsdam is one that many of us will contemplate in coming years. All over the world, in every sector and specialty, education and work are less and less about executing concrete tasks and more and more about engaging in internal thought processes. As Allan Collins observed, these processes are largely inaccessible to *both* novice and expert: the novice doesn't yet know the material well enough, while the expert knows it so well that it has become second nature. This reality means that if we are to extend our thinking with others' expertise, we must find better ways of effecting an accurate transfer of knowledge from one mind to another. Cognitive apprenticeships are one such method; we'll explore several others in this chapter, starting with an approach that boasts both a long historical pedigree and an increasingly robust foundation in scientific research. So what if it makes us a bit uneasy?

INSIDE THE HÔPITAL Universitaire Pitié-Salpêtrière in Paris, a young man stares vacantly into space; a twitch contorts his mouth, and a tremor runs through his body like an electric shock. Nearby, another young man accepts help getting up from his chair. His right arm is bent at an awkward angle, and his right leg drags stiffly behind him. Across the room, a young woman is asked if she can touch her nose with her index finger; she tries, but her finger misses the mark and lands on her cheek.

On this day, the often uncanny symptoms of neurological disease are on vivid display. But these individuals are not patients; they are medical students, doctors in training. Coached by their professors, they are learning how to mimic the symptoms of the ailments they will be called upon to treat. Instructors show the students how to arrange their facial features, how to move their hands, how to sit and stand and walk. Faculty members also guide the responses of another contingent of students, dressed in white lab coats, who are role-playing the physicians they will become. After extensive practice, the "patients" and "doctors" will perform a series of clinical vignettes for their peers from the stage of the hospital's amphitheater.

Jean-Martin Charcot, the nineteenth-century physician known as the father of neurology, practiced and taught at this very institution. Charcot brought his patients onstage with him as he lectured, allowing his students to see firsthand the many forms neurological disease could take. Imitating such forms with one's own face and body is an even more effective means of learning, maintains Emmanuel Roze, who introduced his "mime-based role-play training program" to the students at Pitié-Salpêtrière in 2015. Roze, a consulting neurologist at the hospital and a professor of neurology at Sorbonne University, had become concerned that traditional modes of instruction were not supporting students' acquisition of knowledge, and were not dispelling students' apprehension in the face of neurological illness. He reasoned that actively imitating the distinctive symptoms of such maladies — the tremors of Parkinson's, the jerky movements of chorea, the slurred speech of cerebellar syndrome — could help students learn while defusing their discomfort.

And indeed, a study conducted by Roze and his colleagues found that two and a half years after their neurological rotation, medical students who had participated in the miming program recalled neurological signs and symptoms much better than students who had received only conventional instruction centered on lectures and textbooks. Medical students who had simulated their patients' symptoms also reported that the experience deepened their understanding of neurological illness and increased their motivation to learn about it.

In an earlier chapter on sensations, we saw that our automatic and unconscious mimicry of other people helps us understand them better — aids us in sensing their emotions, for example. The same is true for more deliberate imitation. Researchers have demonstrated, for instance, that intentionally imitating someone's accent allows us to comprehend more easily the words the person is speaking (a finding that might readily be applied to second-language learning). When we copy the accent of our conversation partner — when we produce the sounds that individual is making with our own mouth — we become better able to predict, and thus to make sense of, what he or she is saying. As with the medical students at Pitié-Salpêtrière, it's a matter of attaining understanding from the inside, of taking aspects of the other into ourselves.

We also come to feel more positive about the people whose speech we mimic — an effect that holds true for imitation more generally. Emmanuel Roze has found that the experience of imitating patients makes the young doctors he trains more empathetic, as well as more comfortable with the signs of their patients' disorders. Imitation permits us to extend to the other some of the familiar regard we feel for ourselves, as well as some of the insight we gain from inhabiting the role of dynamic actor in the world, rather than that of passive observer. It is a general purpose strategy with boundless applications in education, in the workplace, and in the learning we do on our own time.

There's just one problem: as a society we are suspicious of imitation, regarding it as juvenile, disreputable, even morally wrong. It's a reaction Roze has come to know well. Despite the demonstrated benefits of mime-based role play, many of his fellow medical school professors have expressed apprehension about implementing the practice. Some of his students, too, initially voiced discomfort at the prospect of imitating patients. Roze is careful to note that those who participate are in no way mocking or making fun of their charges. In fact, he says, the act of imitation is imbued with respect: it's treating patients as the ultimate authority, as the experts on what it's like to have their condition.

The conventional approach to cognition has persuaded us that the only route to more intelligent thinking lies in cultivating our own brain. Imitating the thought of other individuals courts accusations of being derivative, or even of being a plagiarist — a charge that can end a writer's career or a student's tenure at school. But this was not always the case. Greek and Roman thinkers revered imitation as an art in its own right, one that was to be energetically pursued. Imitation occupied a central role in classical education, where it was treated not as lazy cheating but as a rigorous practice of striving for excellence by emulating the masters.

In the Romans' highly structured system of schooling, students would begin by reading and analyzing aloud a model text. Early in pupils' education, this might be a simple fable by Aesop; later on, a complex speech by Cicero or Demosthenes. The students would memorize the text and recite it from memory. Then they would embark on a succession of exercises designed to make them intimately familiar with the work in question. They would paraphrase the model text, putting it in their own words. They would translate the text from Greek to Latin or Latin to Greek. They would turn the text from Latin prose into Latin verse, or even from Latin prose into Greek verse. They would compress the model into fewer words, or elaborate it at greater length; they would alter its tone from plainspoken to grandiloquent, or the other way around. Finally, they would write their own pieces — but in the style of the admired author. Having imitated the model from every angle, students would begin the sequence anew, moving on to a more challenging text.

We know about the Roman system largely from the writings of Quintilian, the "master teacher of Rome." Marcus Fabius Quintilianus, born around the year AD 35, headed a school of rhetoric that enrolled students from the city's most illustrious families, including the emperor Domitian's two heirs. In his masterwork, the *Institutio Oratoria* (subtitle: *Education of an Orator in Twelve Books*), Quintilian unapologetically asserted the value of copying. From authors "worthy of our study," he wrote, "we must draw our stock of words, the variety of our figures, and our methods of composition" so as to "form our minds on the model of every excellence." The educator continued: "There can be no doubt that in art, no small portion of our task lies in imitation, since, although invention came first and is all-important, it is expedient to imitate whatever has been invented with success, and it is a universal rule of life that we should wish to copy what we approve in others."

This system of education, founded on mimicking the masters, was remarkably robust, persisting for centuries and spreading throughout Europe and beyond. Fifteen hundred years after Quintilian, children in Tudor England were still being taught in this fashion. A scholar and teacher of that time, Juan Luis Vives, explained why imitation was necessary. While some basic capacities — such as speaking one's native language — seem to come naturally to humans, "Nature has fashioned man, for the most part, strangely hostile to 'art,'" he observed. "Since she let us be born ignorant and absolutely skill-less of all arts, we require imitation." Vives had intuited a truth that cognitive science would later demonstrate empirically: many of the achievements of human culture do not come "naturally" but must be painstakingly acquired. The most effective way to take possession of these skills, Vives and others of his time believed, was imitation.

Then, as the eighteenth century was drawing to a close, the Romantics arrived on the scene. This band of poets and painters and musicians worshiped originality, venerated authenticity. They rejected all that was old and familiar and timeworn in favor of what was inventive and imaginative and heartfelt. Their insistence on originality came in response to two major developments of the age. The first of these was industrialization. As factories rose brick by brick, an aesthetic countermovement mounted in tandem: *machines* could stamp out identical copies; only humans could come up with one-of-a-kind ideas. A particular sort of machine occasioned the second major development of the time: a flood of texts produced by the newly common printing press. More than any generation before them, the thinkers of the Romantic era bore what literary critic Walter Jackson Bate called "the burden of the past," as masterworks from earlier eras became widely available for the first time. Immersed in a sea of their predecessors' words, they felt an urgent need to create something new and fresh and never before said.

William Blake, the English poet and artist born in 1757, was one of the earliest and most passionately original Romantics. In creating works like *Songs of Innocence and of Experience* and *Visions of the Daughters of Albion,* Blake employed a technique he had invented himself, relief etching, in which he used acid-resistant chemicals to mark a copper plate, then applied acid to etch away the untreated areas. (The mystical-minded Blake claimed that his deceased brother Robert had revealed the technique to him in a vision.) These works took a form that was also newly devised by Blake: illuminated books that combined text and drawings and that were etched, printed, and colored by Blake himself. No two books were the same. And their content was nothing if not original: an elaborate invented cosmology featuring allegorical figures with names like Urizen (representing reason) and Los (imagination). In his illuminated book *Jerusalem*, Los — Blake's alter ego — voiced a sentiment that might have served as the Romantics' motto. "I must create a system," Blake's character declared, or else "be enslav'd by another man's."

Under the Romantics' influence, imitation did not merely become less favored than previously. It came to be actively disdained and disparaged — an attitude that was carried forward into succeeding decades. The naturalists of the late nineteenth century described imitation as the habit of children, women, and "savages," and held up original expression as the preserve of European men. Innovation climbed to the top of the cultural value system, while imitation sank to an unaccustomed low.

This is the cult of originality to which we ourselves subscribe — more so now than ever before. Our society celebrates pioneers and trailblazers — like, for example, the late Steve Jobs, the Apple Computer founder famous for offering dazzling onstage introductions of the company's latest inventions. His company's advertisements glorified those who break the mold, rather than those who allow themselves to be shaped by it. "Here's to the crazy ones," intoned the voice-over of an Apple commercial that aired in 1997. "The misfits. The rebels. The troublemakers. The round pegs in the square holes. The ones who see things differently. They're not fond of rules. And they have no respect for the status quo."

Think different, went the ad's tagline. But at least in some quarters, thinking *the same* — that is, engaging in imitation — is now gaining new respect.

IT WAS THE "proudest moment" of Kevin Laland's professional career, he says: the prestigious journal *Science* published a photograph of Laland mowing his lawn.

In the picture, Laland's three-year-old son is walking just behind him, intently pushing his own toy lawnmower; the photo accompanied a commentary on Laland's research about the importance of imitation in human culture. In the same issue, Laland, a professor of biology at the University of St. Andrews in the UK, reported on the results of a computerized competition he and his collaborators had set up. This was a multi-round tournament in which the contenders — bots that had been programmed to behave in particular ways — battled to victory for a monetary prize. A hundred entrants from around the world had faced off, each designed to act according to one of three strategies (or a combination thereof): applying original ideas, engaging in trial and error, or copying others.

As for which strategy worked best, there was really no contest: copying was far and away the most successful approach. The winning entry exclusively copied others — it *never* innovated. By comparison, a player-bot whose strategy relied almost entirely on innovation finished ninety-fifth out of the one hundred contestants. The result came as a surprise to Laland and to one of his collaborators, Luke Rendell. "We were expecting someone to come up with a really clever way to say, 'In these conditions you should copy, and in these conditions you should learn stuff for yourself,'" says Rendell. "But the winner just copied all the time."

Kevin Laland acknowledges that imitation has a bad reputation. But, he says, researchers like him — in fields from biology to economics to psychology to political science — are discovering how valuable imitation can be as a way of learning new skills and making intelligent decisions. Researchers from these varied disciplines are using models and simulations, as well as historical analyses and real-world case studies, to show that imitation is often the most efficient and effective route to successful performance. And they're elaborating the reasons why this is so — reasons vividly illustrated by examples from the business world.

First on the list: by copying others, imitators allow other individuals to act as filters, efficiently sorting through available options. Finance professors Gerald Martin and John Puthenpurackal examined what would happen if an investor did nothing but copy the moves of celebrated investor Warren Buffett. (Buffett's investment choices are periodically made public, when his company files a report with the Securities and Exchange Commission.) An individual who simply buys what Buffett is buying, the researchers found, will earn an average of more than 10 percent above market returns.

Of course, investors already do take notice of the activity of a high-powered investor like Buffett — but even so, they're missing out on the benefits of imitating his selections more closely. Investors are leaving money on the table, Martin suggests, because everyone likes to think they've got an innovative strategy or an overlooked gem. Although it can feel good to chart our own course, he says, we often perform better when we copy someone more experienced and more knowledgeable than ourselves.

Second: imitators can draw from a wide variety of solutions instead of being tied to just one. They can choose precisely the strategy that is most effective in the current moment, making quick adjustments to changing conditions. That sums up the business model of Zara, a worldwide chain of clothing stores based in the industrial city of Arteixo, Spain.

At the headquarters of its parent company, Inditex, Zara's designers cluster around tables covered with pages ripped from fashion magazines and catalogs; with photographs snapped of stylish people on streets and in airports; and with the deconstructed parts of other designers' garments, fresh from the runway. "Zara is engaged in a permanent quest for inspiration, everywhere and from everybody," says Spanish journalist Enrique Badía, who has written extensively about the company. Zara even copies its own *customers*. Store managers from the chain's hundreds of locations are in frequent touch with its designers, passing along new looks spotted on Zara's fashion-forward clientele.

Kasra Ferdows, a professor of operations and information management at Georgetown University, notes that Zara's adroit use of imitation has helped make Inditex the largest fashion apparel retailer in the world. Its success, he and two coauthors concluded in a company profile written for the *Harvard Business Review,* "depends on a constant exchange of information throughout every part of Zara's supply chain — from customers to store managers, from store managers to market specialists and designers, from designers to production staff." Crucially, the "information" that flows so freely at the company concerns not new ideas, but ideas good enough to copy.

The third advantage of imitation: copiers can evade mistakes by steering clear of the errors made by others who went before them, while innovators have no such guide to potential pitfalls. A case in point: diapers. Among parents who rely on disposable diapers, Pampers is a household name. Less familiar is the brand Chux — and yet Chux was the first to arrive on the market, all the way back in 1935.

The problem: Chux were expensive, costing about 8.5 cents per diaper, at a time when parents could wash cloth diapers at a cost of 1.5 cents each. As a result, parents tended to use the product only when traveling, and Chux accounted for just 1 percent of the overall market for diapers. Procter & Gamble saw an opportunity. It imitated the basic idea behind Chux while intentionally addressing parents' main objection to the product: its high price. When Pampers were rolled out nationwide in 1966 — at a cost of three cents per diaper — P&G's version was enthusiastically welcomed by parents.

Gerard Tellis and Peter Golder, both professors of marketing, conducted a historical analysis of fifty consumer product categories (including diapers, from which the Pampers versus Chux example was taken). Their results showed that the failure rate of "market pioneers" is an alarming 47 percent, while the mean market share they capture is only 10 percent. Far better than being first, Tellis and Golder concluded, is being what some have called a "fast second": an agile imitator. Companies that capitalize on others' innovations have "a minimal failure rate" and "an average market share almost three times that of market pioneers," they found. In this category they include Timex, Gillette, and Ford, firms that are often recalled — wrongly — as being first in their field.

Fourth, imitators are able to avoid being swayed by deception or secrecy: by working directly off of what others do, copiers get access to the best strategies in others' repertoires. Competitors have no choice but to display what social scientists call "honest signals," as they make decisions for themselves based on their own best interests. This is the case in every sort of contest — including sporting events like the America's Cup, the high-profile sailing race.

Jan-Michael Ross and Dmitry Sharapov, both business professors at Imperial College London, studied the competitive interactions among yachts engaged in head-to-head races in the America's Cup World Series. The researchers found that sailors often engaged in "covering," or copying, the moves made by their rivals — especially when their boat was in the lead. It might seem surprising that sailors at the front of the pack would imitate those who are trailing, but Ross notes that such emulation makes sense: as long as the leaders do as their rivals behind them do, their lead will remain locked in place. Says Ross, "Our research challenges the common view that it's only the laggards, the also-rans, who imitate."

Last, and perhaps most important, imitators save time, effort, and resources that would otherwise be invested in originating their own solutions. Research shows that the imitator's costs are typically 60 to 75 percent of those borne by the innovator—and yet it is the imitator who consistently captures the lion's share of financial returns.

Such findings, arriving from many directions, converge on the conclusion that imitation (if we can get past our aversion to it) opens up possibilities far beyond those that dwell inside our own heads. Engaging in effective imitation is like being able to think with other people's brains—like getting a direct download of others' knowledge and experience. But contrary to its reputation as a lazy cop-out, imitating well is not easy. It rarely entails automatic or mindless duplication. Rather, it requires cracking a sophisticated code—solving what social scientists call the "correspondence problem," or the challenge of adapting an imitated solution to the particulars of a new situation. Tackling the correspondence problem involves breaking down an observed solution into its constituent parts, and then reassembling those parts in a different way; it demands a willingness to look past superficial features to the deeper reason why the original solution succeeded, and an ability to apply that underlying principle in a novel setting. It's paradoxical but true: imitating well demands a considerable degree of creativity.

ADAPTING SOLUTION to problem was the tall task facing graduate nursing student Tess Pape in 1999. Pape could see the problem clearly enough: hospital patients were being harmed by medication errors committed by doctors and nurses. In that year, the Institute of Medicine had released a landmark report on patient safety, *To Err Is Human*. The report found that as many as 98,000 Americans were dying each year as a result of preventable medical errors occurring in hospitals—more people than succumbed to car accidents, workplace injuries, or breast cancer. And some significant portion of these deaths involved mistakes in the dispensing of drugs.

But as she investigated ways to address the medication error crisis, Pape didn't rack her brain for an innovative fix. Instead she sought to *imitate* a solution that had been successfully applied in another industry.

That industry was aviation — an enterprise, like health care, in which people's lives depend on professionals' precision and accuracy. While reading up on aviation safety, Pape learned that the moments of highest risk occurred during takeoffs and landings — periods when the plane was under ten thousand feet. She spotted a correspondence in her own field: for hospital patients who are given medication, the riskiest moments happen during the *preparation* of the drug dosage, and during *administration* of the drug to the patient.

Delving deeper, Pape discovered that distractions and interruptions of the pilot by other crew members accounted for a majority of airline "incidents." Another correspondence came into view: interruptions of health care professionals, she knew, were also to blame for many medication mistakes. (Consider this striking incident reported by a team of researchers observing real-life conditions in hospitals: one nurse, dispensing one medication, to one single patient, was interrupted *seventeen* times.) Pape also became aware that aviation experts had devised a solution to the problem of pilot interruption: the "sterile cockpit rule." Instituted by the Federal Aviation Administration in 1981, the rule forbids pilots from engaging in conversation unrelated to the immediate business of flying when the plane is below ten thousand feet.

In her 2002 dissertation, and then in a series of articles published in medical journals, Pape made a case for imitating this practice. "The key to preventing medication errors lies with adopting protocols from other safety-focused industries," Pape wrote in the journal *MEDSURG Nursing* in 2003. "The airline industry, for example, has methods in place that improve pilots' focus and provide a milieu of safety when human life is at stake." Such methods could be adapted to the hospital setting, she argued, by creating a "no-interruptions zone" around medication preparation areas, and by having nurses who are administering medication wear special vests or sashes signaling that they are not to be disturbed. Added Pape, pointedly, "Medication administration should be considered as critical as piloting a plane, because patients place their lives in the hands of healthcare professionals."

Pape wasn't sure if her peers in the health care community would be open to listening to the idea. But listen they did. Hospitals began following the lead of airlines, and the change made a dramatic difference. At Kaiser Permanente South San Francisco Medical Center, for example, the introduction of no-interruption signaling in 2006 led to the "virtual elimination of nurse distractions for those wearing the vests," according to the US government's Agency for Healthcare Research and Quality. Over a six-month period, medication errors at the hospital fell by 47 percent. Nearly two decades after she initiated it, Pape's lifesaving act of imitation has spread all over the country and the world.

Tess Pape figured out the correspondence problem on her own. But what if she had been *taught* how to imitate? Imitating well is a skill, one that Oded Shenkar believes should be deliberately cultivated. Shenkar, a professor of management and human resources at Ohio State University, studies how companies use imitation to gain a strategic edge in the marketplace. He maintains that we are living in a golden "age of imitation," in which access to information about how other people are addressing problems similar to our own has made it more feasible than ever to copy effective solutions. Shenkar would like to see students in business schools and other graduate programs taking courses on effective imitation. He imagines companies opening "imitation departments," devoted to identifying promising opportunities for copying. And he anticipates a day when successful imitators are celebrated and admired just as much as innovators are now.

Shenkar notes that at least one profession has been taking steps in the direction of his push: health care. The urgent need to reduce medical errors, perceived by Tess Pape and many others, has led hospitals to imitate the practices of a host of other industries, including the military, railroads, chemical manufacturing, nuclear power—and, of course, aviation. In addition to the sterile cockpit concept Pape adapted, health care professionals have also borrowed from pilots the onboard "checklist"—a standardized rundown of tasks to be completed. In this case, too, imitation has worked wonders. In 2009, researchers from the Harvard School of Public Health and the World Health Organization reported that after the surgical teams in their study started using a nineteen-item checklist, the average patient death rate fell more than 40 percent, and the rate of complications decreased by about a third.

The medical field has also adopted the "peer-to-peer assessment technique," a common practice in the nuclear power industry. A delegation from one hospital visits another hospital in order to conduct a "structured, confidential, and non-punitive review" of the host institution's safety and quality efforts. Without the threat of sanctions carried by regulators, these peer reviews can surface problems and suggest fixes, making the technique itself a vehicle for constructive copying among organizations.

Even within health care, however, the practice of imitation leaves much room for improvement. It took seventy years for the checklist concept to migrate from aviation to medicine, and twenty years for the sterile cockpit to make the leap. A more structured and intentional approach to imitation could speed up this process considerably. In order to elevate the social value of copying, says Shenkar, we need not only to promote new acts of imitation but also to recognize that imitation is *already* behind the success of many of our most admired individuals and organizations—a group that assuredly includes the famed innovator Steve Jobs.

In 1979, Jobs and his colleagues at the fledgling Apple Computer company were wrestling with how to turn the crude, clunky computers of the day into sleek personal appliances that were easy and even fun to use. In December of that year, he got a glimpse of the solution while on a visit to Xerox PARC, a research facility operated by the photocopier giant in Palo Alto, California. Jobs was shown a series of technological innovations that he knew he could put to use in his own project: a networking platform that allowed computers to connect to and communicate with one another; a set of visually appealing and user-friendly onscreen graphics; a mouse that enabled users to point and click. "This is it!" he shouted to an Apple associate as their car sped away from PARC. "We've got to do it!"

In Oded Shenkar's contemplated academic course on imitation, he might well use Apple as a case study. And he might point out to his class that Jobs had already taken the first of three steps required to solve the all-important correspondence problem. Step one, according to Shenkar: specify one's own problem and identify an analogous problem that has been solved successfully.

Step two: rigorously analyze why the solution is successful. Jobs and his engineers at Apple's headquarters in Cupertino, California, immediately got to work deconstructing the marvels they'd seen at the Xerox facility. Soon they were on to the third and most challenging step: identify how one's own circumstances differ, then figure out how to adapt the original solution to the new setting. Xerox had already brought its own computer to market, but it was awkward and difficult to use. It was designed for the needs of business rather than the individual consumer. And it was prohibitively expensive, costing more than $16,000.

Xerox had found technological solutions that eluded Apple's scientists, but it was Jobs who adapted these solutions to the potential market he saw for personal devices. An example: the mouse he'd seen at PARC had three buttons, which Jobs deemed excessively fussy; it didn't roll easily, even on smooth surfaces; and it cost a whopping $300. Jobs worked with a local design firm to produce a one-button mouse that could be operated on any surface (even his blue jeans, Jobs specified) and that cost only $15.

The rest is history, though not history as it is usually told—a story of solitary geniuses ("the ones who see things differently"). The lesson of this case study is that skilled imitation, and not just brilliant innovation, is behind many of the successes we celebrate.

IMITATION EVEN APPEARS to be behind our success as a *species*. Developmental psychologists are increasingly convinced that infants' and children's facility for imitation is what allows them to absorb so much, so quickly. So efficient is imitation as a method of learning, in fact, that roboticists are studying babies in order to understand how they pull off the trick of observing an adult and then doing as the grown-up does. Imagine if a robot could watch a human perform an action—say, place a silicon chip on a circuit board, or make a repair on a space capsule—and then replicate that movement itself. Elon Musk, founder of Tesla and SpaceX, has invested in research on just such "one-shot imitation learning." But, as University of California, Berkeley, psychologist Alison Gopnik notes, the most sophisticated forms of artificial intelligence "are still far from being able to solve problems that human four-year-olds accomplish with ease."

While it was once regarded as a low-level, "primitive" instinct, researchers are coming to recognize that imitation—at least as practiced by humans, including very young ones—is a complex and sophisticated capacity. Although non-human animals do imitate, their mimicry differs in important ways from ours. For example, young humans' copying is unique in that children are quite selective about *whom* they choose to imitate. Even preschoolers prefer to imitate people who have shown themselves to be knowledgeable and competent. Research shows that while toddlers will choose to copy their mothers rather than a person they've just met, as children grow older they become increasingly willing to copy a stranger if the stranger appears to have special expertise. By the time a child reaches age seven, Mom no longer knows best.

At the same time, children are strikingly unselective about *what* they imitate — another way in which our practice of imitation departs from that of animals. Humans are "high-fidelity" copiers: our young imitate adults to the letter, while other animals will make do with a slapdash approximation. This difference can make apes, monkeys, and even dogs look like the smarter species. Shown a procedure with an extra, unnecessary step — like touching a box with one's forehead before prying it open and retrieving the treat inside — chimps and canines will skip the superfluous move to go right for the goods. Children, however, will faithfully imitate every step.

There is sense behind this seemingly irrational behavior. Humans' tendency to "overimitate" — to reproduce even the gratuitous elements of another's behavior — may operate on a copy now, understand later basis. After all, there might be good reasons for such steps that the novice does not yet grasp, especially since so many human tools and practices are "cognitively opaque": not self-explanatory on their face. Even if there doesn't turn out to be a functional rationale for the actions taken, imitating the customs of one's culture is a smart move for a highly social species like our own. Indeed, researchers have demonstrated that a four-year-old child is *more* likely to overimitate than a two-year-old, indicating a growing sensitivity to social cues. Our tendency to engage in overimitation continues to increase across development, all the way into adulthood. Because so much of human culture is arbitrary in form — why do we clap at the end of a performance, or eat cake at birthday parties, or wear wedding rings on the fourth finger of the left hand? — it depends on imitation for its perpetuation. Imitation is at the root of our social and cultural life; it is, quite literally, what makes us human.

There is evidence that we are born with a predisposition to imitate. Several decades ago, University of Washington psychologist Andrew Meltzoff showed that babies who were days or even hours old responded to his opening his mouth or sticking out his tongue by forming the same facial expressions in return. At the same time, the capacity to imitate, and to learn from observation, can be nurtured: in some present-day cultures, it is cultivated quite deliberately, with impressive results. In studies comparing European American children with Mayan children from Guatemala, psychologists Maricela Correa-Chávez and Barbara Rogoff asked children from each culture to wait while an adult performed a demonstration—folding an origami shape—for another child nearby. The Mayan youth paid far more sustained attention to the demonstration—and therefore learned more—than the American kids, who were often distracted or inattentive. Correa-Chávez and Rogoff note that in Mayan homes, children are encouraged to carefully observe older family members so that they can learn how to carry out the tasks of the household, even at very young ages.

Because our own culture discourages imitation, American children aren't granted similar opportunities to show how competent they can be; they also lack exposure to inspiring examples, or "models," of the kind of work that kids their age are capable of producing. For decades, educator Ron Berger towed around a rolling suitcase filled with hundreds of such models: sketches, poems, and essays created by children, which he would pull out and share with teachers and students at schools around the country. Among his favorite examples is a picture he calls "Austin's Butterfly," drawn by a first-grader in Boise, Idaho. When he displays the drawing—which depicts a tiger swallowtail butterfly in graceful detail—students often murmur in awe. Berger's aim is to inspire his audience with examples of excellent work by their peers, but also to demonstrate to them how such work is made. One by one he shows them the six drafts Austin produced on the way to his finished drawing, and tells them about the constructive critiques Austin received from his classmates at each stage.

The contents of Berger's rolling suitcase are now available in an online archive—but he has found that many teachers and parents object to the use of models, afraid that it will suppress students' creativity and originality. In fact the opposite is true, says Berger, who spent twenty-eight years as a classroom teacher before becoming chief academic officer of the non-profit organization EL Education. Seeing examples of outstanding work motivates students by giving them a vision of the possible. How can we expect students to produce first-rate work, he asks, when they have no idea what first-rate work looks like?

The capacity of models to promote—rather than quash—students' creativity was long recognized by teachers of composition and rhetoric, subjects that were once a central part of the curriculum in American schools. One of the field's leading textbooks was authored by literature scholar Edward P. J. Corbett, who never relinquished the notion that emulating the work of the masters was the first step toward developing one's own distinctive style. *"Imitate, that you may be different!"* Corbett thundered. Even as the use of models has faded from the teaching of English more generally, it is enjoying a resurgence among educators who are training students to write within particular academic genres, an activity sometimes called "disciplinary writing." In this context, the emulation of model texts is valued for its capacity to reduce cognitive load — especially important when students are juggling new concepts and vocabulary while also trying to construct a coherent argument in writing. Following the contours of a prototype provided by the instructor permits students to process more deeply the material they are expected to learn.

Marin Robinson, a chemistry professor at Northern Arizona University, leads an undergraduate course aimed at teaching students to "write like a chemist." Those who enroll practice the writing of four forms central to the discipline: the journal article, the conference abstract, the poster presentation, and the research proposal. In each case, students follow the scientific and linguistic conventions on display in the "authentic texts" they are given: actual articles, abstracts, posters, and proposals. This act of imitation relieves students of some of their mental burden, Robinson notes, allowing them to devote the bulk of their cognitive bandwidth to the content of the assignment. She and her collaborator, English professor Fredricka Stoller, have authored a textbook, *Write Like a Chemist,* that is used at universities around the country.

A similar movement is emerging in teaching legal writing, another endeavor that requires students to assimilate a host of new terms and ideas even as they are learning to write in an unfamiliar genre. In the course of teaching hundreds of first-year law students, Monte Smith, a professor and dean at Ohio State University's law school, grew increasingly puzzled by the seeming inability of his bright, hardworking students to absorb basic tenets of legal thinking and to apply them in writing. He came to believe that the manner of his instruction was demanding more from them than their mental bandwidth would allow. Students were being asked to employ a whole new vocabulary and a whole new suite of concepts, even as they were attempting to write in an unaccustomed style and an unaccustomed form. It was too much, and they had too few mental resources left over to actually learn.

His solution: at the start of the course, Smith provides students with several sample legal memorandums like those written by working lawyers. Guided by a set of instructions and targeted questions, students are expected to detail their responses to various aspects of the memos — thereby relieving them of the burden of having to produce their own memos even as they are laboring to learn what a memo looks and sounds like. Only after several such structured encounters with legal documents are students asked to author their own memorandums.

As Smith notes, the emulation of model texts was once a standard feature of instruction in legal writing; it fell out of favor because of concern that the practice would fail to foster a capacity for independent thinking. The careful observation of how students actually learn, informed by research on the role of cognitive load, may be bringing models back into fashion.

OF COURSE, the richest, deepest, and potentially most useful models are *people*. Yet the very individuals who are most expert are often least able to share what they know. After years of practice, much of experts' knowledge and skill has become "automatized" — so well practiced that they no longer need to think about it. Automatization allows experts to work efficiently and effectively, but it also prevents them from offering to others a full account of *how* they do what they do. Kenneth Koedinger, a professor at Carnegie Mellon University and the director of its Pittsburgh Science of Learning Center, estimates that experts are able to articulate only about 30 percent of what they know. His conclusion is based on research like the following: A study that asked expert trauma surgeons to describe how they insert a shunt into the femoral artery (the large blood vessel in the upper leg) reported that the surgeons neglected to cite nearly 70 percent of the actions they performed during the procedure. A study of expert experimental psychologists found that they omitted or inaccurately characterized an average of 75 percent of the steps they took when designing experiments and analyzing data. And a study of expert computer programmers revealed that they enumerated fewer than half of the tasks they actually carried out when debugging a computer program.

Our systems of academic education and workplace training rely on experts teaching novices, but they rarely take into account the blind spots that experts acquire by virtue of being experts. In the era of knowledge work, it's not only the case that learners and novices must become more assiduous imitators; instructors and experts must also become more legible models. This can be accomplished through what philosopher Karsten Stueber calls "re-enactive empathy": an appreciation of the challenges confronting the novice that is produced by reenacting what it was like to have once been a beginner oneself.

Ting Zhang, an assistant professor of business administration at Harvard Business School, found a clever way to stage such a reenactment among expert musicians. For her experiment, Zhang enlisted a group of experienced guitarists to play their instruments; half of them were asked to play as they normally would, while the other half were instructed to reverse the position of their instrument and play it with their non-dominant hand. All the musicians were then asked to watch a video clip of a beginning guitar student trying to form basic chords, and to offer him advice. The guidance provided by the guitarists in the reversed instrument condition — those who had so recently struggled themselves to play in an unfamiliar manner — was judged to be especially helpful.

But reenacting the experience of being a novice need not be so literal; experts can generate empathy for the beginner through acts of the imagination, changing the way they present information accordingly. An example: experts habitually engage in "chunking," or compressing several tasks into one mental unit. This frees up space in the expert's working memory, but it often baffles the novice, for whom each step is new and still imperfectly understood. A math teacher may speed through an explanation of long division, not remembering or recognizing that the procedures that now seem so obvious were once utterly inscrutable. Math education expert John Mighton has a suggestion: break it down into steps, then break it down again — into *micro*-steps, if necessary.

Though Mighton now holds a PhD in mathematics, he struggled with math as a child. He succeeded by teaching himself how to advance one tiny move at a time — the method he now advocates as the founder of a nonprofit educational organization, JUMP Math (the name stands for Junior Undiscovered Math Prodigies). When instructors make their expertise legible in this way, learners are able to master one small step, then another and another, acquiring a solid understanding and gaining confidence as they go. The approach has allowed many JUMP participants, including some who struggled to grasp the most basic mathematical concepts in school, to achieve proficiency in the subject. An evaluation of the JUMP program, conducted by researchers from the University of Toronto and Toronto's Hospital for Sick Children and published in 2019, provides support for Mighton's method. By the second year of the study, JUMP students in grade three had made greater progress in problem solving, and JUMP students in grade six had made greater gains across a broad range of mathematical skills — calculation, math fluency, and applied problems — than those who received traditional instruction.

Experts have another edge over novices: they know what to attend to and what to ignore. Presented with a professionally relevant scenario, experts will immediately home in on its most salient aspects, while beginners waste their time focusing on unimportant features. But research shows that the expertise of experienced practitioners can be made more accessible by deliberately exaggerating it, even distorting it, such that the pertinent elements "pop out" for the novice as they do for the expert.

A number of years ago, the US Air Force sought the advice of psychologist Itiel Dror, now a senior researcher at University College London. Endeavoring to prevent friendly fire aimed at their own aircraft, air force leaders were looking for ways to improve the ability of pilots in training to instantly recognize the shapes of various planes. Dror observed that the trainees were becoming overwhelmed with details about the many airplanes they were expected to identify. He took a new tack, digitally morphing the outlines of the aircraft diagrams the pilots were given to study. Planes with a wide wingspan became even wider; sharp-angled planes were made pointier; snub-nosed planes appeared more rounded. The differences among the aircraft, once subtle and hard to notice, now leaped out at the pilots — and they continued to recognize these distinguishing features even when they viewed the planes at normal scale.

Dror's method is related to a phenomenon that psychologists call "the caricature advantage": the fact that we recognize a caricatured face even more readily than we recognize a true-to-life depiction. While a caricature does distort its subject's actual appearance, it does so in a *systematic* way, exaggerating what is unique or distinctive about that individual — thereby making him or her even more instantly identifiable. (Think of George W. Bush's prominent ears, or Bill Clinton's bulbous nose, or the late Ruth Bader Ginsburg's oversized glasses.) Experts can leverage the caricature advantage by playing up the unique features that differentiate among a group of examples — examples that, to a novice, may come off as confusingly similar.

A third difference between experts and novices lies in the way they *categorize* what they see: novices sort the entities they encounter according to their superficial features, while experts classify them according to their deep function. A classic experiment by Arizona State University professor Michelene Chi asked eight experts (advanced PhD students from a university physics department) and eight novices (undergraduates who had taken a single semester of physics) to sort two dozen physics problems, each one described on an index card, into categories of their choosing. The classifications devised by the two groups could not have been more different. The undergraduates sorted the problems according to their surface features: whether the problems involved springs, or pulleys, or inclined planes. The graduate students, meanwhile, categorized the problems on the basis of the underlying principles of physics they represented: conservation of energy, the work-energy theorem, conservation of momentum.

There's far more useful information embedded in the categories applied by experts. So why not present novices with information that's *already* organized by function? That notion was behind a new kind of wine store created by Joshua Wesson, an expert sommelier and entrepreneur who founded the chain Best Cellars. "I heard the same questions all the time, and they all reduced to, 'How can I make sense of the world of wine without having to master all the details? How can I deal with all these choices when all I want is a wine that goes with pizza?'" says Wesson. In most stores selling wine, he observes, bottles are arranged by grape (Chardonnay, Cabernet Sauvignon) or by region (California, France). Such classifications communicate little to the uneducated wine consumer.

Wine experts, meanwhile, know about surface-level characteristics like grapes and regions — but they *think* about wine in terms of function: wines that are luscious and fruity, good for pairing with spicy food; wines that are big and bold and can stand up to a hearty meal; wines that are fizzy and festive, fit for a celebration. "Luscious," "Big," and "Fizzy" are, in fact, three of the eight categories Wesson devised for his stores (the others are "Soft," "Fresh," "Juicy," "Smooth," and "Sweet"). Foregrounding these features is like giving customers a shortcut to thinking the way a sommelier does. Following Wesson's lead, experts more generally can make themselves into accessible models for imitation by communicating the categories they use to organize information, categories that are themselves packed with meaning about how experts' thinking operates.

These strategies — breaking down agglomerated steps, exaggerating salient features, supplying categories based on function — help pry open the black box of experts' automatized knowledge and skill. Extended technology may offer an even more direct probe into the mind of the expert. For example: the nature of expertise is now being studied using eye tracking, the automated monitoring of where the expert's gaze falls, when, and for how long. Research has shown that, across disciplines, experts look in ways different from novices: they take in the big picture more rapidly and completely, while focusing on the most important aspects of the scene; they're less distracted by visual "noise," and they shift more easily among visual fields, avoiding getting stuck. Within any occupation — among surgeons, pilots, programmers, architects, even high school teachers — experts' gaze patterns are highly similar, while beginners' are widely divergent and idiosyncratic.

Yet experts are not aware of how they engage in looking; their gaze patterns are not available for conscious inspection. Eye-tracking technology can capture this aspect of their expertise and make it available for use by novices, guiding their gaze with unobtrusive cues about where to look. It's a way of "cheating experience," as one researcher puts it — a shortcut around hours of observation and practice that could potentially make learning much more efficient and effective.

Researchers are also experimenting with "haptic" signals: physical nudges delivered via special gloves or tools that help mold a novice's movement patterns into those of an expert. While brainbound approaches to education and training convey information almost exclusively through visual and auditory channels, haptic technologies supply guidance and feedback directly to the body. Preliminary results suggest that their use can reduce cognitive load and improve performance for many kinds of learners, from students learning to play the violin to medical residents learning to perform laparoscopic surgery.

In a sense, these innovations represent a technologically enhanced take on the teaching that has unfolded within apprenticeships for centuries —

the twenty-first-century version of a master craftsman's pointing finger or guiding hand. The indenture contracts for those old-time apprenticeships often proposed to provide the apprentice's labor in exchange for instruction in "the trade, art, and mystery" of the craft, whether it be carpentry or blacksmithing or shipbuilding. In an age of knowledge work, the "mystery" of expertise is even more enshrouded, hidden by the scrim of automatization. Pulling this curtain aside requires experts to forgo the familiar conventions of brainbound instruction — to think outside the brain, where their cognition can be seen.

8

■

Thinking with Peers

CARL WIEMAN KNOWS his way around a perplexing problem. A professor of physics at Stanford University, Wieman was awarded the Nobel Prize in 2001 for figuring out (along with his colleague Eric Cornell) how to create, in the laboratory, an extreme state of matter known as the Bose-Einstein condensate. But Wieman's mastery in the lab did not extend to the classroom, as he would be the first to admit. For years he wrestled with what would seem to be a straightforward task: how to get undergraduates to understand physics in the way *he* understood physics. Laying it out for them—describing, explaining, even demonstrating the core concepts of the discipline—was not working. No matter how clearly he elucidated these ideas, no matter how energetically he communicated them, his students' ability to solve the problems he posed to them remained rudimentary.

This failure to "think like a physicist" is more the rule than the exception. Decades of research have found that high school and college students who are taught physics in the conventional manner, via lectures and textbooks, typically don't learn the subject in any depth. So it was for Wieman and his students. Wieman knew how to cool and trap atoms using light from a powerful laser. From work carried out in his lab, he understood how atoms interact at ultracold temperatures, more than four hundred degrees below zero. He had discovered how to cause atoms to oscillate at the same frequency, or "sing in unison," as the Royal Swedish Academy of Sciences put it in announcing his Nobel. But he could not seem to pin down the process by which a halting thinker turned into a dexterous one.

Wieman ultimately found the key to his conundrum in an unexpected place: not in his undergraduate classes but among the graduate students who came to work in his laboratory. When they first arrived at the lab, Wieman noticed, his PhD candidates were more like the undergrads than not. They knew plenty about physics, but their habits of thought were narrow and rigid. Within just a year or two, however, these same graduate students had grown into models of the kind of supple, flexible thinker Wieman was trying so earnestly, and unsuccessfully, to mold. "It was clear to me that there was some kind of intellectual process present in the research lab that was sorely missing from the traditional education process," Wieman recounts.

A major factor in the grad students' transformation, he concluded, was their experience of intense *social engagement* around a body of knowledge — the hours they spent advising, debating with, and recounting anecdotes to one another. A 2019 study published in the *Proceedings of the National Academy of Sciences* supports Wieman's hunch. Tracking the intellectual advancement of several hundred graduate students in the sciences over the course of four years, its authors found that the development of crucial skills such as generating hypotheses, designing experiments, and analyzing data was closely related to the students' engagement with their peers in the lab, and *not* to the guidance they received from their faculty mentors.

Social interaction appeared to be an essential facilitator of intelligent thought — but, Wieman realized, such exchanges were almost completely absent from traditional undergraduate lecture courses. The way he'd been teaching, students sat listening to him talk, hardly uttering a word to one another. Wieman set out to change that, aiming to generate within his college classes the very sort of "intellectual process" that turned the graduate students in his lab into first-rate thinkers. Students no longer sat silently in rows; instead they huddled together in clusters, debating the solution to a challenging physics problem Wieman had posed. While the deliberations went on, Wieman and his teaching assistants circulated around the room, listening for misconceptions and offering feedback. When Wieman retook the podium, it was to reveal and explain the right answer, and to offer a commentary on where alternative responses went wrong. By initiating what he called "multiple brief small-group discussions" among the students, and by asking them to venture a judgment—one they were expected to defend against challenges from classmates who held a different view—Wieman was creating the conditions in which undergraduates could learn to think like expert physicists.

Wieman is one of a growing number of professors in the STEM disciplines who are bringing this "active learning" approach to their courses. Research demonstrates that students who engage in active learning acquire a deeper understanding of the material, score higher on exams, and are less likely to fail or drop out. Wieman, who holds an appointment in Stanford's school of education as well as in its physics department, spends most of his time evangelizing for more effective ways of teaching science; he donated his Nobel Prize winnings ("this big pot of money that fell from the sky") to improving physics instruction. His aspiration is to move science education away from the lecture format, toward a model that is more active and more engaged.

Wieman is working to achieve wider recognition of an often overlooked truth: the development of intelligent thinking is fundamentally a social process. We *can* engage in thinking on our own, of course, and at times solitary cognition is what's called for by a particular problem or project. Even then, solo thinking is rooted in our lifelong experience of social interaction; linguists and cognitive scientists theorize that the constant patter we carry on in our heads is a kind of internalized conversation. Our brains evolved to think *with* people: to teach them, to argue with them, to exchange stories with them. Human thought is exquisitely sensitive to context, and one of the most powerful contexts of all is the presence of other people. As a consequence, when we think socially, we think differently—and often better—than when we think non-socially.

To offer just one example: the brain stores social information differently than it stores information that is non-social. Social memories are encoded in a distinct region of the brain. What's more, we remember social information more accurately, a phenomenon that psychologists call the "social encoding advantage." If findings like this feel unexpected, that's because our culture largely excludes social interaction from the realm of the intellect. Social exchanges with others might be enjoyable or entertaining, this attitude holds, but they're no more than a diversion, what we do around the edges of school or work. Serious thinking, *real* thinking, is done on one's own, sequestered from others.

Science has not infrequently served to reinforce this notion. All of us have seen the images generated by fMRI, or functional magnetic resonance imaging: the gray mass of the brain, enlivened with patches of color denoting regions that are actively engaged in thought. The way technologies like fMRI are applied is a *product* of our brainbound orientation; it has not seemed odd or unusual to examine the individual brain on its own, unconnected to others. And the ubiquitous depictions generated by fMRI in turn *perpetuate* that very orientation: the scans offer a vivid visual affirmation of the assumption that everything worth observing happens within the bounds of a single skull. Scientists who might have wished to investigate the role of social interaction on cognition have until recently been hampered by technical constraints; for many years following the introduction of fMRI, researchers were all but required to examine the individual in seclusion, shut inside the solitary bore of the MRI machine. Thus the neuroscientific study of how people think has been, for decades, the study of people thinking alone.

Now that is changing, as a growing interest in the social dimension of cognition has arrived alongside a new generation of more flexible and adaptable tools. Technologies such as electroencephalography (EEG) and functional near-infrared spectroscopy (fNIRS) are allowing scientists to scan multiple people's brains as they interact in naturalistic settings — making deals, playing games, or simply talking to one another. Using these tools, researchers have found persuasive evidence for what is known as the "interactive brain hypothesis": the premise that when people interact socially, their brains engage different neural and cognitive processes than when those same people are thinking or acting on their own.

A representative example of this research emerges out of the study of how the brain comprehends and produces language. As far back as the nineteenth century, two bundles of gray matter — Broca's area and Wernicke's area — have been regarded as the brain's "canonical" language regions. They are named after Paul Broca and Carl Wernicke, scientists who discovered the regions' language-related function through their studies of brain-damaged patients (including autopsies of the patients following their deaths). Confirming the scientists' hunches a century after they were advanced, Broca's area and Wernicke's area were the very regions that lit up in fMRI scans of study participants who were asked to read or listen to words. But the long-established canon elaborating the brain's functional anatomy is now being revised by a new wave of research, carried out with a new array of tools.

In experiments that track brain activity while subjects are not reading or passively listening but actually *talking* to other people, a third and heretofore unknown language-related neural circuit has been identified. Studies using fNIRS—a brain-scanning technology that works via a flexible band encircling the head—demonstrate that this newly recognized network, called the subcentral area, is specialized for predicting and responding to language as it is used moment by moment in conversation. This discovery adds to accumulating evidence showing that engaging in real-time conversation involves much more nimble and nuanced cognitive processes than does the simple recognition of discrete words. It requires us to anticipate the language our conversational partner will use in speaking to us, and to improvise the language we ourselves will muster in response.

A related finding emerged when scientists, again employing fNIRS, compared the brain scans of people playing poker with a human partner to those of people playing the same game with a computer. The areas of the brain involved in generating a "theory of mind"—inferring the mental state of another individual—were active in competing with a human but dormant in matching wits with a machine. In a sense, it was not the "same game" at all; play against a human partner produced a distinctively different pattern of brain activity. A larger number of brain regions were activated, and these regions manifested a higher degree of connectivity with one another. Playing against another human produced a richer experience, neurologically speaking, than playing against a computer. Other studies have found that areas of the brain involved in planning and anticipation, and in feeling empathy, are more active when we are playing against a human as compared to a computer. Brain regions associated with reward also show stronger stimulation when we play—and especially when we *win*—against a human opponent.

The tools employed in such research are so unobtrusive they can even be used with babies and toddlers, allowing scientists to explore how social interaction shapes the thinking of children as they grow and develop. EEG is a technology that tracks and records brain wave patterns via a cap of electrodes placed on a subject's scalp; it was applied in a study conducted by Patricia Kuhl, a psychologist at the University of Washington. Kuhl and her collaborators observed nine-month-old infants from English-speaking homes as they interacted with a Spanish-speaking tutor, counting how many times the babies shifted their gaze between the tutor and the toys the tutor pointed to while speaking the names of the toys in Spanish. Such eye movements, Kuhl explains, provide an indication of the degree to which the children were bringing their social capacities to bear on their learning of a new language.

After twelve of these tutoring sessions, the researchers obtained a neural measure of the babies' second-language learning, using EEG to gauge how strongly their brains reacted to hearing Spanish spoken aloud. The infants who had engaged in the most social interaction — frequently looking back and forth between the tutor and the toys the tutor was talking about — also showed the most evidence of having learned Spanish, as indicated by their brain activity in response to Spanish sounds. Such neuroscientific findings join a larger body of evidence generated by psychology and cognitive science, all pointing to a striking conclusion: we think *best* when we think *socially*.

Yet even as scientific evidence of the link between social interaction and intelligent thought accumulates, our society remains mired in a brainbound approach to cognition; our activities at school and at work still treat thinking as the manipulation of abstract symbols inside individual heads. We are asked to produce facts (on tests, in reports) without the presence of a person to edify. We make arguments (write essays, author memos) without the presence of a person to debate. We are asked to set information out (log entries in a knowledge management system), or take information in (read manuals and instructions), without the presence of a person with whom to trade stories.

We are, that is, continually expected to think about abstract symbols for the benefit of an abstract audience, an expectation that overlooks our actual strength. Humans are not especially good at thinking about concepts; our ability to think about *people,* however, is superlative. Consider the Wason Selection Task, a test of reasoning widely used in experimental psychology. Introduced by psychologist Peter Wason in 1966, the task seems straightforward enough. One version of it goes something like this: "Take a look at the cards shown here. Each card has a vowel or a consonant on one side and an even or an odd number on the other. Which card or cards must be turned over in order to determine whether it is true that *If a card has a vowel on one side, it has an even number on the other?*" Four cards are displayed; the first is marked with an "E," the second with a "K," the third with the number 3, and the fourth with the number 6.

People's performance on this task is abysmal. Studies by many researchers over many years have shown that only about 10 percent of subjects given the task complete it correctly. Even when the language of the task is rephrased to pose a familiar-sounding problem — like which train a subway rider should take to get to her destination — people's performance remains strikingly poor. Change one particular aspect of the task, however, and the percentage of participants getting it right shoots up to 75 percent. What is that change? Make it *social.*

In the social version of the task, participants are told: "You are serving at a bar and have to enforce the rule that if a person is drinking beer, they must be 21 years of age or older. The four cards shown here have information about people sitting at a table. One side of the card tells you what a person is drinking, and the other side tells their age. Which card or cards must you turn over to see if the rule is being broken?" The puzzle, once so befuddling, now seems easily solved.

Some evolutionary psychologists have speculated that people do so much better on the social version of the Wason Selection Task because natural selection has furnished our brains with a dedicated "cheater-detection module," specialized for the crucial task of spotting transgressors who are breaking the rules of the community. It's more likely, however, that people so readily succeed at solving the puzzle in this form simply because it is social in nature, and we are past masters at thinking about social relations.

Indeed, scientists have theorized that we humans developed our oversized brains in order to deal with the complexity of our own social groups. As a result of evolutionary pressures, each of us alive today possesses a specialized "social brain" that is immensely powerful, says Matthew Lieberman, a psychologist at the University of California, Los Angeles. The social brain with its "superpowers," as Lieberman calls them, starts developing in early childhood; in adolescence, it kicks into high gear.

"HERE. THIS MAP is going to be your guide to North Shore."

With these words, Janis Ian thrusts a piece of paper into the hands of a wide-eyed Cady Heron. In the 2004 film *Mean Girls,* Cady (played by Lindsay Lohan) is the new girl at North Shore High School. The first person she meets there is the wisecracking Janis, played by Lizzy Caplan. Janis takes it upon herself to show Cady the lay of the land, as mapped out in her detailed diagram of the lunchroom.

"Now, where you sit in the cafeteria is crucial," Janis explains, jabbing at the map in Cady's hand. "Because you've got *everybody* there." The camera pans over the groups gathered around their cafeteria tables as Janis catalogs them, in not quite politically correct fashion: "You've got your freshmen . . . ROTC guys . . . preps . . . JV jocks . . . Asian nerds . . . cool Asians . . . varsity jocks . . . unfriendly black hotties . . . girls who eat their feelings . . . girls who don't eat *anything* . . . desperate wannabes . . . burnouts . . . sexually active band geeks" — and, Janis concludes with a note of scorn, the well-groomed girls she calls "the Plastics": "*Beware. Of. The Plastics.*"

Almost every adolescent maintains a mental flowchart like the one Janis put down on paper. Teens may not remember how to find the square root of a fraction or recall all the elements on the periodic table, but they can effortlessly explain and analyze the complicated social hierarchy that prevails at their high school. Starting with the launch of puberty, young people become powerfully driven to form bonds with, and establish a place among, their peers — an activity that entails an almost obsessive focus on the intricacies of relationships. They can't help it: structural and hormonal changes taking place in the brains of teenagers persistently orient them toward the social sphere.

During adolescence, teens' brains become more sensitive to social and emotional cues — responding more strongly to pictures of faces, for example, than do the brains of children or adults. Teenagers' brains also become more attuned to reward as puberty prompts an increase in the activity of neural circuits involving the feel-good chemical dopamine. And the sweetest reward for a teen is to be accepted and liked by her peers. In the service of navigating a newly complex and valued interpersonal ecosystem, the social brain appears to be "turned on" in teenagers nearly all the time. Says Matthew Lieberman of UCLA, "What the brain really wants to do, particularly during adolescence, is explore and master the social world."

Yet at this very moment in their development, we tell teenagers to *turn off* their social brains when they arrive at school, and to focus instead on abstract information devoid of social meaning or context. Teachers, parents, and other adults treat social life as an unwelcome diversion from the real work at hand, and in so doing, they set up a struggle for students' attention and effort. The results are predictable: boredom, distraction, disengagement, even acting out. Of course, we can't simply allow adolescents to attend to their social lives all day. But we *can* leverage their burgeoning sociability in the service of learning the material they need to learn. How to do this? One effective technique is to involve them in highly social relationships in which academic content is also front and center: that is, engage them in *teaching others.*

Given their well-advertised ambivalence about school, deputizing teenagers to act as teachers may seem like a questionable prescription. But that's just it: While our species and its young did not evolve to care about the Pythagorean theorem or the War of 1812, we *did* evolve to educate others about the vital arcana of our particular tribe. (Think of how much informal "teaching" of adolescent social norms goes on in the cafeteria or the student lounge.) As teachers, human beings are naturals; we are born to instruct others and to learn from them. Evidence of teaching has been found in the archaeological record reaching back hundreds of thousands of years, and the act of teaching has been observed in every human culture around the world, including the hunter-gatherer tribes that live today in a fashion similar to that of our ancient forebears.

The "teaching instinct" manifests just as reliably among modern people like us. In the course of everyday interactions, we unconsciously offer cues to others — eye contact, a change in our tone of voice — that signal our intent to instruct; these cues in turn induce our social partners to become more receptive to the information we have to convey. Such signaling begins at birth: mothers and fathers of newborn babies immediately start speaking to their infants in "parentese," a distinctively high-pitched, slowed-down, exaggerated way of talking. Research has found that hearing parentese helps infants and toddlers learn new words more readily than hearing ordinary speech. Before too long, children them-selves are engaging in instruction; teaching behavior has been observed among toddlers as young as three and a half years of age.

Across the lifespan, engagement with other people orients us toward taking in new information — but this reflexive adjustment may happen only if we encounter those people in the flesh. In a study using fNIRS brain-scanning technology, a team of researchers at Yale University found that an area of the social brain was activated when adult partici-pants looked directly into one another's eyes, but not when they gazed at the eyes of others recorded on video. "Eye contact opens the gate be-tween the perceptual systems of two individuals, and information flows," says Joy Hirsch, the Yale neuroscientist who led the study. Another fac-tor that seems to "gate," or initiate, the process of learning is *contingent* communication: social exchanges in which the utterances of one partner are directly responsive to what the other has said. When contingent com-munication is absent, learning may simply fail to occur. A particularly striking example: toddlers under the age of two and a half readily learn new words and actions from a responsive adult but pick up almost noth-ing from prerecorded instruction delivered on a screen — a phenomenon that researchers call the "video deficit."

Humans learn best from other (live) humans. Perhaps more surpris-ing, people learn from *teaching* other people — often more than the pu-pils themselves absorb. Consider this finding: firstborn children have an IQ that is on average 2.3 points higher than that of their younger brothers and sisters. After disconfirming several potential explanations, such as better nutrition or differential parental treatment, researchers concluded that firstborn children's higher IQs stem from a simple fact of family life: older siblings engage in teaching younger ones. Outside the family, lab-oratory research and real-world programs consistently show that en-gaging students in tutoring their peers has benefits for all involved, and especially for the ones doing the teaching. Why would the act of teach-ing produce learning — for the teacher? The answer is that teaching is a deeply social act, one that initiates a set of powerful cognitive, attentional, and motivational processes that have the effect of changing the way the teacher thinks.

One such process kicks in even before the tutoring session begins: students who learn information in preparation for teaching someone else review the material more intensively and organize it more thoroughly in their own minds than do students who are learning the same information in order to take a test. For social creatures like us, the prospect of engaging in an interpersonal interaction — with all of its potential for feeling admired or embarrassed — is far more motivating than the relatively anonymous activity of supplying written answers on an exam. Likewise, social interactions with other people alter our physiological state in ways that enhance learning, generating a state of energized alertness that sharpens attention and reinforces memory. Students who are studying on their own experience no such boost in physiological arousal, and so easily become bored or distracted; they may turn on music or open up Instagram to give themselves a dose of the human emotion and social stimulation they're missing.

More learning happens for the tutor in the course of teaching. When explaining academic content, the tutor is forced to make explicit the details she might herself have glossed over; the gaps in her own knowledge and understanding become visible. When directing the tutee to the most important aspects of the subject, and drawing connections among these features, the tutor is herself led to engage in a deeper level of mental processing. When fielding the pupil's questions, and posing questions of her own, the tutor is obliged to adopt a "metacognitive" stance toward the material, consciously monitoring what her pupil knows and what she herself knows. Researchers have found that while students often possess the mental tools required to understand challenging academic content, they simply don't apply them when studying on their own. When placed in the role of teacher, however, students are compelled to put those tools to use, with previously unrealized benefits for their own learning.

So powerful is the teacher role, in fact, that some of its cognitive effects can be evoked even when one's "students" don't exist. Vincent Hoogerheide, an assistant professor of education at Utrecht University in the Netherlands, has conducted several studies in which participants are asked to explain academic content on camera, to an imagined audience. After studying the material themselves, participants create a short video lesson (on the calculation of probability, for example, or on syllogistic reasoning, among other subjects). No tutee is in attendance, and there is no tutor-tutee interaction — and yet, Hoogerheide has found, the act of teaching on video enhances the teacher's own learning, improves her test performance, and enhances her ability to "transfer" the learned information to new situations. Writing out an explanation of the same material for an imagined tutee does not generate the same gains. Hoogerheide theorizes that teaching on camera generates persuasive feelings of "social presence" — the sense that there is someone watching and listening. Explaining oneself while being recorded, he notes, measurably increases the explainers' physiological arousal — a state that is associated with enhanced memory, attention, and alertness.

Face-to-face interaction between teacher and student remains the ideal, however, and it produces benefits that go beyond academics. The act of teaching can positively affect students' identity and self-image, as demonstrated by a number of real-world peer-tutoring programs — for example, a nonprofit initiative called the Valued Youth Partnership. Though we might imagine that tutors should be drawn from the ranks of the most accomplished students, Valued Youth does just the opposite: it deliberately recruits struggling students and assigns them to teach younger kids. Evaluations of the program show that students who engage in tutoring earn higher grades, attend school more consistently, and stay enrolled at higher rates than similar students who do not participate. Such outcomes may be due, in part, to the experience of what psychologists call "productive agency": the sense that one's own actions are affecting another person in a beneficial way. Actually seeing the fruits of one's labor is especially gratifying; research finds that tutors learn more, and derive more motivation, from a tutoring session when they have the opportunity to watch their tutees answer questions about what they've learned.

The experience of teaching others can also help tutors become more fully integrated into an academic or professional community. The Summer Premed Program, operated out of the medical school at the University of California, Irvine, enlists African American and Latino medical students to teach college students who are themselves members of minority groups. The undergraduates, in turn, teach students from Irvine's predominantly black and Latino public high schools. Started in 2010, the program has been shown to enhance the self-confidence and motivation of all three tiers of students.

This "cascading mentorship" model, in which participants both teach and are taught, shows promise in many settings, including the workplace. Just as students benefit from teaching their classmates, professionals gain from advising their colleagues. Holly Chiu, an associate professor of business management at Brooklyn College, reported in a study published in 2018 that employees who engage in sharing job-related knowledge with their co-workers enlarge their own expertise in the bargain. By "systematically going through the knowledge, examining it, understanding it, integrating it and presenting it," Chiu notes, these workers increased the depth and breadth of their knowledge, and subsequently turned in job performances that were rated more highly by their supervisors.

Far from being frivolous or unserious, social interaction is a vital complement to intellectual activity, activating aptitudes and capabilities that might otherwise remain unused. But because the brainbound approach to cognition regards information as information, no matter how it is encountered, the social element of thinking is often sacrificed in the name of efficiency and convenience. The spread of technology into education and the workplace has reinforced this tendency, as students are asked to learn mathematical operations from Khan Academy videos and employees are expected to train themselves using online resources. But technology could be used in another fashion: to promote the kind of in-person social exchanges that do so much to extend our mental capacities.

For example: Family Playlists, a tool developed by the education nonprofit PowerMyLearning. After being introduced to a concept at school, students are directed to take their new knowledge home and teach it to their parents or other relatives or caregivers. Family Playlists provides support by sending the "family partners" a link via a text message; the link takes them to a Web page describing the "collaborative learning activity" in which they are to participate. Family members use this same platform to provide feedback to their child's teacher about how well the child understood and explained the lesson. PowerMyLearning has now implemented Family Playlists in more than one hundred schools nationwide; the organization's CEO, Elisabeth Stock, notes that an internal research study found that students using the tool made gains in math equivalent to four months of additional learning. Even more important, she adds, teachers report that their relationships with students' families have improved, and students themselves are more engaged and enthusiastic about learning.

Teaching is a mode of social interaction we can deliberately deploy in order to think more intelligently. There's another form of social exchange that we can use to our advantage, one that comes just as naturally to the human animal: arguing.

THE STUDY was positively devilish in its design.

Participants were asked, first, to solve a series of logic puzzles: "A produce shop sells a variety of fruits and vegetables, some of which are organic and some of which are not. The apples sold by this shop are not organic. Which of the following statements about the shop's wares are true? Provide a reason for each answer. 1) All the fruits are organic; 2) None of the fruits are organic; 3) Some of the fruits are organic; 4) Some of the fruits are not organic; 5) We cannot tell anything for sure about whether the fruits in this shop are organic." After solving the puzzles, study participants were then asked to evaluate the responses provided by *other* participants — that is, to judge whether the reasons given by others seemed valid or not.

The trick: one of the answers presented in this second round did *not* issue from someone else but rather was an answer the participant herself had supplied in the first round. Some participants recognized their own response, but many others did not. What happened next was fascinating. More than half of those who believed they were evaluating someone else's response *rejected* as invalid the answer they themselves had put forth! They were especially likely to reject their own response when they had, in fact, offered a logically invalid answer originally. In other words, they applied more critical analysis to (what they thought were) other people's arguments than to their own — and this scrutiny made them more accurate.

There was a purpose behind the deviousness of the researchers who designed this study. Hugo Mercier, a cognitive scientist at the National Center for Scientific Research (CNRS) in Paris, and his coauthors were out to expose the peculiar nature of human reason. As we've seen, people often perform poorly when asked to think in a logical fashion. Recall that fewer than 10 percent of people who take the standard (non-social) form of the Wason Selection Task complete it correctly; performance on other standardized measures of reasoning, like the Thinking Skills Assessment and the Cognitive Reflection Test, is similarly mediocre, even among people who are generally well educated and even among people who have been expressly trained in argumentation and rhetoric.

An entire academic field is devoted to cataloguing the cognitive biases and other mental distortions that interfere with rational thinking. There is our well-documented confirmation bias, for example — the tendency to selectively seek out and believe evidence that supports our prior beliefs. Originally named by none other than Peter Wason, confirmation bias has been further elaborated by the psychologist Daniel Kahneman. In his 2011 book *Thinking, Fast and Slow,* Kahneman observed, "Contrary to the rules of philosophers of science, who advise testing hypotheses by trying to refute them, people (and scientists, quite often) seek data that are likely to be compatible with the beliefs they currently hold." The human mind, he lamented, is "a machine for jumping to conclusions."

But why should this be so? Why would the most intelligent creatures on the planet be hobbled by these built-in mental defects? Kahneman and others who study cognitive biases have no convincing answer to this question, according to Hugo Mercier; they treat human reason as if it were a "flawed superpower," at once impressively capable and strangely prone to breaking down. In the view of these psychologists, such glitches in the mind's ability to reason are inherent and unavoidable; the most we can do, they say, is to remain alert for the emergence of bias, and then endeavor to correct it.

Mercier begs to differ. With his collaborator Dan Sperber, also a cognitive scientist at CNRS, he has proposed a provocative alternative — different in its explanation for reason's afflictions, and different in its recommended remedy. We did not evolve to solve tricky logic puzzles on our own, they point out, and so we shouldn't be surprised by the fact that we're no good at it, any more than by the fact that we're no good at breathing underwater. What we *did* evolve to do is persuade other people of our views, and to guard against being misled by others. Reasoning is a social activity, in other words, and should be practiced as such.

Mercier's and Sperber's premise, which they advanced in their 2017 book *The Enigma of Reason,* makes coherent sense of the very aspects of human thought that have seemed so confounding: the fact that people are capable of stringently evaluating the validity of arguments, along with the fact that they so often fail to do so when the arguments are their own. Both tendencies are fully predicted by the authors' "argumentative theory of reasoning." We have every incentive to closely examine the arguments of others — who might be out to exploit or manipulate us for their own ends — but few inducements to scrutinize the arguments we make ourselves. After all, being completely convinced of the merits of our case can only make us more credible to others. And expending a lot of effort on picking apart our own argument isn't necessary, not when we can rely on our sparring partner to conduct the audit for us.

The argumentative theory also makes specific predictions about the conditions in which reason will function best, such as: the weaknesses of our reasoning faculty will be most evident when we use it outside the context in which it evolved. That context is raucously, noisily *social*. When we reason alone, inside our own heads, we will be dangerously vulnerable to confirmation bias — constructing the strongest case for our own point of view, and fooling ourselves in the process. Of course, in our brainbound culture, thinking alone is how thinking is usually done, with predictably disappointing results. Mercier and Sperber urge a different approach: *arguing together*, with the aim of arriving jointly at something close to the truth.

Arguing together is something Brad Bird and his frequent collaborator John Walker have turned into an art form. Bird is the Academy Award–winning director of Pixar movies like *Ratatouille* and *The Incredibles;* Walker is the producer who helped to manage the making of these and other films. The two are "famous for fighting openly," Bird has acknowledged, "because he's got to get it done and I've got to make it as good as it can be before it gets done." Some of the arguments they had while creating *The Incredibles* were so epic that they made it into the bonus materials included on the movie's DVD. "Look, I'm just trying to get us across the line," yells Walker in one moment captured by the camera. Bird hollers back, "I'm trying to get us across the line *in first place!*"

In an interview that took place after the movie's release, Bird explained that he counts on Walker to push back against the arguments he makes, saying of his producer: "I don't want him to tell me, 'Whatever you want, Brad' . . . I love working with John because he'll give me the bad news straight to my face. Ultimately, we both win. If you ask within Pixar, we are known as being efficient. Our movies aren't cheap, but the money gets on the screen because we're open in our conflict."

Stanford University business school professor Robert Sutton conducted the interview with Bird, whom he calls "a vigorous practitioner of creative abrasion." Bird is on the right track with his approach, says Sutton: "A pile of studies show that when people fight over ideas, and do so with mutual respect, they are more productive and creative." Indeed, research has consistently found that argument — when conducted in the right way — produces deeper learning, sounder decisions, and more innovative solutions (not to mention better movies).

Why does arguing help us think better? Hugo Mercier and Dan Sperber have their theory: Engaging in active debate puts us in the position of evaluating others' arguments, not simply constructing (and promoting) our own. Such objective analysis, unclouded by self-interested confirmation bias, makes the most of humans' discriminating intelligence. But there are additional reasons why confrontations enhance our cognition, reasons that are likewise rooted deep in human nature.

For example, there's the simple fact that conflict irresistibly seizes our attention and motivates us to learn more. We would probably put down a novel or switch off a movie that didn't introduce conflict early on — whether that conflict centers on a resolute hero battling the odds, two lovers separated by fate, or a looming disaster that might yet be averted. The drama inherent in conflict is what keeps us reading or watching. Yet we expect students and employees to attend to information that's been drained of conflict, blandly presented as the established account or consensus view. In fact, almost every topic can be cast in terms that highlight opposing perspectives — and should be, according to David Johnson, a psychologist at the University of Minnesota. It is "a general rule of teaching," he has written, "that if an instructor does not create an intellectual conflict within the first few minutes of class, students won't engage with the lesson." Johnson has spent decades investigating the uses of what he calls "constructive controversy," or the open-minded exploration of diverging ideas and beliefs. In his studies, Johnson has found that students who are drawn into an intellectual dispute read more library books, review more classroom materials, and seek out more information from others in the know. Conflict creates uncertainty — who's wrong? who's right? — an ambiguity that we feel compelled to resolve by acquiring more facts.

Intellectual clashes can also generate what psychologists call "the accountability effect." Just as students prepare more assiduously when they know they'll be teaching the material to others, people who know they'll be called upon to defend their views marshal stronger points, and support them with more and better evidence, than people who anticipate merely presenting their opinions in writing. Once the debate has begun, the act of arguing enhances thinking in an additional way: it relieves cognitive load by effectively distributing intellectual positions among the disputants. While an individual reasoning alone must keep in mind the details of each claim she contemplates, the person who argues with others can divide that task among her fellow debaters, allowing each person to stand in for a particular point of view. Relieved of the burden of carrying on a debate inside her own head, she has more mental resources to devote to evaluating the arguments on their merits.

The ability to argue emerges early in life, as any parent knows. Children as young as two or three are capable of producing justifications and constructing arguments when they find themselves at odds with their parents or siblings. "As they acquire more language, cognitive skills, and social knowledge about rules and rights," children become increasingly effective advocates for their own points of view, notes Nancy Stein, a psychologist at the University of Chicago who studies the development of argumentative thinking. The ability to critically evaluate *others'* arguments — to distinguish strong claims from weak ones — also emerges early in childhood.

We are "natural-born arguers," in Hugo Mercier's phrase — and we can deliberately deploy that innate capacity to correct our mistakes, clarify our thinking, and reach sounder decisions. The key is to approach the act of arguing with the aim not of winning at all costs but of reaching the truth through a vigorous process of advancing claims and evaluating counter-claims. We use argument to its full advantage when we make the best case for our own position while granting the points lodged against it; when we energetically critique our partner's position while remaining open to its potential virtues. According to Robert Sutton, the Stanford business school professor, we should endeavor to offer "strong opinions, weakly held"; put another way, he says, "People should fight as if they are right, and listen as if they are wrong."

Listening — and telling — is at the heart of one more way we can use social interaction to enhance our thinking: through the exchange of stories.

ALL OF THE seventh- and eighth-grade students enrolled in a 2012 study of educational methods were learning about the science of radioactive elements. The manner in which they encountered the subject, however, was strikingly different. One group was given an account written in the soporifically dull style of a textbook: "Elements are individual pieces of matter that combine with each other and make up everything we see around us. Most of what we see and use in the world, like air and water, is not made up of one single element. For example, sodium and chlorine are two different elements that make up the salt that we use for cooking . . ." — and on it droned, adding, "We now know of 92 elements that are a natural part of the Earth."

A second group of students learned the same material, but with a twist. Their version of the account picked up the thread this way: "By the late 1800s, scientists had already found most of these elements, but some were yet to be discovered. At this time, Polish-born Marie Curie and her French husband, Pierre, were two chemists living in France, trying to find all of Earth's natural elements. Although it was very hard detective work, Marie and Pierre loved solving the mysteries of elements. One day, a fellow scientist named Henri Becquerel showed Marie and Pierre a special kind of rock called pitchblende. When Henri took this pitchblende rock into a dark room, Marie could see that it gave off a light blue glow."

This second account continued: "Henri explained that the pitchblende had a lot of the element called uranium, and that he believed that the glow came from the uranium. Certainly, this was one of the strangest rocks that Marie and Pierre had ever seen, which is why they wanted to learn as much as possible about the mysterious blue glow and whether it came from the uranium." The second group of students went on to hear about how Marie and Pierre crushed the rock into tiny pieces, how they burned it at different temperatures and added different kinds of acid to see what would happen. They read about how the two scientists discovered that the uranium in the rock was emitting energetic particles, a property they named "radioactivity." And, the students were told, "while working with excitement and hope for developing a brand new element, Pierre and Marie noticed that they were beginning to feel tired and sick" — the effects of radiation poisoning.

Study author Diana Arya, an assistant professor of education at the University of California, Santa Barbara, wanted to see whether the difference in presentation would produce a difference in learning. It did. Students understood the material more thoroughly, and remembered it more accurately, when it was given to them in the form of a *story* — in particular, a story that captured the human motives and choices that lay behind the creation of what is now well-established knowledge. Arya notes that it's not the case that the second version was artificially imbued with narrative drama; rather, it's the conventional text that has been stripped of "a sense of the feelings of importance and intrigue that originally inspired the discovery."

Alas, such conventional texts — devoid of human stories and human sentiment — make up the bulk of the information students encounter in school, and, for that matter, of the information employees encounter in the workplace. This "depersonalized" approach, as other educational psychologists have called it, fails to take advantage of the distinctive power wielded by narrative. Cognitive scientists refer to stories as "psychologically privileged," meaning they are granted special treatment by our brains. Compared to other informational formats, we *attend* to stories more closely. We *understand* them more readily. And we *remember* them more accurately. Research has found that we recall as much as 50 percent more information from stories than from expository passages.

Why do stories exert these effects on us? One reason is that stories shape the way information is shared in cognitively congenial ways. The human brain has evolved to seek out evidence of causal relationships: *this* happened because of *that*. Stories are, by their nature, all about causal relationships; Event A leads to Event B, which in turn causes Event C, and so on. If a speaker were to relate a story in which the first part of the tale had no bearing on the second part, listeners would justifiably protest that this so-called "story" made no sense. At the same time, stories don't spell everything out for us either. If a storyteller were to laboriously connect every narrative dot, listeners would again rightly object: *Okay, we get it!* When stories are told well, only the highlights are included, leaving listeners to fill in the causal inferences that lend the story its full meaning. Such inferences require some mental effort, though not too much, making stories enjoyable to listen to and think about. But precisely because we *do* have to think about stories in order to understand them — do have to maintain a mental chain of events that links beginning, middle, and end — we're more likely to remember stories than to remember information that doesn't require such cognitive processing.

There's another reason why stories affect us more deeply than nonnarrative forms of information: when we listen to a story, our brains experience the action as if it were happening to us. Brain-scanning studies show that when we hear about characters emoting, the emotional areas of our brains become active; when we hear about characters moving vigorously, the motor regions of our brains are roused. We even tend to remember what characters in a story are said to remember and forget what the characters forget. On the basis of such evidence, researchers have concluded that we understand stories by running a simulation of them in our minds. Because stories by their nature feature human actors carrying out observable actions, our brains generate a mental movie of the events — an imaginary film strip that doesn't unfurl when we're reading a set of facts or instructions. Such simulations offer a kind of practice by proxy; the experiences we hear about in stories didn't happen to us, but thanks to the mental dress rehearsal we conduct as we listen, we'll be better prepared when they do.

Christopher Myers witnessed this phenomenon firsthand in the course of conducting an unusual form of academic research. Myers, an assistant professor of management and organization at Johns Hopkins University's Carey Business School, has logged many hours in the air, watching medical transport teams at work. These nurses and paramedics travel by helicopter to pick up patients from the scene of an accident, or from small community hospitals, ferrying them to larger facilities for advanced care. On the way, they administer treatment for a staggeringly wide range of illnesses and injuries. No single member of the medical transport team could possibly claim firsthand experience of every condition for which the team must provide care — forcing them to rely on the accumulated expertise of their teammates. And the way this expertise is shared, Myers discovered, is largely through narrative.

During months of fly-alongs, he observed that much of the knowledge held by flight nurses was acquired not in formal training sessions, nor from guidebooks or manuals, but through informal storytelling in the downtime between missions. "I don't want to read about Toxic Shock Syndrome in a book," one nurse said to Myers. "Tell me about the case you've just flown. What symptoms did he present? What did it look like? What did you do for him? We have protocols, but what if you guys added something that wasn't in the protocol? Tell me why. Did it work?" Team members regularly related stories to one another about technical problems they had encountered with the helicopter's equipment, about interpersonal issues they had confronted in taking over patient care from the staff at various hospitals — and, of course, about medical procedures they had performed or witnessed.

There was, for example, the tale of a patient who had fallen off a balcony at a wedding and impaled herself on the wedding singer's microphone stand. The story of how the medical transport team successfully treated her injuries reoccurred to a flight nurse years later when her own team was called upon to help a bicyclist whose torso was impaled on one of his handlebars. "I had never actually seen something like that before," the nurse told Myers about the bicycle injury. "But I'd heard the microphone-stand story, and so when we showed up on the scene it just kind of kicked in, like — 'Well, this is what they did with her, so this is a good place to start.'"

As Myers points out, such vicarious learning is increasingly necessary across any number of industries. The variety of unanticipated scenarios that may arise at a given moment is too great for any individual to have had direct experience with them all. Pressed by unfamiliar circumstances, workers may have no time to page through a procedural manual, or even to search for answers online; a trial-and-error approach is also too time-consuming, and too risky. But the professional who is in the habit of exchanging stories with co-workers has a deep well of vicarious experience on which to draw. The medical transport teams studied by Myers fly more than sixteen hundred missions a year; an individual nurse typically serves on a small fraction of these, perhaps two hundred. As one of them told Myers, listening to his colleagues' stories gave him access to "fourteen hundred experiences a year that I don't have personally. The more you know about those other patients, the more you're ready for the next one."

Narratives emerge organically in our communications with others; the role for leaders and managers lies in offering supports for, and removing barriers to, the storytelling in which their people would naturally engage. Two of the most important allowances that higher-ups can provide are *time* and *space*. In his research with transport nurses, Christopher Myers learned that stories weren't usually shared during the crush of a work shift. A nurse he interviewed told him: "There's too much going on for there to be a 'Hey, listen to this story' or 'This happened, and this . . .' It seems to be a more informal [thing that happens when you're] sitting around sharing war stories."

Some supervisors may look askance at such "sitting around," but research shows it's time well spent. One study found, for example, that a 1 percent reduction in efficiency, allowing time for "unstructured employee interaction," produces a threefold increase in group performance over the long term. During such interaction, it may seem as though employees are simply exchanging gossip. "But what is gossip?" asks Sandy Pentland, a computational scientist and MIT professor who has conducted many studies demonstrating the benefits of workplace interaction. "Gossip is stories about what happened and what you did" in response. He adds: "If you think about what needs to happen for a healthy organization, people need to know the rules of the road. They need to know how things are done. Which means they have to hear the stories."

The space where such interactions unfold is also important. In the case of the medical transport teams Myers studied, the designated story-telling locale was a ten-by-fifteen-foot area near the door to the helipad, just outside a supply room. Over time, this unassuming spot became the unofficial site for trading job-related anecdotes. The informality of the space was part of its appeal and part of its value. Myers notes that there was another, more formal space set aside for the transport nurses to share stories: the weekly doctor-supervised meetings known as "grand rounds." The patient case studies presented at these meetings were "cleaner," he reports — more focused and concise — than the stories recounted outside the supply room. But in polishing up their narratives, the nurses often omitted just those details that would be most useful to their colleagues should they encounter a similar situation in the future.

Such nitty-gritty details constitute what psychologists call "tacit knowledge": information about how things are done, when, and under which circumstances. It's what gets left out of the depersonalized information employees encounter in more formal meetings and training sessions. It's also where the "knowledge management systems" in which so many firms have invested go wrong: the information such systems make available is devoid of context, stripped of detail, and thereby rendered all but useless. "Much of the knowledge needed for employees to learn and thrive at work is not the kind of formal, codified information that is typically documented in online repositories or knowledge management systems," Myers notes. "Instead, what is often critical for success is mastery of the tacit knowledge of the organization — the complex, often subtle interpretive knowledge that is difficult to capture or write down."

The disappointing track record of such repositories makes Myers think of another interview he conducted, this one with a professional at a large tech company. This employee's organization had invested millions of dollars in a sophisticated knowledge management system, intended to codify the expertise held inside the heads of the company's workforce. "I use the knowledge management system all the time," he assured Myers — but not in the way the company's leaders intended: "I just scroll down to the bottom of the entry to see who wrote it, and then I call them on the phone." What this individual is seeking is richly contextualized information, full of detail and nuance; what he's looking for, in short, is a story.

9

.

Thinking with Groups

A FTER SEVERAL DAYS conducting military drills off the coast
of California, the USS *Palau* was headed home. The massive
aircraft carrier, large enough to transport twenty-five helicop-
ters, was steaming into San Diego Harbor at a brisk clip. Inside the pilothouse — located on the navigation bridge, two levels up from the flight
deck — the mood was buoyant. Members of the crew would soon be dis-
embarking and enjoying themselves on shore. Conversation turned to
where they would go for dinner that night. Then, suddenly, the intercom
erupted with the voice of the ship's engineer.

"Bridge, Main Control," he barked. "I am losing steam drum pressure.
No apparent cause. I'm shutting my throttles."

A junior officer, working under the supervision of the ship's navigator,
moved quickly to the intercom and spoke into it, acknowledging, "Shutting throttles, aye." The navigator himself turned to the captain, seated on
the port side of the pilothouse. "Captain, the engineer is losing steam on
the boiler for no apparent cause," he repeated.

Everyone present knew the message was urgent. Losing steam pressure effectively meant losing power throughout the ship. The consequences of this unexpected development soon made themselves evident.
Just forty seconds after the engineer's report, the steam drum had emp-
tied, and all steam-operated systems ground to a halt. A high-pitched
alarm sounded for a few seconds; then the bridge fell eerily quiet, as the
electric motors in the radars and other devices spun down and stopped.

But losing electrical power was not the full extent of the emergency.

A lack of steam meant the crew had no ability to slow the ship's rate of speed. The ship was moving too fast to drop anchor. The only way to reduce its momentum would have been to reverse the ship's propeller — operated, of course, by steam. On top of that, loss of steam hobbled the crew's ability to steer the ship, another consequence that soon became painfully evident. Gazing anxiously out over the bow of the ship, the navigator told the helmsman to turn the rudder to the right ten degrees. The helmsman spun the wheel, but to no effect.

"Sir, I have no helm, sir!" he exclaimed.

The helm did have a manual backup system: two men sweating in a compartment in the stern of the ship, exerting all their might to move the unyielding rudder even an inch. The navigator, still gazing out over the bow, whispered, "Come on, damn it, swing!" But the seventeen-thousand-ton ship sailed on — now veering far off its original course and headed for the crowded San Diego Harbor.

Watching all of this unfold in real time was Edwin Hutchins. Hutchins was a psychologist employed by the Naval Personnel Research and Development Center in San Diego. He had boarded the *Palau* as an observer conducting a study, taking notes and tape-recording conversations. Now the ship was roiled by a crisis — a "casualty," in the crew's lingo — and Hutchins was along for the ride.

From his corner of the pilothouse, Hutchins looked over at the crew's leader. The captain, he noted, was acting calm, as if all this were routine. In fact, Hutchins knew, "the situation was anything but routine": "The occasional cracking voice, a muttered curse, the removal of a jacket that revealed a perspiration-soaked shirt on this cool spring afternoon, told the real story: the *Palau* was not fully under control, and careers, and possibly lives, were in jeopardy."

Hutchins used his time aboard the ship to study a phenomenon he calls "socially distributed cognition," or the way people think with the minds of others. His aim, he later wrote, was to "move the boundaries of the cognitive unit of analysis out beyond the skin of the individual person and treat the navigation team as a cognitive and computational system." Such systems, Hutchins added, "may have interesting cognitive properties of their own." Faced with a predicament that no single mind could resolve, the socially distributed cognition of the *Palau*'s crew was about to be put to the test.

Among the downstream effects of the steam-engine malfunction was the failure of the gyrocompass, the principal tool relied upon by the *Palau*'s navigation team. Without the gyrocompass, the team had to manually ascertain the position of the ship, calculating the relationship among bearings taken from multiple landmarks on shore. And because the *Palau*'s position was a moving target, this calculation had to be generated once every minute. The ship's quartermaster chief, a man named Richards, got down to work at the chart table in the pilothouse—but it soon became clear that the job was too much for one brain to handle.

At first, Hutchins observed, Richards reached for ways to spread the burden of the task across his own body and across the tools he had at hand. He "subvocally rehearsed" the numbers he was computing, repeating the digits under his breath—using his voice and his auditory sense to expand the capacity of his working memory. He traced the columns of numbers being added with his fingertip, using his hand to help keep track of the masses of information he was managing. With a pencil, he jotted down intermediate sums in the margin of the navigation chart, fixing in place a kind of "external memory," in Hutchins's phrase. And he pulled out a calculator, using it to relieve his brain of the burden of carrying out mathematical operations. Still, laboring on his own, Richards began to fall behind. He recruited yet one more resource: the mental ability of his teammate, Quartermaster Second Class Silver. The addition of another mind created a new challenge, however: how to figure out, on the fly, the best way to divide up the complex and fast-paced task.

All the while, the ship kept moving, and now a new emergency arose: the *Palau* was bearing down on a sailboat, a small craft whose occupants were oblivious to the bigger ship's dire condition.

"Normally the *Palau* would have sounded five blasts with its enormous horn," Hutchins noted. But the *Palau*'s whistle was a steam whistle, and without steam pressure it was mute. Onboard the ship was a small manual foghorn, "basically a bicycle pump with a reed and a bell," in Hutchins's description. A junior officer—the keeper of the deck log—was sent running to find the foghorn, take it out to the bow, and let it sound. Meanwhile, the captain gripped the microphone for the flight deck's public-address system and spoke into it: "Sailboat crossing *Palau*'s bow, be advised that I have no power. You cross at your own risk. I have no power."

By this time, the sailboat had disappeared under the *Palau*'s bow; only the tip of its sail was visible from the pilothouse. The crew braced for the impending collision. The keeper of the deck log reached the bow at last and let out five feeble honks, surely too late to do any good. But a few seconds later, the sailboat emerged, still sailing, from under the starboard bow—one casualty, at least, averted.

Back inside the pilothouse, Richards and Silver were still huddled over the chart table, struggling to apportion the task between them. According to Hutchins's scrupulous observations, the pair made thirty-two attempts before "a consistent pattern of action appeared" and an effective division of labor between the two men was established. On try thirty-three, he noted, "they perform what will be the stable configuration for the first time."

Once this configuration was in place, the teammates settled into a rhythm, taking in new bearing data and churning out new position calculations. With their coordinated efforts, and those of the rest of the crew, the huge ship was guided to safety. "Twenty-five minutes after the engineering casualty and more than two miles from where the wild ride had begun, the *Palau* was brought to anchor at the intended location in ample water just outside the bounds of the navigation channel," Hutchins reported.

"The safe arrival of the *Palau* at anchor was due in large part to the exceptional seamanship of the bridge crew," he continued. "But no single individual in the bridge acting alone—neither the captain nor the navigator nor the quartermaster chief supervising the navigation team—could have kept control of the ship and brought it safely to anchor."

A psychologist on the lookout for "socially distributed cognition" could hardly have chanced upon a better example. Too often, however, we're not alert to such instances of collective thought. Our culture and our institutions tend to fixate on the individual—on his uniqueness, his distinctiveness, his independence from others. In business and education, in public and private life, we emphasize individual competition over joint cooperation. We resist what we consider conformity (at least in its overt, organized form), and we look with suspicion on what we call "groupthink."

In some measure, this wariness may be justified. Uncritical group thinking can lead to foolish and even disastrous decisions. But the limitations of excessive "cognitive individualism" are becoming increasingly clear as well. Individual cognition is simply not sufficient to meet the challenges of a world in which information is so abundant, expertise is so specialized, and issues are so complex. In this milieu, a single mind laboring on its own is at a distinct disadvantage in solving problems or generating new ideas. Something beyond solo thinking is required — the generation of a state that is entirely natural to us as a species, and yet one that has come to seem quite strange and exotic: the *group mind*.

HOW DOES A group of minds think as one? It can seem mysterious or even magical. Indeed, the study of the group mind by Western science got off to a dubious start in this regard — a relatively recent historical interlude that, on top of our culture's long-standing ideological commitment to individualism, helps to explain the unease with which the group mind is often regarded. The episode in question got under way in the late nineteenth and early twentieth centuries, a period in which the day's intellectuals — such as the French physician Gustave Le Bon and the British psychologist William McDougall — conceived a fascination with the way crowds of people seemed to have minds of their own. The group mind was believed to be powerful but also dangerous: primitive, irrational, incipiently violent. Significantly, it was also assumed that the group was *less intelligent* than the individual. Complex ideas are "only accessible to crowds after having assumed a very simple shape," asserted Le Bon in *The Crowd: A Study of the Popular Mind*, first published in 1895. "It is especially when we are dealing with somewhat lofty philosophic or scientific ideas that we see how far-reaching are the modifications they require in order to lower them to the level of the intelligence of crowds," he wrote. McDougall sounded a similar note in *The Group Mind: A Sketch of the Principles of Collective Psychology*, published in 1920. "Not only mobs or simple crowds, but such bodies as juries, committees, corporations of all sorts, which are partially organized groups, are notoriously liable to pass judgments, to form decisions, to enact rules or laws, so obviously erroneous, unwise, or defective that anyone, even the least intelligent member of the group concerned, might have been expected to produce a better result," he averred.

This conception of the group mind was hugely influential; its echoes linger today in our prevailing distrust and even disparagement of group thinking. But the field rested on shaky empirical foundations. Without a way to explain how the group mind operated, its theorists turned to vague, unscientific, and even supernatural speculation. Le Bon conjectured about a "magnetic influence" at work within crowds. McDougall mused about the possibility of "telepathic communication." Even the psychoanalyst Carl Jung got into the act, advancing the notion of a shared "genetic ectoplasm" that bound a group of people as one. Ultimately the entire field collapsed under its own imprecision and incoherence. The notion of a group mind "slipped ignominiously into the history of social psychology," writes one observer. It was "banished from the realm of respectable scientific discourse," notes another. Social scientists took as their near exclusive focus the individual, thinking and acting on his own.

But the serious study of the group mind is now staging a surprising comeback. It owes its resurgence to sheer necessity: contemporary conditions demand it. Knowledge is more abundant; expertise is more specialized; problems are more complex. The activation of the group mind — in which factual knowledge, skilled expertise, and mental effort are distributed across multiple individuals — is the only adequate response to these developments. As group thinking has become more imperative, interest has grown in learning how to do it well. At the same time, reimagined theories and novel investigative methods have granted researchers new insight into how the group mind actually operates, placing the field on a genuinely scientific footing. Neither senseless nor supernatural, group thinking is a sophisticated human ability based on a few fundamental mechanisms. We'll begin with this one: *synchrony.*

EVERY MORNING AT 6:30, the program starts up with the jaunty plinking of a piano.

"Nobinobi to senobi no undoo kara!" announces the narrator of *Radio Taiso,* a three-minute calisthenics routine broadcast daily in Japan for decades: "Stand tall and stretch your whole body!" On cue, millions of Japanese — gathered in office buildings, factories, construction sites, community centers, and public parks — begin working through a series of exercises they've known by heart since childhood.

"Ichi, ni, ushiro ni sorasete; tsugi wa ude, to ashi no undoo!" — "One, two, and back down; stretch your back, next arms and legs!" Classmates, co-workers, groups of young mothers and senior citizens reach, flex, twist,

and hop in unison. "Now, forward bends, with rhythmic bounces! Bend three times, then hands on your hips, bend backwards." Arms swinging, knees dipping, they move as if one body, right up to the program's closing lines: *"Fukaku iki o suimasuu, yukkuri sutte yukkuri haite; go, roku, moo ikkai!"*—"And we end with deep breaths, slowly in and slowly out; five, six, and one more time!"

The benefits of this activity—practiced by everyone from the youngest schoolchildren to top executives at Sony and Toyota—may go well beyond fitness and flexibility for those who take part. A substantial body of research shows that *behavioral* synchrony—coordinating our actions, including our physical movements, so that they are like the actions of others—primes us for what we might call *cognitive* synchrony: multiple people thinking together efficiently and effectively.

A study by psychologists at the University of Washington, for example, asked pairs of four-year-old children to play on a swing set apparatus installed in their lab; the researchers then discreetly manipulated whether the kids swung in unison or out of sync. After getting down from the swings, the preschoolers who had swung in time with their partners were more likely to cooperate with those same partners on a subsequent set of tasks. Comparable results were found among eight-year-olds who experienced synchronized play on a computer game; afterwards they reported feeling a greater sense of similarity and closeness to their partners than did participants who also played the game but did so out of sync with their peers. Studies conducted with adults show the same: moving in sync makes us better collaborators.

Why would this be? On the most basic level, synchrony sends a tangible signal to others that we are *open* to cooperation, as well as *capable* of cooperation. Synchronized movement acts as an invitation to work together, along with an assurance that such work will be productive. In addition to this signaling function, synchrony appears to initiate a cascade of changes in the way we view ourselves and others. The recognition that we're moving in the same way at the same time as other people heightens our awareness of being part of a group, leading us to focus less on ourselves as individuals. Because these others are making motions similar to our own, we're able to interpret and predict their actions more easily. Research shows that we are more apt to "mentalize" about them, forming a notion of what's going on in their heads. Synchrony even alters the nature of our perception, making our visual system more sensitive to the occurrence of movement. As a result of such changes, we form more accurate memories of people we have synchronized with, including the way they look, the moves they make, and the words they say. We learn from them more readily. We communicate with them more fluidly. And we pursue shared goals with them more effectively.

On an emotional level, synchrony has the effect of making others, even strangers, seem a bit like friends and family. We feel more warmly toward those with whom we have experienced synchrony; we're more willing to help them out, and to make sacrifices on their behalf. We may experience a blurring of the boundaries between ourselves and others—but rather than feeling that our individual selves have shrunk, we feel personally enlarged and empowered, as if all the resources of the group are now at our disposal. Studies of athletes and dancers have even found that moving in unison increases endurance and reduces the perception of physical pain. Synchronization sweeps us up into what one researcher calls a "social eddy," in which the press of our individual interests is diminished and the performance of the group becomes paramount. When we are carried along by the social eddy, cooperation with others feels smooth, almost effortless.

The popularity of synchronized exercises in Japan—a society famous for its communal spirit and internal cohesion—would appear to be firmly grounded in research and in the workings of human nature. In every culture and every era, armies, churches, and other institutions have used synchronous movement to bond disparate individuals into a unified whole. Picture a group of American citizens placing their right hands over their hearts and reciting the Pledge of Allegiance, for example, or a Catholic congregation kneeling and bowing their heads, speaking aloud the same words at the same time from the missal. Synchrony is a highly effective "biotechnology of group formation," as neuroscientist Walter Freeman put it—but why would such a technology be necessary?

Because, says Jonathan Haidt, "human nature is 90 percent chimp and 10 percent bee." Haidt, a psychologist at NYU's Stern School of Business, notes that in the main, we are competitive, self-interested animals intent on pursuing our own ends. That's the chimp part. But we can also be like bees—"ultrasocial" creatures who are able to think and act as one for the good of the group. Haidt argues for the existence in humans of a psychological trigger he calls the "hive switch." When the hive switch is flipped, our minds shift from an individual focus to a group focus — from "I" mode to "we" mode. Getting this switch to turn on is the key to thinking together to get things done, to extending our individual minds with the groups to which we belong.

Synchronous movement is one way of flipping this switch; it reliably produces what the late historian William McNeill called "muscular bonding." He maintained that the long-standing dominance of European armies over other fighting forces was due in part to the psychological effect of close-order drill, a practice that took root in the Netherlands in the sixteenth century before spreading to other European nations. Soldiers spent hours marching in formation, their movements tightly coordinated — thereby creating a mental and emotional bond that elevated their performance on the battlefield.

McNeill, a distinguished military scholar, wrote about the transformative effect of martial drills not only from his erudition but from his personal experience as well. As a young man, he was drafted into the US Army and sent to basic training in Texas. There he and his fellow recruits were ordered to march, "hour after hour, moving in unison and by the numbers in response to shouted commands, sweating in the hot sun, and, every so often, counting out the cadence as we marched: Hut! Hup! Hip! Four!" McNeill recounted. "A more useless exercise would be hard to imagine," he notes wryly, and yet as the hours wore on, he found himself entering "a state of generalized emotional exaltation."

"Words are inadequate to describe the emotion aroused by the prolonged movement in unison that drilling involved," he wrote. "A sense of pervasive well-being is what I recall; more specifically, a strange sense of personal enlargement; a sort of swelling out, becoming bigger than life, thanks to participation in collective ritual." McNeill continued, "Obviously, something visceral was at work; something, I later concluded, far older than language and critically important in human history, because the emotion it arouses constitutes an indefinitely expansible basis for social cohesion among any and every group that keeps together in time, moving big muscles together and chanting, singing, or shouting rhythmically."

What happened to McNeill and his comrades on that "dusty, graveled patch of the Texas plain" was surely the product of behavioral synchrony, of moving together in a coordinated manner. But there was likely another factor affecting them as well: not just shared *movement* but shared *arousal*. Their bodies' common response to the physical exertion of marching, the heat of the sun, the shouted commands of their superiors—this too supported the emergence of a group mind.

The significance of shared arousal was demonstrated in an ingenious experiment designed by researcher Joshua Conrad Jackson and published in the journal *Scientific Reports* in 2018. Jackson and his colleagues set out "to simulate conditions found in actual marching rituals"—which, they noted, "required the use of a larger venue than a traditional psychology laboratory." They chose as the setting for their study a professional sports stadium, with a high-definition camera mounted twenty-five meters above the action. After gathering 172 participants in the stadium and dividing them into groups, the experimenters manipulated their experience of both synchrony and arousal: one group was directed to walk with their fellow members in rank formation, while a second group walked in a loose and uncoordinated fashion; a third group speed-walked around the stadium, boosting their physiological arousal, while a fourth group strolled at a leisurely pace. Jackson and his collaborators then had each group engage in the same set of activities, asking them to gather themselves into cliques, to disperse themselves as they wished across the stadium's playing field, and finally to cooperate in a joint task (collecting five hundred metal washers scattered across the field).

The result: when participants had synchronized with one another, *and* when they had experienced arousal together, they then behaved in a distinctive way—forming more inclusive groups, standing closer to one another, and working together more efficiently (observations made possible by analyzing footage recorded by the roof-mounted camera). The findings suggest that "behavioral synchrony and shared physiological arousal in small groups independently increase social cohesion and cooperation," the researchers write; they help us understand "why synchrony and arousal often co-occur in rituals around the world."

As demonstrated in this study, physical exertion is a dependable method of producing physiological arousal—but it's not the only one. An experience of heightened emotion will also do the trick. Whether hearts are racing as a result of running laps or because of hearing an exciting story, such shared arousal is another way of getting a group of individuals to cohere. In behavioral synchrony, group members are moving their arms and legs as if they were one being; in physiological synchrony, their hearts are beating and their skin is perspiring as if they were one body. Both behavioral and physiological synchrony, in turn, generate greater cognitive synchrony. Emerging research even points to the existence of "neural synchrony" — the intriguing finding that when a group of individuals are thinking well together, their patterns of brain activity come to resemble one other's. Though we may imagine ourselves as separate beings, our minds and bodies have many ways of bridging the gaps.

A HOST OF LABORATORY experiments, as well as countless instances of real-world rituals, show that it's possible to activate the group mind — to flip the hive switch, as it were — by "hacking" behavioral synchrony and physiological arousal. The key lies in creating a certain kind of group experience: real-time encounters in which people act and feel together in close physical proximity. Yet our schools and companies are increasingly doing just the opposite. Aided by technology, we are creating individual, asynchronous, atomized experiences for students and employees — from personalized "playlists" of academic lessons to go-at-your-own-pace online training modules. Then we wonder why our groups don't cohere, why group work is often frustrating and disappointing, and why thinking with groups doesn't extend our intelligence.

Why is our current approach so wrongheaded? It assumes that information is information, however it is encountered; that tasks are tasks, no matter how we take them on. But in fact, the new science of the group mind is demonstrating that we think differently — and often better — when we think as part of a close-knit group rather than as individuals. This is particularly the case regarding our *attention* and our *motivation*. The nature of these two states is altered in meaningful ways when we enter them collectively instead of alone.

First, attention: the phenomenon that psychologists call "shared attention" occurs when we focus on the same objects or information at the same time as others. The awareness that we are focusing on a particular stimulus along with other people leads our brains to endow that stimulus with special significance, tagging it as especially important. We then allocate more mental bandwidth to that material, processing it more deeply; in scientists' terms, we award it "cognitive prioritization." In a world of too much information, we use shared attention to help us figure out what to focus on, then direct our mental resources toward the object that the spotlight of shared attention has illuminated. As a result of these (mostly automatic) processes, we *learn* things better when we attend to them with other people. We *remember* things better when we attend to them with other people. And we're more likely to *act upon* information that has been attended to along with other people.

The practice of engaging in shared attention starts in infancy. By nine months, a baby begins to look in the direction in which an adult turns her head. Infants will gaze longer at what the grownups around them appear to be looking at, and they are more likely to recognize objects that they earlier jointly attended to with a caregiver than objects they attended to alone. In this subtle and mostly unconscious way, parents are continually instructing their offspring on what is important, what merits attention, and what can be safely ignored.

By one year of age, a baby will reliably look in the direction of an adult's gaze, even absent the turning of the adult's head. Such gaze-following is made easier by the fact that people have visible whites of the eyes. Humans are the only primates so outfitted, an exceptional status that has led scientists to propose the "cooperative eye hypothesis" — the theory that our eyes evolved to support cooperative social interactions. "Our eyes see, but they are also meant to be seen," notes science writer Ker Than.

The ability to experience the world from a shared perspective is an evolved adaptation that grants humans an unequalled capacity for co-ordinating thought and behavior with other members of their species. Shared attention, and the increased cognitive resources devoted to information that is mutually attended to, produces greater overlap in group members' "mental models" of a problem, and therefore smoother cooperation while solving it. It is, in a sense, what makes all human achievements possible — from jointly maneuvering a piece of furniture through a narrow doorway to collaboratively designing and launching a rocket sent to the moon. And it starts with babies following the direction of our eyes.

Shared attention remains important among adults, though it plays a different role than in the interactions of caregivers and children. Here the function of shared attention is not so much the expert instruction of a novice but rather the maintenance of a mutual store of information and impressions. We feel compelled to continuously monitor what our peers are paying attention to, and to direct our own attention to those same objects. (When the face of everyone on the street is turned skyward, we look up too.) In this way, our mental models of the world remain in sync with those of the people around us.

The common ground established by shared attention is especially crucial for teams working together to solve a problem. Studies of groups laboring on a shared task — from students programming a robot to surgeons performing an operation — show that the members of effective teams tend to synchronize their gaze, looking at the same areas at the same time. More of these "moments of joint attention" are associated with more successful outcomes. Research suggests that the ability to coordinate such moments can be acquired with practice. One study of physician teams performing surgery on a simulator found that the gaze of experienced surgeons overlapped at a rate of about 70 percent, while the gaze of novices overlapped only about 30 percent of the time. But effective collaborators aren't *always* looking at the same place at the same time; rather, they cycle between looking on their own, then looking together.

If attention is different when we experience it as a member of a group, so too is *motivation*. Common conceptualizations of motivation — such as "grit," the notion popularized by University of Pennsylvania psychologist Angela Duckworth — are based on the assumption that engagement and persistence are individual matters, individually willed. What this understanding omits is that our willingness to persevere can be enhanced when our efforts are made on behalf of a group we care about. Membership in a group can be a potent source of motivation — if we feel a genuine sense of belonging to the group, and if our personal identity feels firmly tied to the group and its success. When these conditions are met, group membership acts as a form of *intrinsic* motivation: that is, our behavior becomes driven by factors internal to the task, such as the satisfaction we get from contributing to a collective effort, rather than by external rewards such as money or public recognition. And as psychologists have amply documented, intrinsic motivation is more powerful, more enduring, and more easily maintained than the extrinsic sort; it leads us to experience the work as more enjoyable, and to perform it more capably.

Experiencing ourselves as part of a collective "we," rather than as a singular "I," changes the way we direct our focus and the way we allocate our energies — often in felicitous fashion. Yet so much in our every-man-for-himself society conspires against the creation of a robust sense of "we." Our emphasis on individual achievement, and our neglect of group cohesion, means that we are failing to reap the rich benefits of shared attention and shared motivation. Even when groups do exist in name, they are often weak and dilute in their bonds. Psychologists have found that groups differ widely on what they call "entitativity" — or, in a catchier formulation, their "groupiness." Some portion of the time and effort we devote to cultivating our individual talents could more productively be spent on forming teams that are genuinely groupy.

In order to foster a sense of groupiness, there are a few deliberate steps we can take. First, people who need to think together should *learn* together — in person, at the same time. The omnipresence of our digital devices can make it difficult to ensure that shared learning takes place, even among students gathered in a single classroom. Some years back, high school teacher Paul Barnwell realized that many of his students were physically present, but mentally absent, during class. "They were pecking away at their smartphones under their desks, checking their Facebook feeds and texts," recalls Barnwell, who teaches English at Fern Creek Traditional High School in Louisville, Kentucky.

Moreover, once he got their attention and directed them to engage in a group assignment, he discovered that his students didn't know how to carry on an academic discussion. They were so used to the stutter-stop rhythm of asynchronous text exchanges that holding a substantive conversation in real time was an unfamiliar and unpracticed activity. (Notably, research has found that asynchronous communication — of the kind that is now common not only among teenagers but among adult professionals as well — reduces the efficiency and effectiveness of group work.) In a clever jujitsu move, Barnwell redirected his students' use of technology: he asked them to record one another with their smartphones and then analyze their own and their partners' conversational patterns. Before long, his students were holding lively class-wide conversations — thinking and acting more like a group, and reaping the cognitive benefits that only a group can generate.

A second principle for engendering groupiness would go like this: people who need to think together should *train* together — in person, at the same time. Research shows that teams that trained as a group collaborate more effectively, commit fewer errors, and perform at a higher level than teams made up of people who were trained separately. Training together can also reduce the "silo effect," a common phenomenon in which co-workers fail to communicate or collaborate across different departments and disciplines. Yet training together is not the norm in many industries. In medicine, for example, health care providers representing various specialties — surgeons, nurses, anesthesiologists, pharmacists — must collaborate closely when caring for patients. But traditionally, their training occurs in isolation from one another, in different departments and even different institutions.

Some medical schools and hospitals are now experimenting with group training across disciplinary lines. The University of Minnesota has found an especially engaging way to do so: creating an "escape room." In this activity (modeled on an adventure game), UMN students studying nursing, pharmacy, physical therapy, and social work, among other disciplines, are invited into a simulated hospital room. There they are given the case study of a fictional patient — for example: "A 55-year-old male with a past medical history of bipolar disorder and type I diabetes presents to the emergency room with diabetic ketoacidosis, triggered by a recent manic episode." Acting under the pressure of a one-hour time limit, the students must work together to develop a discharge plan for the patient by solving a series of puzzles, making use of the objects and information available in the room — and drawing on the participants' varied areas of expertise. The game is followed by a guided debriefing session in which students reflect on the challenges of collaborating across fields. The "interprofessional escape room" is now part of the formal curriculum for students studying the health sciences at the University of Minnesota; similar activities have been introduced at hospitals and medical schools located in Philadelphia, Pennsylvania; Buffalo, New York; Tucson, Arizona; and Lubbock, Texas.

A third principle for generating groupiness would hold that people who need to think together should *feel* together — in person, at the same time. Laboratory research, as well as research conducted with survivors of battlefield conflicts and natural disasters, has found that emotionally distressing or physically painful events can act as a kind of "social glue" that bonds the people who experienced them together. But the emotions that unite a group need not be so harrowing. Studies have also determined that simply asking members to candidly share their thoughts and feelings with one another leads to improvements in group cohesion and performance.

The Energy Project, a training and consulting firm based in New York, holds a companywide "community meeting" every Wednesday. Each of the organization's employees is asked a series of simple questions, starting with "How are you feeling?" "That's a very different question than the standard 'How are you?' we all ask each other every day," notes Tony Schwartz, the company's founder and CEO. "When people stop and reflect, and then say, one at a time, how each of them are really feeling, it opens up a deeper level of dialogue." At times, he recounts, his colleagues' answers have been searching or even wrenching, reflecting a personal crisis or family tragedy. But even when responses are more run-of-the-mill, the members of his close-knit staff have shared an emotional experience with one another, one that is fleshed out by the remaining questions in the series: "What's the most important thing you learned last week?" "What's your goal for this week?" "What are you feeling most grateful for?"

The fourth and final mandate for eliciting groupiness is this: people who need to think together should *engage in rituals* together — in person, at the same time. For this purpose, a ritual can be any meaningful organized activity in which members of a group take part together. If the rituals involve synchronized movement or shared physiological arousal, all the better. Both of these switches are flipped at Clearview Elementary School in Sherburne County, Minnesota, where each weekday starts with the "Morning Mile." Students in every grade spend twenty minutes walking briskly before class, usually outside. The physical exertion involved means that shared physiological arousal is a given; teachers report that students arrive at their desks with cheeks flushed (especially during the cold Minnesota winters). But the Morning Mile generates synchronous movement as well. Research shows that when people walk or run together, they automatically and unconsciously match up their bodily movements.

Even so ordinary a ritual as sharing a meal can make a difference in how well a group thinks together. Lakshmi Balachandra, an assistant professor of entrepreneurship at Babson College in Massachusetts, asked 132 MBA students to role-play executives negotiating a complex joint venture agreement between two companies. In the simulation she arranged, the greatest possible profits would be created by parties who were able to discern the other side's preferences and then work collectively to maximize profits for the venture as a whole, rather than merely considering their own company's interests. Balachandra found that participants who dined together while negotiating—at a restaurant, or over food brought into a conference room—generated 12 percent higher profits, on average, than those who bargained while not eating.

The explanation may go back, again, to synchrony. Balachandra notes that when we eat together, we end up mirroring one another's movements: lifting the food to our mouths, chewing, swallowing. "This unconscious mimicking of each other may induce positive feelings towards both the other party and the matter under discussion," she writes. Other research has found that the positive effect of shared meals on cooperation is heightened if participants dine "family style"—eating the same food, served from communal dishes. It may also be enhanced if very spicy entrees are on the menu, since consuming such food increases body temperature and perspiration, raises blood pressure, speeds up heart rate, and prompts the release of adrenaline, all hallmarks of physiological arousal. A group of Australian researchers reported greater economic cooperation among people who together had eaten bird's eye chilies, a painfully hot pepper.

In addition to incorporating the now familiar factors of behavioral synchrony and physiological arousal, consuming food with others is in itself uniquely meaningful: our very survival depends on this elemental sharing of resources. "Eating together is a more intimate act than looking over an Excel spreadsheet together," observes Kevin Kniffin, an assistant professor of management at Cornell University. "That intimacy spills back over into work." In a study published in the journal *Human Performance,* Kniffin and his coauthors reported that teams of firefighters who eat their meals together perform better than firefighters who dine on their own. He believes that our focus on individual achievement—and individual rewards—leads us to overlook the performance-enhancing effects of group rituals. "Coworkers who eat together tend to perform at a higher level than their peers, yet cafeterias are often undervalued by companies," he notes. And those tech companies that do offer luxurious cafeterias as a perk to their employees? The key may not be the freshness of the sushi or the deliciousness of the vegan grain bowls but whether the firm's workers consume such delicacies *together.*

All of these approaches to generating groupiness are firmly grounded in our nature as embodied, situated, social beings; their effectiveness depends on people moving, talking, and working together, so closely that their brains and bodies fall into a joint rhythm. This marks a difference from notions such as "crowdsourcing" and the "hive mind," which have enjoyed a sustained surge of popularity. In theory and in practice, these concepts are highly brainbound: a bunch of disembodied minds bouncing ideas around, usually online. Technology more generally has often served to isolate us from one another, sealing us within our individual digital bubbles. But this need not be the case; promising models of extended technology—that is, extended by the age-old resource of the human group—are now emerging.

For example: scientists at Germany's Max Planck Institute and elsewhere are experimenting with automatic "rapport detection" within groups. Sensors embedded in a conference room or in video-conferencing equipment unobtrusively monitor group members' nonverbal behavior (their facial expressions, hand motions, gaze direction, and so on); these data are analyzed in real time to yield a measure of how well a group is cooperating. When rapport falls below a critical level, nudges can be applied to move the group toward greater cohesion: the system might alert the group's leader that a shared coffee break is in order, or it might suggest to him, via a pop-up message, that he engage in more mirroring of his co-workers. Inside wired-up "smart meeting rooms," it may even elect to raise the temperature by a few degrees, or introduce some soothing white noise.

Another technology-assisted tactic for generating groupiness are activities in which members of a group are challenged to synchronize their movements while dancing with one another or moving in time to music. Body-worn sensors compute the degree of synchrony group members achieve, and real-time feedback allows participants to fine-tune their movements such that they become ever more closely matched to those of their peers. "We set out to design a mobile-based play experience that enhanced in-person social interaction and connection," explains Katherine Isbister, a professor of computational media at the University of California, Santa Cruz, who says she was inspired by research that shows "how being physically 'in sync' brings people together emotionally and builds trust." Isbister notes that her game, called Yamove!, encourages looking not at a screen but at the other players. "The more players look at each other, the better results they achieve in coordination and the stronger the lingering positive social effects," she says. Used as an icebreaker or team-building activity, games such as Yamove! may strike some as embarrassing or ridiculous. But unlike those we typically endure, these digital nudges in the direction of greater synchrony might actually work.

And—potential embarrassment aside—what's striking about the group experiences explored here is how very *positive* they are. Military historian William McNeill entered "a state of generalized emotional exaltation" while marching with his fellow recruits in basic training. Tony Schwartz, the founder of the consulting company The Energy Project, says that he and his staff members find their weekly community meetings "powerful" and "liberating," even "transformational." On surveys, participants in educational escape room activities describe the experience as "engaging," "motivating," and even "fun."

This is, to put it mildly, not the perspective most of us bring to group projects. Group work is widely disliked, even despised, in both educational and professional settings; it is commonly viewed as inefficient, unfair, and just plain annoying. The research literature has even given this phenomenon a name: "grouphate," defined as "a feeling of dread that arises when facing the possibility of having to work in a group." What explains the disparity between the productive, invigorating, even ecstatic ideal of group thought and action—which, as we've seen, human beings *evolved* to do well—and the dispiriting reality as most of us experience it? The answer may lie in a profound mismatch between the present-day demands of knowledge work and a set of ideas about such endeavors that is rooted deep in the past.

THE LETTER TO Albert Einstein, dated June 4, 1924, began on a deferential note. "Respected Sir: I have ventured to send you the accompanying article for your perusal and opinion. I am anxious to know what you think of it." The author of the letter, Satyendra Nath Bose, was an obscure academic at a university in East Bengal. The paper he was sending to Einstein had already been submitted to—and rejected by—a professional journal. And the letter's addressee "was not just the most famous scientist of his time; he was one of the best-known individuals on the entire planet," notes Yale University physics professor A. Douglas Stone. But

Bose felt at ease reaching out to Einstein, he explained in the letter, "because we are all your pupils." And so, with "some combination of veneration and chutzpah," in Stone's words, Bose went on to make an astonishing request.

"I do not know sufficient German to translate the paper. If you think the paper worth publication, I shall be grateful if you arrange for its publication in *Zeitschrift für Physik*," he wrote, naming Germany's leading physics journal. Even more surprising, Einstein agreed to his entreaty. Reading the paper, he saw that Bose had solved a problem that Einstein had labored on without success: how a law of radiation, formulated by German physicist Max Planck some twenty-four years earlier, could be deduced from the theory that light is a particle as well as a wave (a theory that Einstein himself had proposed in 1905). It was, as Einstein wrote to Bose, "a beautiful step forward." Bose took this step on his own, powered only by his own curiosity. It was simple, he later explained: "I wanted to know how to grapple with the difficulty in my own way." In 1925 Bose's paper was published in the journal he had specified, *Zeitschrift für Physik*, along with a commentary from Einstein. It is not an exaggeration to say that the course of scientific history was altered by one man thinking alone.

Ninety years later, another paper was published. It documented a successive step forward in the discovery process to which Bose had made a signal contribution, reporting a newly precise measurement of the Higgs boson mass. (Bosons, which were named in honor of Bose, are a type of particle that obeys the rule of what is called, in physics, "Bose-Einstein statistics"; their mass is measured by means of a particle accelerator, a gargantuan machine that propels charged particles to extremely high speeds.) The author of *this* paper was Georges Aad — and Brad Abbott, and Jalal Abdallah, and Ovsat Abdinov, and Rosemarie Aben, and Maris Abolins, and Ossama AbouZeid, and Halina Abromowicz, and Henso Abreu ... on and on, for a total of 5,154 authors. The publication, in the journal *Physical Review Letters*, is but an extreme example of a trend now ascendant in every industry and occupation: in order to carry out the intensely complex work demanded by the modern world, people must think together in groups.

The shift is easiest to see — and to measure — in the social and physical sciences, where contributions by single individuals were once the norm. Today, fewer than 10 percent of journal articles in science and technology

are authored by just one person. An analysis of book chapters and journal articles written across the social sciences likewise found "a sharp decline in single-author publishing." In economics, solo-authored articles
once predominated; now they account for only about 25 percent of publications in the discipline. In the legal field, a 2014 survey of law reviews
concluded that nowadays, "team authors dominate solo authors in the
production of legal knowledge." Even the familiar archetype of the solo
inventor (think Thomas Edison or Alexander Graham Bell) is no longer
representative. A 2011 report found that over the previous forty years, the
number of individuals listed on each US patent application had steadily
increased; nearly 70 percent of applications now named multiple inventors.

This development is more than an academic fad, says Brian Uzzi, a
professor of management at Northwestern University who conducted
some of this research: "It suggests that the process of knowledge creation
has fundamentally changed." More broadly, Uzzi notes, "almost everything that human beings do today, in terms of generation of value, is no
longer done by individuals. It's done by teams." What has *not* changed is
our model of how intelligent thinking happens. We're still convinced that
good ideas and new insights and ingenious solutions come from a single
brain; we're a bunch of pencil-wielding Satyendra Boses in an era of particle accelerators and mega-collaborations. This fundamental mismatch
lies at the root of many of our struggles with group work.

It's time we tossed out that individual model and replaced it with one
better suited to the world in which we actually live. We can begin by
identifying those ways in which thinking with a group is different from
thinking on our own, and by instituting new practices that support the
smooth operation of the group mind. Once these practices are put into
place, research shows, a group can think more efficiently and more effectively than any one of its members — a phenomenon that psychologists
call "collective intelligence."

The ways in which group thinking differs from individual thinking
are obvious and yet almost always overlooked. The first of these: When
we think on our own, all of our thoughts get a hearing. But when we think
as part of a team, it takes intentional effort to ensure that everyone speaks
up and that everyone shares what they know. Research on group dynamics reveals that this rarely happens. Instead, very few people — and some-

times just one — dominate the conversation; in addition, group members often neglect to contribute their "uniquely-held information," gravitating instead to discussion of information that everyone present already knows. Thus do less than optimal patterns of communication produce the inefficiency and ill will associated with group work, without generating any of the potential benefits.

This outcome isn't inevitable, however; simple changes in the way communication is carried out can steer teams toward the group mind. Steven Rogelberg, a professor of management at the University of North Carolina at Charlotte, notes that group members "often hold back in meetings, waiting to hear what others say and what their boss might say out of fear of being perceived as difficult, out of touch, or off the mark." Asking attendees to write out their contributions instead of speaking them, he says, "can be a solution to this problem, allowing space for unique knowledge and novel ideas to emerge." Participants jot down their thoughts on index cards, which the group's leader then reads aloud. Or they write them on sheets of paper posted around the room, after which participants circulate again — this time marking down comments on their colleagues' ideas, which the group as a whole then discusses.

Another potential change to communication patterns centers on the behavior of group leaders. Cass Sunstein is a professor at Harvard Law School who also served as administrator of the White House Office of Information and Regulatory Affairs under Barack Obama. Upon assuming this role, Sunstein learned a valuable lesson in group leadership: if he began a meeting by stating his own views, he discovered, the ensuing discussion was far less expansive and open than if he started out by saying, "What do you all think? This is a tough one." As soon as a leader makes his preferences known, says Sunstein, many who work for him will choose to engage in "self-silencing" rather than rock the boat with a dissenting view. And, he notes, "some people are more likely to silence themselves than others"; these may include women and members of minority groups, as well as individuals with less status, less experience, or less education. Yet it's just this range of voices that must be heard if the group mind is to exert its unique power. One solution, says Sunstein, is for leaders to silence *themselves;* the manager or administrator who adopts an "inquisitive and self-silencing" stance, he maintains, has the best chance of hearing more than his own views reflected back to him.

The second way in which group thinking differs from individual thinking is this: when thinking as part of a collective, we need to make our thought processes *visible* to others on our team. While we do leave "traces" for ourselves when engaging in private thought — underlining, jotting notes in margins, moving papers from the "unread" pile to the "have gone through" pile — these traces must be far more specific and explicit if they are to be used productively by others. Philosopher Andy Clark, observing the progressive delegation of our mental operations to our devices, has noted that "the mind is just less and less in the head" these days. More than that, the mind *must* be less and less in the head, and more and more emblazoned on the world, if we are to extend our minds with the minds of others.

Once again, verbal communication is key — but not of the unstructured type that too often conforms to and confirms our individual-oriented model of thinking. Rather, researchers recommend that we implement a specific sequence of actions in response to our teammates' contributions: we should *acknowledge, repeat, rephrase,* and *elaborate* on what other group members say. Studies show that engaging in this kind of communication elicits more complete and comprehensive information. It re-exposes the entire group to the information that was shared initially, improving group members' understanding of and memory for that information. And it increases the accuracy of the information that is shared, a process that psychologists call "error pruning." Although it may seem cumbersome or redundant, research suggests that this kind of enhanced communication is part of what makes expert teamwork so effective. A study of airplane pilots, for example, found that experienced aviators regularly repeated, restated, and elaborated on what their fellow pilots said, while novice pilots failed to do so — and as a result, the less experienced pilots formed sparser and less accurate memories of their time in the air.

Another way to make our thinking visible to others is to collaborate on the creation of what Gary Olson and Judith Olson call "shared artifacts." The Olsons, both professors of informatics at the University of California, Irvine, have spent more than three decades studying how people think and work together. One major contributor to the success of group cognition, they have discovered, is the effective use of such artifacts, or tangible representations of the task to be completed — which are, ideally, *large, complex, persistent,* and *revisable.* Over the course of their long careers, the Olsons have often evaluated the effectiveness of workplace technology such as video-conferencing software and digital collaboration platforms. But the baseline against which they compare these tools — the working arrangement they regard as the best of all worlds — is decidedly analog: a group of people gathered together in a room that is dedicated to their current project, with plenty of space on the walls to tack up those shared artifacts (which may take the form of lists, graphs, charts, or sketches).

It matters above all that these artifacts are, in fact, shared. At one design meeting they observed, all the participants were handed individual copies of a system diagram. "As they discussed and agreed to things, they took notes on their own copy of the diagram, adding things and crossing things out," the Olsons recounted in one of their academic articles. "We noted at the end of the meeting that different people had made different marks, implying different understandings of what they had agreed to." Without the capacity to refer to a single shared artifact, the Olsons concluded, these co-workers "ended up not 'singing from the same sheet of music.'"

In addition to being shared, it's beneficial for group artifacts to be large and complex. The Olsons have found that people often gesture at large artifacts, enhancing their own thinking and that of the people who observe them; meanwhile, a complex artifact (as opposed to a simple or schematic one) allows more of the group's thinking to be explicitly represented for all to see, rather than remaining concealed inside individuals' heads. Finally, shared artifacts are most effective when they are persistent — preserved, retained, and kept continuously visible — but also revisable, able to be changed as new information or insight emerges. Describing another team at work, the Olsons noted that this group's artifacts "were often put up in the order in which they were produced. People knew where to look for something because they knew when it was produced, and they could tell something about another person's attention by seeing where that person was looking." Lamenting the "inherent invisibility of much of today's computing artifacts" — we can't see what's inside our colleague's laptop, any more than we can see what's inside her head — the Olsons report that the best materials with which to create these representations are simple ones: a felt-tip pen, used on large sheets of paper.

There's a third way in which group thinking diverges from individual thinking: when engaged in the latter, we of course have access to the full depth of our own knowledge and skill. That's not the case when we're thinking collectively—and that's a good thing. One of the great advantages of the group mind is its capacity to bring together many and varied areas of proficiency, ultimately encompassing far more expertise than could ever be held in a single mind. We *couldn't* know all that our fellow group members know, nor should we want to; our mental bandwidth would quickly become overloaded. We do, however, need to know that *they* know it, in order to call upon it when it's needed. The process by which we leverage an awareness of the knowledge other people possess is called "transactive memory."

It could be said that the study of transactive memory began on the day when Daniel Wegner and Toni Giuliano were wed. The groom later wrote: "Toni and I noticed not long after we were married that we were sharing memory duties. I remembered where car and yard things were, she remembered where house things were, and we could each depend on the other to be an expert in domains we didn't need to master." As social psychologists, Wegner and Giuliano quickly identified their experience as not only an interesting feature of newlywed life but also a promising target of scientific investigation. A year later the couple (along with colleague Paula Hertel) published a paper introducing what Wegner called a new way "to understand the group mind." As he observed: "Nobody remembers everything. Instead, each of us in a couple or group remembers some things personally—and then can remember much more by knowing who else might know what we don't. In this way, we become part of a transactive memory system."

Over the past few decades, psychologists have confirmed Wegner's claim that a robust transactive memory system effectively multiplies the amount of information each group member has. Members of such groups are able to work on deepening their own areas of expertise while still remaining in contact, through their colleagues, with a broader range of relevant information. Their cognitive load is reduced, as they need only to attend in the moment to the portion of incoming information that concerns them, secure in the knowledge that their teammates are doing the same. And members of these groups are able to engage in smooth and efficient coordination, directing tasks to those members who are best suited to carry them out. As a result, research finds, teams that build a strong transactive memory structure perform better than teams for which that structure is less defined.

Within a group of any size, a transactive memory system will spontaneously assemble itself, as it did for Daniel Wegner and Toni Giuliano once they married and began living together. But because such systems usually aren't cultivated in an intentional way, much of their potential to extend the group's intelligence is lost. The goal of such cultivation is to make the group aware of what its members know, without requiring each member to take on the full burden of her teammates' specialized knowledge. Notice how this model differs from early notions of the "group mind," in which members of a group were believed to be thinking the same thoughts at the same time. By contrast, the value of a transactive memory system lies in its members thinking *different* thoughts while also remaining aware of the contents of their fellow members' minds. The struggle to cope with information overload has led many of us to turn to technological filters — smartphone alerts and email applications that offer to sort for us the information that must be attended to from the information that can be ignored. Research suggests, however, that *other people* can function as the most sensitive and discriminating filters of all — as long as we're aware of what they know and can access their knowledge when we need it.

All of us maintain a set of mental markers that help us locate information we don't currently possess; we may not recall every detail contained in a report, but we know the folder (physical or digital) where the report can be found. Such markers also point us toward the people who possess information we do not; the aim in building a robust transactive memory system is to make these pointers as explicit and as accurate as possible. The process of setting out these markers should begin early in a team's work together. It's important to establish from the outset not only who's responsible for *doing* what but also who's responsible for *knowing* what. Group members should be explicitly informed about their colleagues' distinctive talents or spheres of specialization, and clear protocols should be established for directing questions and tasks to the appropriate individual. Research shows that groups perform best when each member is clearly in charge of maintaining a particular body of expertise — when each topic has its designated "knowledge champion," as it were. Studies further suggest that it can be useful to appoint a *meta*-knowledge champion: an individual who is responsible for keeping track of what others in the group know and making sure that group members' mental "directory" of who knows what stays up to date.

There's one more way in which solo thought and collective thought differ. For an individual thinking on her own, applying her mental effort toward the advancement of her own interests is a straightforward matter. When a group of people think together, however, their potentially divergent interests need somehow to be directed toward the accomplishment of a collective goal. Incentives must thus be engineered such that, instead of pursuing their own ends, group members are inspired by a sense of "shared fate": the results achieved by one member are felt to benefit them all. Psychological research, and some still relevant history, show that such reengineering can be dramatically effective — even in the most volatile of situations.

THE PUBLIC SCHOOLS in Austin, Texas, were in crisis. It was 1971, and the education system was undergoing court-ordered desegregation, bringing white students, African American students, and Latino students together in classrooms for the first time. The schools were roiling with conflict and even physical violence. Matthew Snapp, assistant superintendent of schools, turned for help to his former academic mentor, a social psychologist and University of Texas professor named Elliot Aronson.

"The first step was to find out what the hell was going on in those classrooms," Aronson recalled. He and his graduate students sat in the back and watched, and what they saw was in many ways typical of middle school education, both then and now: "The teacher stands in front of the room, asks a question, and waits for the students to indicate that they know the answer," Aronson recounted. "Most frequently, six to ten youngsters strain in their seats and raise their hands — some waving them vigorously in an attempt to attract the teacher's attention. Several other students sit quietly with their eyes averted, as if trying to make themselves invisible."

Through daily experience with such scenarios, he noted, "students learn more than the content of the material explicitly taught in the classroom. The medium is the message; they learn implicit lessons from the process as well." And what they learned was that "there is no payoff for consulting with their peers"; rather, the payoff lay solely in "giving the correct answer—the one which the teacher had in her head."

In the best of times, such single-minded pursuit of one's own interests was incompatible with cooperation and collaboration; in the uneasy environment that then prevailed, it was exacerbating tensions and reinforcing stereotypes. Aronson and his team sought to foster a more communal spirit among the students, but they knew that simply encouraging them to work together wasn't the answer. Instead, they shifted the incentives to which students were responding—by creating, in Aronson's words, "a situation where they needed to cooperate with one another in order to understand the material." They called their procedure the "jigsaw classroom."

This is how it worked: Students were divided into groups of five or six. When a class began a new unit—say, on the life of Eleanor Roosevelt—each student in the group was assigned one section of the material: Roosevelt's childhood and young adulthood, or her role as first lady, or her work on behalf of causes such as civil rights and world peace. The students' task was to master their own section, then rejoin the group and report to the others on what they had learned. "Each student has possession of a unique and vital part of the information, which, like the pieces of a jigsaw puzzle, must be put together before anyone can learn the whole picture," Aronson explained. By arranging instruction in this manner, he was effectively creating a transactive memory system on the spot, turning each student into an expert on a particular facet of the subject under study. "In this situation," Aronson added, "the only way a child can be a good learner is to begin to be a good listener and interviewer"; the jigsaw structure "demands that the students utilize one another as resources."

The effects of the new approach were immediately apparent: the same students who had been straining to be recognized for their individual brilliance, or endeavoring to retreat into a cloak of invisibility, were now focused on collaborating with one another. In studies carried out by Aronson and his graduate students that compared the jigsaw method to traditional modes of instruction, they identified longer-term effects as well. Students learned the material faster and performed better on exams when they participated in the jigsaw exercise; they also developed greater empathy and respect for their classmates. In the Austin-area schools where the jigsaw classroom was implemented, racial tensions diminished, absenteeism declined, and students reported more favorable attitudes toward school.

In an effort to collect as much objective evidence as possible, Elliot Aronson asked one of his graduate students to climb to the roofs of the schools in which they were working and take pictures of the playgrounds at recess. At the start, these photographs showed a dispiriting reality: students were tightly clustered in groups defined by race, ethnicity, and gender. As the jigsaw experiment progressed, however, the pictures documented a striking shift, as these clenched formations began to loosen and disperse. Pupils were mingling and mixing more freely, their play reflecting their new experiences in the classroom. From several stories up, an observer could see the change in Austin's students: they were at last getting outside their own heads.

Conclusion

SOME FIFTEEN YEARS after Elliot Aronson ventured into the restive classrooms of Austin, Texas, his twenty-five-year-old son set out to follow his father's professional path. The younger man arrived on the campus of Princeton University to pursue his PhD in 1986. But Joshua Aronson soon encountered an unanticipated obstacle to achieving his aim of becoming a social psychologist: he found himself struck dumb whenever he met with his graduate school adviser, a distinguished scholar named Edward Ellsworth Jones. "I was totally intimidated by him," Aronson recounts. "I'd walk into his office as prepared as I could be, but without fail, I would lose ten or fifteen IQ points the minute I came through the door. They would get sucked out of my head just by being in the presence of this person."

The humiliation of standing, dull-witted and tongue-tied, in Professor Jones's office affected him so deeply that it shaped the course of his burgeoning career. Less than a decade later, as a junior professor at the University of Texas at Austin, Aronson helped design a study that became one of the most influential ever conducted in psychology — "a modern classic," as it's been called. The paper described for the first time a phenomenon that he and his coauthor, Claude Steele, named "stereotype threat": a temporary condition that saps the brainpower of those affected, rendering them effectively less intelligent. Aronson and Steele's experiments demonstrated that members of groups stereotyped as academically inferior — such as female students enrolled in math and science courses, or African American and Latino students attending college — score lower on tests of intellectual ability when made conspicuously aware of their gender or ethnicity.

Stereotype threat has since become a crucial concept within psychology, guiding researchers' inquiries into why women are underrepresented in STEM fields, for example, and why well-prepared minority high school graduates may still struggle in college. These investigations are rooted in a more general truth, says Aronson, one that that applies to every one of us. Intelligence is not "a fixed lump of something that's in our heads," he explains. Rather, "it's a transaction": a fluid interaction among our brains, our bodies, our spaces, and our relationships. The capacity to think intelligently emerges from the skillful orchestration of these internal and external elements. And indeed, studies have shown that such mental extensions can help us think more effectively when confronted with a challenge like stereotype threat. Using "cognitive reappraisal" to reinterpret bodily signals, as we learned to do in chapter 1, can head off the performance-suppressing effects of anxiety. Adding "cues of belonging" to the physical environment, of the kind we explored in chapter 5, can generate a sense of psychological ease that's conducive to intelligent thought. And carefully structuring the expert feedback offered to a "cognitive apprentice," as we learned about in chapter 7, can instill the confidence necessary to overcome self-doubt.

With a wry smile, Joshua Aronson (currently an associate professor of psychology at New York University) refers to his stammering state when encountering his adviser as "conditional stupidity." Knowing what we do about mental extensions and how they work, we are now able to assemble the conditions for intelligence, even brilliance. In this book we've looked intently at one extension at a time: interoceptive signals, movements, and gestures; natural spaces, built spaces, and the "space of ideas"; experts, peers, and groups. But evidence suggests that extensions are most powerful when they are employed in combination, incorporated into mental routines that draw on the full range of extra-neural resources we have at hand.

The skilled use of extensions is a proficiency that has gone largely unrecognized and uncultivated by our schools and workplaces, and it was long ignored by researchers in psychology, education, and management. But some general principles of effective extending are now clearly discernible, implicit in the more recent research we've covered in previous chapters. Let's take up, in turn, three sets of such principles — three lenses through which to view the project of extending the mind.

The first set of principles lays out some habits of mind we would do well to adopt, starting with this one: whenever possible, we should *offload* information, externalize it, move it out of our heads and into the world. Throughout this book we've encountered many examples of offloading and have become familiar with its manifold benefits. It relieves us of the burden of keeping a host of details "in mind," thereby freeing up mental resources for more demanding tasks, like problem solving and idea generation. It also produces for us the "detachment gain," whereby we can inspect with our senses, and often perceive anew, an image or idea that once existed only in the imagination.

In its most straightforward form, offloading is the simple act of putting our thoughts down on paper — simple, but often skipped over in a world that values doing things in our heads. As we learned from the story of Charles Darwin and the ship's log he kept on the HMS *Beagle,* a habit of *continuous* offloading — through the use of a daily journal or field notebook — can extend our ability to make fresh observations and synthesize new ideas. And as we saw in the example of historian Robert Caro, offloading information onto a space that's big enough for us to *physically* navigate (wall-sized outlines, oversized concept maps, multiple-monitor workstations) allows us to apply to that material our powers of spatial reasoning and spatial memory.

Externalizing information takes a more involved form: it may entail carefully designing a task such that one part of the task is offloaded even as another part absorbs our full attention. This was the practice adopted by law professor Monte Smith, who had his students offload the task of structuring a legal memo onto a model while they focused their efforts on understanding and articulating their newly acquired knowledge of the law. Offloading need not require written language, either. At times, offloading may be *embodied:* when we gesture, for example, we permit our hands to "hold" some of the thoughts we would otherwise have to maintain in our head. Likewise, when we use our hands to move objects around, we offload the task of visualizing new configurations onto the world itself, where those configurations take tangible shape before our eyes. (Picture an interior designer manipulating a model as she tries out new groupings of furniture, for example, or a Scrabble player rearranging the tiles on his tray to form new words.)

At other times, offloading may be *social:* we've seen how engaging in argument allows us to distribute among human debaters the task of tallying points for and against a given proposition; we've learned how constructing a transactive memory system offloads onto our colleagues the task of monitoring and remembering incoming information. Offloading also occurs in an interpersonal context when we externalize "traces" of our own thinking processes for the benefit of our teammates; in this case, we're offloading not to unburden our own minds but to facilitate collaboration with others.

Onward to the second principle: whenever possible, we should endeavor to transform information into an artifact, to make data into something *real*—and then proceed to interact with it, labeling it, mapping it, feeling it, tweaking it, showing it to others. Humans evolved to handle the concrete, not to contemplate the abstract. We extend our intelligence when we give our minds something to grab onto: when we experience a concept from physics as a bicycle wheel spinning in our hands, for example, or when we turn a foreign language vocabulary word into a gesture we can see and sense and demonstrate to others. Vague impressions of what constitutes "excellent work" can usefully take form as a display of actual models to which to aspire (remember "Austin's Butterfly"?); dry intellectual deliberations can acquire a rooted, embodied dimension when we closely attend to, and label and track, the internal signals that arise in our bodies. Our days are now spent processing an endless stream of symbols; with a bit of ingenuity, we can find ways to turn these abstract symbols into tangible objects and sensory experiences, and thereby think about them in new ways.

In a related vein, the third principle: whenever possible, we should seek to productively alter *our own state* when engaging in mental labor. We've repeatedly confronted the limits of the brain-as-computer analogy, and here we come up against perhaps its most conspicuous flaw. When fed a chunk of information, a computer processes it in the same way on each occasion — whether it's been at work for five minutes or five hours, whether it's located in a fluorescent-lit office or positioned next to a sunny window, whether it's near other computers or is the only computer in the room. This is how computers operate, but the same doesn't hold for human beings. The way we're able to think about information is dramatically affected by the state we're in when we encounter it.

Effective mental extension, then, requires us to think carefully about inducing in ourselves the state that is best suited for the task at hand. We might engage in a bout of brisk exercise before sitting down to learn something new, for example; we might seek out an opportunity to engage in group synchrony and shared physical arousal (spicy food, anyone?) when we're expected to work together as a team. We might get up from our desk and get our hands and bodies moving when we're seeking to understand a spatial concept; we might plan a three-day trip into the wilderness when we're in need of a creative boost. Deliberately altering our own state could entail taking a walk in a nearby park when our frazzled attention requires restoration, or seeking out a sparring partner with whom to argue when we want to make sure our ideas are sound. Instead of heedlessly driving the brain like a machine, we'll think more intelligently when we treat it as the context-sensitive organ it is.

The second set of principles offers a higher-level view of how mental extension works, in accordance with an understanding of what the brain evolved to do. The brain is well adapted to sensing and moving the body, to navigating through physical space, and to interacting with other members of our species. On top of this basic suite of human competencies, civilization has built a vast edifice of abstraction, engaging our brains in acts of symbolic processing and conceptual cognition that don't come as naturally. These abstractions have, of course, allowed us to expand our powers exponentially — but now, paradoxically, further progress may depend on running this process in reverse. In order to succeed at the increasingly complex thinking modern life demands, we will find ourselves needing to translate abstractions *back into* the corporeal, spatial, and social forms from which they sprang — forms with which the brain is still most at ease.

We can begin to understand what this means by taking up the fourth principle: whenever possible, we should take measures to *re-embody* the information we think about. The pursuit of knowledge has frequently sought to disengage thinking from the body, to elevate ideas to a cerebral sphere separate from our grubby animal anatomy. Research on the extended mind counsels the opposite approach: we should be seeking to draw the body back into the thinking process. That may take the form of allowing our choices to be influenced by our interoceptive signals—a source of guidance we've often ignored in our focus on data-driven decisions. It might take the form of enacting, with bodily movements, the academic concepts that have become abstracted, detached from their origin in the physical world. Or it might take the form of attending to our own and others' gestures, tuning back in to what was humanity's first language, present long before speech. As we've seen from research on embodied cognition, at a deep level the brain still understands abstract concepts in terms of physical action, a fact reflected in the words we use ("reaching for a goal," "running behind schedule"); we can assist the brain in its efforts by bringing the literal body back into the act of thinking.

The fifth principle emphasizes another human strength: whenever possible, we should take measures to *re-spatialize* the information we think about. We inherited "a mind on the hoof," as Andy Clark puts it: a brain that was built to pick a path through a landscape and to find the way back home. Neuroscientific research indicates that our brains process and store information—even, or especially, abstract information—in the form of mental maps. We can work in concert with the brain's natural spatial orientation by placing the information we encounter into expressly spatial formats: creating memory palaces, for example, or designing concept maps. In the realm of education research, experts now speak of "spatializing the curriculum"—that is, simultaneously drawing on and strengthening students' spatial capacities by having them employ spatial language and gestures, engage in sketching and mapmaking, and learn to interpret and create charts, tables, and diagrams. The spatialized curriculum has obvious applications to subjects like geometry, but researchers report that learning in a spatial mode can also help students think in more advanced ways about topics including chemistry, biology, and history. Nor should spatial reasoning be restricted to schools; the workplace offers abundant opportunities for reconceiving information in spatial terms —terms that put us back in touch with our natural talent for navigation.

The sixth principle rounds out the roster of our innate aptitudes: whenever possible, we should take measures to *re-socialize* the information we think about. We learned earlier in this book that the continual patter we carry on in our heads is in fact a kind of internalized conversation. Likewise, many of the written forms we encounter at school and at work—from exams and evaluations, to profiles and case studies, to essays and proposals—are really social exchanges (questions, stories, arguments) put on paper and addressed to some imagined listener or interlocutor. As we've seen, there are significant advantages to turning such interactions at a remove back into actual social encounters. Research we've reviewed demonstrates that the brain processes the "same" information differently, and often more effectively, when other human beings are involved—whether we're imitating them, debating them, exchanging stories with them, synchronizing and cooperating with them, teaching or being taught by them. We are inherently social creatures, and our thinking benefits from bringing other people into our train of thought.

The final set of principles of mental extension steps back for a still wider view, taking up a rather profound question: What *kind* of creatures are we? We can't design effective protocols for extension without a nuanced understanding of our highly particular, intriguingly eccentric human nature. A clear-eyed acknowledgment of our quirks can lead us to create new kinds of mental routines, such as the one encapsulated in the seventh principle: whenever possible, we should manage our thinking by *generating cognitive loops.*

As Andy Clark has pointed out, when computer scientists develop artificial intelligence systems, they don't design machines that compute for a while, print out the results, inspect what they have produced, add some marks in the margin, circulate copies among colleagues, and then start the process again. That's not how computers work—but it is how *we* work; we are "intrinsically loopy creatures," as Clark likes to say. Something about our biological intelligence benefits from being rotated in and out of internal and external modes of cognition, from being passed among brain, body, and world. This means we should resist the urge to shunt our thinking along the linear path appropriate to a computer—input, output, done—and instead allow it to take a more winding route.

We can pass our thoughts through the portal of our bodies: seeking the verdict of our interoception, seeing what our gestures have to show us, acting out our ideas in movement, observing the inspirations that arise during or after vigorous exercise. We can spread out our thoughts in space, treating the contents of the mind as a territory to be mapped and navigated, surveyed and explored. And we can run our thoughts through the brains of the people we know, gathering from the lot of them the insights no single mind could generate. Most felicitous of all, we can loop our thoughts through all three of these realms. What we *shouldn't* do is keep our thoughts inside our heads, inert, unchanged by encounters with the world beyond the skull.

We are loopy creatures — and we are also situationally sensitive ones, responsive to the immediate conditions and circumstances in which we find ourselves. Hence, the eighth principle: whenever possible, we should manage our thinking by *creating cognitively congenial situations.* We often regard the brain as an organ of awesome and almost unfathomable power. But we're also apt to treat it with high-handed imperiousness, expecting it to do our bidding as if it were a docile servant. Pay attention to *this,* we tell it; remember *that;* buckle down *now* and get the job done. Alas, we often find that the brain is an unreliable and even impertinent attendant: fickle in its focus, porous in its memory, and inconstant in its efforts. The problem lies in our attempt to command it. We'll elicit improved performance from the brain when we approach it with the aim not of issuing orders but of creating situations that draw out the desired result.

Instead of dictating to a student the information she needs to learn, for example, have her explain it in front of a group of her peers; the gestures she makes will generate a deeper level of understanding. Instead of handing an employee a manual packed with guidelines, create spaces and occasions where stories — full of the tacit knowledge manuals can't convey — will be shared among his co-workers. Instead of instructing a team to cooperate and work together, plan an event (a shared meal, a group hike, karaoke!) where synchronized movement and mutual physiological arousal are bound to take place. The art of creating intelligence-extending situations is one that every parent, teacher, and manager needs to master.

The final principle of extension doubles back on itself with a self-referential observation. What kind of creatures are we? The kind who *extend,* eagerly and energetically, when given the chance. Consider: research from neuroscience and cognitive psychology indicates that when we begin using a tool, our "body schema" — our sense of the body's shape, size, and position — rapidly expands to encompass it, as if the tool we're grasping in our hand has effectively become an extension of our arm. Something similar occurs in the case of mental extensions. As long as extensions are available — and especially when they are reliably, persistently available — we humans will incorporate them into our thinking. Accordingly, the ninth principle: whenever possible, we should manage our thinking by *embedding extensions* in our everyday environments.

Think of the cues of belonging and identity, for example, that bolster our motivation and improve our performance when displayed in our study and work spaces. Recall the transactive memory system we construct with a group of colleagues over time, in which the burden of attending to and remembering information is distributed across group members. Picture, even, the indoor plants and "green" walls and roofs that help restore our attention by providing regular glimpses of nature. Once securely embedded, such extensions can function as seamless adjuncts to our neural capacity, supporting and augmenting our ability to think intelligently.

It's worth noting that this principle bears a bias toward stability: enduring cues of belonging and identity are hard to sustain in an office where "hot-desking," or unassigned work space, is the norm; a transactive memory system is difficult to build in a work environment where turnover is high or team composition is constantly changing. In a dynamic and fast-changing society that celebrates novelty and flexibility, the maintenance and preservation of valued mental extensions also deserve our respect. We may not know how much they bolster our intelligence until they're gone.

THIS NESTED SET of principles, what we might call a "curriculum of the extended mind," is not currently taught in any school or addressed in any workplace training. That ought to change; learning to extend the mind should be an element of everyone's education. At present, to the degree that people know how to extend their minds, it's something they've figured out on their own. Strikingly, we now have evidence that individuals do differ in how fully they have developed their capacity to extend. Furthermore, scientists have found that this competence can be accurately and precisely measured, using a variation on conventional IQ tests (which, in their unaltered form, deliberately exclude all kinds of mental extensions: test takers are not permitted to use tools such as calculators or the Internet, nor are they allowed to move their bodies, rearrange their environments, or talk to their neighbors). Most intriguing, results from these studies show that skill at employing extensions, as assessed by a test, corresponds to real-world performance: empirical evidence that individuals who can extend their minds more fully can solve problems more effectively in everyday life.

In February 2019, a group of psychologists from the Netherlands — plus philosopher Andy Clark — published a study in the journal *Nature Human Behaviour*. The researchers set out, they wrote, "to quantitatively assess a powerful, although understudied, feature of human intelligence: our ability to use external objects, props and aids to solve complex problems." They started with a conventional test of intelligence, the Raven Advanced Progressive Matrices; this IQ test, used millions of times all over the world since it was first introduced in 1938, presents users with a series of geometric puzzles, each of which is missing a piece. Test takers are asked to select, from a number of options provided, the piece that correctly completes each pattern. (The test is available in a paper-and-pencil version, but today it is most often administered on a computer.)

In the standard version of "the Raven," as it is known, test takers are expected to carry out the required operations in their heads, imagining how each potential choice might or might not fit. The rules of the test don't permit them to extend their minds with extra-neural resources; they must rely on their internal reasoning processes alone. In the version of the test designed by Clark and his colleagues, by contrast, test takers are able to digitally manipulate the potential solution pieces, moving them around the screen to create new configurations. To assess the validity of the new test they had created, the researchers recruited 495 students from Leiden University and Erasmus University, both located in the Netherlands. Half the students were randomly assigned to take the conventional version of the Raven; the other half were given the extended-mind variant of the test. In the case of that second group, the researchers monitored how actively test takers engaged in manipulating the layout on the screen.

A suggestive finding soon surfaced: test takers who took full advantage of the new interactive feature were often able to identify patterns that had not been apparent to them before they began shifting the pieces around. An analysis of the moves they made while taking the test showed that these active extenders seemed to be running their thinking processes through successive loops — switching between external actions, which altered the problem-solving space in helpful ways, and internal evaluations of the new configurations thus created. "Our study showed very clearly the relationship between the amount of interaction participants engaged in and how well they solved the problems," says Bruno Bocanegra, an assistant professor of psychology at Erasmus University and the lead author of the paper. "We saw people interacting with the pieces, reflecting on the new configurations, reassessing their strategy, and then reaching out to interact again. These loops are what allowed them to solve the problems effectively."

Final results demonstrated that the more test takers extended their minds using the movable pieces, the more successful they were at solving the complex visual puzzles. What's more, the researchers found, the extended-mind version of the test was better than the standard "static" Raven at predicting students' intellectual performance outside the lab — in the form of the grades they received in their college courses. The test that measured the students' skill at mental extension, the authors wrote, "might be tapping into an additional behavioral aspect of intelligence that is not currently measured" by conventional IQ tests. Says Bruno Bocanegra: "People are applying an underappreciated wealth of strategies to solve problems — underappreciated, in part, because people are not good at describing their own thought processes. They often don't have conscious access to their strategies — but they are using them nonetheless. We're interested in studying people over time to see if they can develop more sophisticated strategies."

Bocanegra's publication is just a start, but it's easy to envision a broad expansion of similar efforts. Imagine a test that would evaluate how well an individual is able to use interoception, movement, and gesture to think; how adept she is at soaking up natural settings, designing built environments, and exploiting the space of ideas to enhance her cognition; how skillfully she manages thinking with experts, thinking with peers, and thinking with groups. Such an assessment could represent a new kind of IQ test, measuring a new sort of intelligence. ("New" in the sense of newly admitted into our society's definition of smart; as we've seen throughout this book, humans have been extending their minds since time immemorial.)

"Much more than we usually recognize, humans use their environment to solve problems — an environment that is both material and social," says Bocanegra. "When you see things that way, it starts to seem very silly to think that we can measure intelligence as some internal, intrinsic, individual quality." Of course, it's possible that such a test would be misused, as IQ tests have so often been misused — employed to rank, divide, and exclude people instead of helping them to develop. But such misuse need not be inevitable. Once we make mental extensions visible, what we do with our new awareness is up to us.

We might begin by applying it to an issue now roiling our society: America's pervasive inequality, a state of affairs that many are finding ever less justifiable or tolerable. Defenders of the status quo have long argued that social and economic inequality merely reflects a kind of organic inequality, determined by nature, in the talents and abilities with which individuals are born. That argument appears less plausible when viewed through the lens of the extended mind. If our ability to think intelligently is shaped so profoundly by the availability of extra-neural resources, how then can we continue to justify their extraordinarily inequitable distribution?

In a famous thought experiment, the contemporary philosopher John Rawls imagined designing the ideal society — but doing so from behind a "veil of ignorance" regarding how the designers themselves would fare in the new world they are creating. While engaged in deciding how society's affluence and opportunities are to be distributed, Rawls writes, "no one knows his place in society, his class position or social status, nor does anyone know his fortune in the distribution of natural assets and abilities, his intelligence, strength, and the like."

Intriguing though it is to contemplate, Rawls's scenario has always been hard to enter into fully — so closely are we identified with what we take to be our "natural assets and abilities," our "intelligence" foremost among them. The theory of the extended mind is a tool with which we might begin to pry loose this instinctive identification. Unlike innate intelligence, which we imagine to be an inseparable part of who we "are," access to mental extensions is more readily understood as a matter of chance or luck. This radically new conceptual theory harbors within it an old and humble moral sentiment: "There but for the grace of God go I." Acknowledging the reality of the extended mind might well lead us to embrace the extended heart.

Acknowledgments

The cognitive psychologist Stephen Kosslyn has proposed a notion that is a close cousin to the theory of the extended mind: he submits that one individual can engage another as a "social prosthetic." As he has written (with G. Wayne Miller): "The crucial idea is that when you are interacting with another person in this way, he or she has the capacity to make you more effective . . . He or she fills in for your lack. And in the process, at that moment you become a different person, transformed by your interactions with the other."

Two people functioned in this way for me during the writing of this book: Tina Bennett and Eamon Dolan. At moments when I doubted that I could pull off this ambitious project, Tina and Eamon filled in for my lack, supplying me with all the hope and confidence and inspiration I could want. Thank you, Tina, and thank you, Eamon.

Many other people worked hard to make this book a reality. I'm grateful to the editorial and production team at Houghton Mifflin Harcourt: Deb Brody, Ivy Givens, Bruce Nichols (now at Little, Brown), and Beth Burleigh Fuller. Amanda Heller provided a scrupulous copyedit. Mark Robinson designed a gorgeous cover. At WME, Jay Mandel was a surefooted guide, assisted by the extremely capable Sian-Ashleigh Edwards. Joseph Hinson and his team at Out:think Group built me a wonderful website; Stephanie Anestis cheerfully took my author photo on a cold and windy New Haven day. Mika Yamaguchi, of Yale University's Department of East Asian Languages and Literatures, helped me with the section on Radio Taiso.

During the long period of researching and writing this book, I was fortunate enough to be supported by three fellowships granted by New America: the Bernard L. Schwartz Fellowship, the Future Tense Fellowship, and the Learning Sciences Exchange Fellowship (offered in conjunction with the Jacobs Foundation). My sincere thanks are due to Anne-Marie Slaughter, Lisa Guernsey, Andrés Martinez, Peter Bergen, Awista Ayub, Torie Bosch, and Elise Franchino. Work on this book was also supported by the Spencer Foundation. As a recipient of the Spencer Education Journalism Fellowship, I'm grateful to the Spencer Foundation's president emeritus Michael McPherson; to LynNell Hancock and Marguerite Holloway of the Columbia University Graduate School of Journalism; and to my fellow fellows Jamaal Abdul-Alim and Lauren Smith Camera. I'd also like to recognize my erstwhile colleagues at the Poorvu Center for Teaching and Learning at Yale University: Jennifer Frederick, Mark Graham, Tracie Addy, and Beth Luoma.

The content of this book was enriched by many conversations with researchers, educators, writers, and editors. I'm indebted to the scientists who made time to answer my questions about their work. My deepest appreciation as well to a cadre of informal advisers who helped me think through the issues explored in this book: Paul Bloom, David Daniel, David Dockterman, Dan Willingham, and Carl Zimmer. I'd also like to acknowledge the insights gleaned from interactions with Mahzarin Banaji, Tina Barseghian, Emily Bazelon, Ulrich Boser, Bill Brown, Adam Cohen, Nicky Dawidoff, Jacob Hacker, Jake Halpern, Carla Horwitz, Scott Barry Kaufman, Holly Korbey, Doug Lemov, Steve Levingston, Daniel Markovits, Kyle Pedersen, Robert Pondiscio, Jessica Sager, Jason Stanley, Bror Saxberg, Liz Taylor, Greg Toppo, Liz Willen, and Aubrey White.

Several friends were kind enough to read the manuscript of this book and offer their comments: Kathryn Bowers, Ben Dattner, Mark Oppenheimer, Wendy Paris, Gretchen Rubin, and Amy Sudmyer. Other friends offered much-appreciated moral support: Victor Agran and Cassandra Albinson, Suzie Alderman, Sarah Bilston, Jenny Brown and James Feighny, Alison Burns, Andy, Emma, and Jack Bowers, Susie Cain, Julie Cooper, Lisa Damour, Rachel Doft, Emily Gordon, Marla Geha and Matt Polly, Alison Illick, Jessica Kaufman, Chris Kenneally, Oz Kus, Nadya Labi, Marguerite Lamb, Alison MacKeen and Scott Shapiro, Michele Orecklin, Pamela Paul, Eeva Pelkonen and Turner Brooks, Dan Perkins, Emily Rubin,

Emma Seppälä, Katherine Stewart, Bob Sullivan, Shel Swanson, Bron Tamulis, Abby Tucker, Jenny Watts and the Deverells (Bill, John, Helen, Linus, Petal, and Biggie), Karen Wynn, and Grace Zimmer.

For more than fifteen years the Invisible Institute — a writers' group I started with Alissa Quart — has provided invaluable advice and enjoyable camaraderie. My warm thanks go to those members already mentioned and to Gary Bass, Susan Burton, Ada Calhoun, KJ Dell'Antonia, Lydia Denworth, Elizabeth Devita-Raeburn, Rebecca Donner, Abby Ellin, Randi Epstein, Sheri Fink, Lindsay Gellman, Anya Kamenetz, Reem Kassis, Maria Konnikova, Brendan Koerner, Jess Lahey, Katherine Lanpher, Ron Lieber, Judith Matloff, Katie Orenstein, Kaja Perina, Andrea Peterson, Mary Pilon, Josh Prager, Paul Raeburn, Alexander Russo, Catherine St. Louis, Debbie Stier, Maia Szalavitz, Lauren Sandler, Debbie Siegel, Rebecca Skloot, Stacy Sullivan, Harriet Washington, and Tom Zoellner.

Closer to home, I'm deeply grateful for the support and insight of Miryam Welbourne; the calm presence and good humor of Darcy Chase; and the devoted fathering of my co-parent, John Witt. I honor the work of my children's teachers: at the Calvin Hill Day Care Center and Kitty Lustman-Findling Kindergarten, the Foote School, and Hopkins School. Finally, I give thanks for my family: my mother, Nancy Paul; the memory of my father, Tim Paul; my sister, Sally Paul; Sally's partner, Billy Fowks; and their son, Frankie Fowks.

And to my beloved sons, Teddy Witt and Gus Witt: more than anyone else, you two have led me to become a different person — transformed, in the best way, by my interactions with you.

Notes

PROLOGUE

page

xi   *"How quickly we guess"*: Friedrich Nietzsche, quoted in Frédéric Gros, *A Philosophy of Walking*, trans. John Howe (London: Verso, 2014), 19.

xii   *"Think of the scribe's body"*: Gros, *A Philosophy of Walking*, 19.

    "the walking body": Gros, *A Philosophy of Walking*, 19.

    "sit as little as possible": Friedrich Nietzsche, quoted in Gros, *A Philosophy of Walking*, 11.

    "retains and expresses": Gros, *A Philosophy of Walking*, 20.

xiii   *"I did the best thinking"*: Daniel Kahneman, *Thinking, Fast and Slow* (New York: Farrar, Straus and Giroux, 2011), 40.

    "brain in a vat": Evan Thompson and Diego Cosmelli, "Brain in a Vat or Body in a World? Brainbound Versus Enactive Views of Experience," *Philosophical Topics* 39 (Spring 2011): 163–80.

xv   *"Where does the mind stop"*: Andy Clark and David Chalmers, "The Extended Mind," *Analysis* 58 (January 1998): 7–19. The phrase "extended mind" was originally proposed by David Chalmers.

    Ned Block likes to say: Cited in David J. Chalmers, "Extended Cognition and Extended Consciousness," in *Andy Clark and His Critics,* ed. Matteo Colombo, Elizabeth Irvine, and Mog Stapleton (New York: Oxford University Press, 2019), 12.

    "What about socially extended cognition?": Clark and Chalmers, "The Extended Mind."

    "an important role": Andy Clark, *Supersizing the Mind: Embodiment, Action, and Cognitive Extension* (New York: Oxford University Press, 2011), 131.

    "designer environments": Andy Clark, "Embodied, Situated, and Distributed Cognition," in *A Companion to Cognitive Science,* ed. William Bechtel and George Graham (Malden, MA: Blackwell Publishing, 1998), 510.

    the "brainbound" perspective: Clark, *Supersizing the Mind,* xxv.

xvii   *"quest for a different light"*: Gros, *A Philosophy of Walking*, 19.

INTRODUCTION: THINKING OUTSIDE THE BRAIN

2 *"the most complex structure"*: Alexis Madrigal, "Mapping the Most Complex Structure in the Universe: Your Brain," *Wired*, January 24, 2000.

3 *"the will trying to do the work"*: William Butler Yeats, "Emotion of Multitude," in *The Collected Works of W. B. Yeats*, vol. 4, *Early Essays*, ed. George Bernstein and Richard J. Finneran (New York: Scribner, 2007), 159.

"By my faith!": Molière [Jean-Baptiste Poquelin], *The Middle Class Gentleman*, trans. Philip Dwight Jones, act 2, posted on the website of Project Gutenberg, https://www .gutenberg.org/files/2992/2992-h/2992-h.htm.

4 *"neurocentric bias"*: Andy Clark, *Natural-Born Cyborgs: Minds, Technologies, and the Future of Human Intelligence* (New York: Oxford University Press, 2003), 4–5.

"I used to think": Emo Philips, *Emo Philips Live! At the Hasty Pudding Theatre* (New York: HBO Video, 1987).

5 *Electronic Numerical Integrator and Computer:* My discussion of the ENIAC draws on the following sources: Thomas Haigh, Mark Priestley, and Crispin Rope, *ENIAC in Action: Making and Remaking the Modern Computer* (Cambridge: MIT Press, 2016); Scott McCartney, *ENIAC: The Triumphs and Tragedies of the World's First Computer* (New York: Walker Publishing, 1999); Jean Jennings Bartik, *Pioneer Programmer: Jean Jennings Bartik and the Computer That Changed the World* (Kirksville, MO: Truman State University Press, 2013); Walter Isaacson, *The Innovators: How a Group of Hackers, Geniuses, and Geeks Created the Digital Revolution* (New York: Simon & Schuster, 2014); Annie Jacobsen, *The Pentagon's Brain: An Uncensored History of DARPA* (New York: Little, Brown, 2015); C. Dianne Martin, "ENIAC: The Press Conference That Shook the World," *IEEE Technology and Society Magazine*, December 1995; Steven Levy, "The Brief History of the ENIAC Computer," *Smithsonian Magazine*, November 2013; David K. Allison, "Using the Computer: Episodes Across 50 Years," paper presented at the 24th Annual ACM Computer Science Conference, February 1996; T. R. Kennedy, "Electronic Computer Flashes Answers, May Speed Engineering," *New York Times*, February 14, 1946; National Museum of American History, "ENIAC Accumulator #2," post on the website of NMAH, https://americanhistory.si.edu/collections/ search/object/nmah_334742; Penn Engineering, "ENIAC at Penn Engineering," post on the website of the University of Pennsylvania, https://www.seas.upenn.edu/about/ history-heritage/eniac/.

6 *"if made to take place"*: Kennedy, "Electronic Computer Flashes Answers, May Speed Engineering."

"It was unheard of": Bartik, *Pioneer Programmer*, 25.

"One of the war's top secrets": Kennedy, "Electronic Computer Flashes Answers, May Speed Engineering."

compared to a human brain: C. Dianne Martin, "Digital Dreams: Public Perceptions About Computers," *ACM Inroads* 4 (September 2013): 34–35.

"giant electronic brain": Cited in Haigh, Priestley, and Rope, *ENIAC in Action.*

"robot brain": Cited in C. Dianne Martin, "The Myth of the Awesome Thinking Machine," *Communications of the ACM* 36 (April 1993): 120–33.

7 *"took seriously the idea"*: Steven Sloman and Philip Fernbach, *The Knowledge Illusion: Why We Never Think Alone* (New York: Riverhead Books, 2017), 24–25.

"New Research Shows": Lisa S. Blackwell, "Psychological Mediators of Student

Achievement During the Transition to Junior High School: The Role of Implicit Theories" (PhD diss., Columbia University, 2002).

8 *"the incremental theory"*: Lisa S. Blackwell, Kali H. Trzesniewski, and Carol Sorich Dweck, "Implicit Theories of Intelligence Predict Achievement Across an Adolescent Transition: A Longitudinal Study and an Intervention," *Child Development* 78 (January–February 2007): 246–63.

"growth mindset": Carol S. Dweck, *Mindset: The New Psychology of Success* (New York: Random House, 2006), 7.

"The key message": Blackwell, Trzesniewski, and Dweck, "Implicit Theories of Intelligence Predict Achievement Across an Adolescent Transition."

"perseverance and passion": Angela L. Duckworth, Christopher Peterson, Michael D. Matthews, and Dennis R. Kelly, "Grit: Perseverance and Passion for Long-Term Goals," *Journal of Personality and Social Psychology* 92 (June 2007): 1087–1101.

"Like a muscle": Angela Duckworth, *Grit: The Power of Passion and Perseverance* (New York: Scribner, 2016), 192.

9 *"neuromyths"*: Elena Pasquinelli, "Neuromyths: Why Do They Exist and Persist?," *Mind, Brain, and Education* 6 (May 2012): 89–96.

likened to machines: Matthew Cobb, *The Idea of the Brain: The Past and Future of Neuroscience* (New York: Basic Books, 2020).

"Because we do not understand": John R. Searle, *Minds, Brains and Science* (Cambridge: Harvard University Press, 1984), 44.

"What do we do": John Harvey Kellogg, *First Book in Physiology and Hygiene* (New York: Harper & Brothers, 1888), 106.

"the oldest issues": Stephen Jay Gould, *The Mismeasure of Man* (New York: W. W. Norton, 1996), 27.

10 *"Essentialism shows up"*: Paul Bloom, *Descartes' Baby: How the Science of Child Development Explains What Makes Us Human* (New York: Norton, 2004), 48.

a power outage: Park Sae-jin, "Power Firm Seeks Every Method to Keep Magpies Off Power Poles," *Aju Business Daily,* July 3, 2018.

12 *"biologically primary"*: David C. Geary, "Reflections of Evolution and Culture in Children's Cognition: Implications for Mathematical Development and Instruction," *American Psychologist* 50 (January 1995): 24–37.

a century-long climb: James R. Flynn, "Massive IQ Gains in 14 Nations: What IQ Tests Really Measure," *Psychological Bulletin* 101 (March 1987): 171–91.

scores have stopped rising: James R. Flynn and Michael Shayer, "IQ Decline and Piaget: Does the Rot Start at the Top?," *Intelligence* 66 (January 2018): 112–21.

13 *"our brains are already working"*: Nicholas S. Fitz and Peter B. Reiner, "Time to Expand the Mind," *Nature* 531 (March 2016): S9.

Lumosity alone claims: Lumos Labs, "Press Resources," https://www.lumosity.com/en/resources/.

little more than disappointment: Bobby Stojanoski et al., "Targeted Training: Converging Evidence Against the Transferable Benefits of Online Brain Training on Cognitive Function," *Neuropsychologia* 117 (August 2018): 541–50.

"little evidence": Daniel J. Simons et al., "Do 'Brain-Training' Programs Work?," *Psychological Science in the Public Interest* 17 (October 2016): 103–86.

"is rare, or possibly inexistent": N. Deniz Aksayli, Giovanni Sala, and Fernand Gobet,

"The Cognitive and Academic Benefits of Cogmed: A Meta-Analysis," *Educational Research Review* 27 (June 2019): 229–43.

"appears to have no benefits": Joseph W. Kable et al., "No Effect of Commercial Cognitive Training on Brain Activity, Choice Behavior, or Cognitive Performance," *Journal of Neuroscience* 37 (August 2017): 7390–402.

reported for older individuals: Giovanni Sala et al., "Working Memory Training Does Not Enhance Older Adults' Cognitive Skills: A Comprehensive Meta-Analysis," *Intelligence* 77 (November–December 2019): 1–13.

a $2 million fine: Federal Trade Commission, "Lumosity to Pay $2 Million to Settle FTC Deceptive Advertising Charges for Its 'Brain Training' Program," press release posted on the FTC website, January 5, 2016, https://www.ftc.gov/news-events/press -releases/2016/01/lumosity-pay-2-million-settle-ftc-deceptive-advertising-charges.

haven't fared much better: Masud Husain and Mitul A Mehta, "Cognitive Enhancement by Drugs in Health and Disease," *Trends in Cognitive Sciences* 15 (January 2011): 28–36.

one "nootropic" drug: Christina Farr, "This Start-Up Raised Millions to Sell 'Brain Hacking' Pills, but Its Own Study Found Coffee Works Better," CNBC, November 30, 2017.

actually enhance intelligence: Martha J. Farah, "The Unknowns of Cognitive Enhancement: Can Science and Policy Catch Up with Practice?," *Science* 350 (October 2015), 379–80.

14 *left his laptop*: Andy Clark interviewed by Natasha Mitchell, "Natural Born Cyborgs: Minds, Technologies, and the Future of Human Intelligence," *All in the Mind*, Australian Broadcasting Company, May 18, 2003.

"like a sudden and somewhat vicious": Clark, *Natural-Born Cyborgs*, 4, 10.

"Where does the mind": Clark and Chalmers, "The Extended Mind."

three other journals: Julian Baggini, "A Piece of iMe: An Interview with David Chalmers," *The Philosophers' Magazine*, 2008.

16 *"10,000 hours"*: K. Anders Ericsson, Ralf T. Krampe, and Clemens Tesch-Römer, "The Role of Deliberate Practice in the Acquisition of Expert Performance," *Psychological Review* 100 (July 1993): 363–406.

economy, efficiency, and optimality: Paul P. Maglio, Michael J. Wenger, and Angelina M. Copeland, "The Benefits of Epistemic Action Outweigh the Costs," paper presented at the 25th Annual Conference of the Cognitive Science Society, July–August 2003.

1. THINKING WITH SENSATIONS

21 *working as a financial trader*: John Coates, *The Hour Between Dog and Wolf: Risk Taking, Gut Feelings, and the Biology of Boom and Bust* (New York: Penguin Press, 2012), 119, 98. My discussion of John Coates's years on Wall Street is drawn from this source.

22 *a group of financial traders*: Narayanan Kandasamy et al., "Interoceptive Ability Predicts Survival on a London Trading Floor," *Scientific Reports* 6 (September 2016).

awareness of the inner state: For an overview, see A. D. (Bud) Craig, "Interoception: The Sense of the Physiological Condition of the Body," *Current Opinion in Neurobiology* 13 (August 2003): 500–505.

23 *feel them more keenly*: Vivien Ainley et al., "'Bodily Precision': A Predictive Coding

Account of Individual Differences in Interoceptive Accuracy," *Philosophical Transactions of the Royal Society B: Biological Sciences* 371 (November 2016).

heartbeat detection test: Sarah N. Garfinkel et al., "Knowing Your Own Heart: Distinguishing Interoceptive Accuracy from Interoceptive Awareness," *Biological Psychology* 104 (January 2015): 65–74.

"How on earth": Vivien Ainley, quoted in Jyoti Madhusoodanan, "Inner Selfie," *Science Notes* (2014).

"He had always been able": Author interview with Vivien Ainley.

visible to scientists: Hugo D. Critchley et al., "Neural Systems Supporting Interoceptive Awareness," *Nature Neuroscience* 7 (January 2004): 189–95. See also A. D. Craig, "Human Feelings: Why Are Some More Aware Than Others?," *Trends in Cognitive Science* 8 (June 2004): 239–41.

24 *interoceptive capacities already operating:* Jennifer Murphy et al., "Interoception and Psychopathology: A Developmental Neuroscience Perspective," *Developmental Cognitive Neuroscience* 23 (February 2017): 45–56.

Differences in sensitivity: Jennifer Murphy et al., "Estimating the Stability of Heartbeat Counting in Middle Childhood: A Twin Study," *Biological Psychology* 148 (November 2019). See also Kristina Oldroyd, Monisha Pasupathi, and Cecilia Wainryb, "Social Antecedents to the Development of Interoception: Attachment Related Processes Are Associated with Interoception," *Frontiers in Psychology* 10 (April 2019).

communications we receive: A. D. (Bud) Craig, *How Do You Feel?: An Interoceptive Moment with Your Neurobiological Self* (Princeton: Princeton University Press, 2014), 7–8.

this process in microcosm: Pawel Lewicki, Maria Czyzewska, and Hunter Hoffman, "Unconscious Acquisition of Complex Procedural Knowledge," *Journal of Experimental Psychology: Learning, Memory, and Cognition* 13 (1987): 523–30.

25 *"Nonconscious information acquisition":* Pawel Lewicki, Thomas Hill, and Maria Czyzewska, "Nonconscious Acquisition of Information," *American Psychologist* 47 (June 1992): 796–801.

scrolling through our mental archive: Moshe Bar, "The Proactive Brain: Memory for Predictions," *Philosophical Transactions of the Royal Society B: Biological Sciences* 364 (May 2009): 1235–43.

"The human cognitive system": Lewicki, Hill, and Czyzewska, "Nonconscious Acquisition of Information."

"somatic rudder": Guy Claxton, "Corporal Thinking," *Chronicle of Higher Education* 62 (September 2015): 19.

a gambling game: Antoine Bechara et al., "Deciding Advantageously Before Knowing the Advantageous Strategy," *Science* 275 (1997): 1293–95. See also Tasha Poppa and Antoine Bechara, "The Somatic Marker Hypothesis: Revisiting the Role of the 'Body-Loop' in Decision-Making," *Current Opinion in Behavioral Sciences* 19 (February 2018): 61–66.

26 *more interoceptively aware:* Natalie S. Werner et al., "Enhanced Cardiac Perception Is Associated with Benefits in Decision-Making," *Psychophysiology* 46 (November 2009): 1123–29. See also Barnaby D. Dunn et al., "Listening to Your Heart: How Interoception Shapes Emotion Experience and Intuitive Decision Making," *Psychological Science* 21 (December 2010): 1835–44.

shows us something important: Antonio R. Damasio, *Descartes' Error* (New York: G. P. Putnam's Sons, 1994), 165–204.

27 *Mindfulness meditation:* Dana Fischer, Matthias Messner, and Olga Pollatos, "Improvement of Interoceptive Processes After an 8-Week Body Scan Intervention," *Frontiers in Human Neuroscience* 11 (September 2017). See also Bethany E. Kok and Tania Singer, "Phenomenological Fingerprints of Four Meditations: Differential State Changes in Affect, Mind-Wandering, Meta-Cognition, and Interoception Before and After Daily Practice Across 9 Months of Training," *Mindfulness* 8 (August 2017): 218–31.

alter the size and activity: Paul B. Sharp et al., "Mindfulness Training Induces Structural Connectome Changes in Insula Networks," *Scientific Reports* 8 (May 2018). See also Norman A. S. Farb, Zindel V. Segal, and Adam K. Anderson, "Mindfulness Meditation Training Alters Cortical Representations of Interoceptive Attention," *Social Cognitive and Affective Neuroscience* 8 (January 2013): 15–26.

"People find the body scan": Jon Kabat-Zinn, *Full Catastrophe Living: Using the Wisdom of Your Body and Mind to Face Stress, Pain, and Illness* (New York: Bantam Books, 2013), 88, 95–97.

28 *a profound effect:* Matthew D. Lieberman et al., "Subjective Responses to Emotional Stimuli During Labeling, Reappraisal, and Distraction," *Emotion* 11 (June 2011): 468–80.

a series of impromptu speeches: Andrea N. Niles et al., "Affect Labeling Enhances Exposure Effectiveness for Public Speaking Anxiety," *Behaviour Research and Therapy* 68 (May 2015): 27–36.

reduces activity in the amygdala: Uwe Herwig et al., "Self-Related Awareness and Emotion Regulation," *Neuroimage* 50 (April 2010): 734–41. See also Matthew D. Lieberman et al., "Putting Feelings into Words: Affect Labeling Disrupts Amygdala Activity in Response to Affective Stimuli," *Psychological Science* 18 (May 2007): 421–28.

produces greater *activity:* Herwig et al., "Self-Related Awareness and Emotion Regulation."

a larger number of terms: Niles et al., "Affect Labeling Enhances Exposure Effectiveness for Public Speaking Anxiety."

29 *as granular as possible:* Todd B. Kashdan, Lisa Feldman Barrett, and Patrick E. McKnight, "Unpacking Emotion Differentiation: Transforming Unpleasant Experience by Perceiving Distinctions in Negativity," *Current Directions in Psychological Science* 24 (February 2015): 10–16.

making sounder decisions: Jeremy A. Yip et al., "Follow Your Gut? Emotional Intelligence Moderates the Association Between Physiologically Measured Somatic Markers and Risk-Taking," *Emotion* 20 (April 2019): 462–72.

acting less impulsively: Aleksandra Herman et al., "Interoceptive Accuracy Predicts Nonplanning Trait Impulsivity," *Psychophysiology* 56 (June 2019).

decline low offers: Gary E. Bolton and Rami Zwick, "Anonymity Versus Punishment in Ultimatum Bargaining," *Games and Economic Behavior* 10 (July 1995): 95–121.

scanned the brains: Ulrich Kirk, Jonathan Downar, and P. Read Montague, "Interoception Drives Increased Rational Decision-Making in Meditators Playing the Ultimatum Game," *Frontiers in Neuroscience* 5 (April 2011).

30 *"in this study, we identified":* Kirk, Downar, and Montague, "Interoception Drives Increased Rational Decision-Making in Meditators Playing the Ultimatum Game."

popularized by psychologist: Kahneman, *Thinking, Fast and Slow.*

a series of interviews: Mark Fenton-O'Creevy et al., "Thinking, Feeling and Deciding: The Influence of Emotions on the Decision Making and Performance of Traders," *Journal of Organizational Behavior* 32 (November 2011): 1044–61. See also Mark Fenton-O'Creevy, "'The Heart Has Its Reasons': Emotions and Cognition in the World of Finance," *Financial World*, February 2015.

"You have to trust": Financial trader, quoted in Shalini Vohra and Mark Fenton-O'Creevy, "Intuition, Expertise and Emotion in the Decision Making of Investment Bank Traders," in *Handbook of Research Methods on Intuition*, ed. Marta Sinclair (Northampton: Edward Elgar Publishing, 2014), 88–98.

31 *in that moment*: Mark Fenton-O'Creevy et al., "Emotion Regulation and Trader Expertise: Heart Rate Variability on the Trading Floor," *Journal of Neuroscience, Psychology, and Economics* 5 (November 2012): 227–37.

fast-paced decisions: Vohra and Fenton-O'Creevy, "Intuition, Expertise and Emotion in the Decision Making of Investment Bank Traders."

"De-biasing approaches": Mark Fenton-O'Creevy et al., "A Learning Design to Support the Emotion Regulation of Investors," paper presented at the OECD-SEBI International Conference on Investor Education, February 2012.

a specially designed video game: Mark Fenton-O'Creevy et al., "A Game Based Approach to Improve Traders' Decision Making," paper presented at the International Gamification for Business Conference, September 2015. See also Gilbert Peffer et al., "xDelia Final Report: Emotion-Centred Financial Decision Making and Learning," report produced by Open University, January 2012.

"interoceptive learning": Sahib S. Khalsa et al., "Interoception and Mental Health: A Roadmap," *Biological Psychiatry: Cognitive Neuroscience and Neuroimaging* 3 (June 2018): 501–13.

32 *"éminence grise"*: Coates, *The Hour Between Dog and Wolf*, 53.

33 *scarce, precious energy*: For an overview, see Lisa Feldman Barrett, *How Emotions Are Made: The Secret Life of the Brain* (New York: Houghton Mifflin Harcourt, 2017), 56–83.

monitor and manage: Johann D. Kruschwitz et al., "Self-Control Is Linked to Interoceptive Inference: Craving Regulation and the Prediction of Aversive Interoceptive States Induced with Inspiratory Breathing Load," *Cognition* 193 (December 2019). See also Hayley A. Young et al., "Interoceptive Accuracy Moderates the Response to a Glucose Load: A Test of the Predictive Coding Framework," *Proceedings of the Royal Society B: Biological Sciences* 286 (March 2019).

In a study he conducted: Lori Haase et al., "When the Brain Does Not Adequately Feel the Body: Links Between Low Resilience and Interoception," *Biological Psychology* 113 (January 2016): 37–45.

a list of statements: Kathryn M. Connor and Jonathan R. T. Davidson, "Development of a New Resilience Scale: The Connor-Davidson Resilience Scale (CD-RISC)," *Depression and Anxiety* (September 2003): 76–82.

34 *champion swimmer*: Diana Nyad, *Find a Way* (New York: Alfred A. Knopf, 2015), 223–24. See also Ariel Levy, "Breaking the Waves," *The New Yorker*, February 3, 2014.

"Of course I'm competitive": Nyad, *Find a Way*, 224.

"Clumped at the bottom": Nyad, *Find a Way*, 224.

35 *found the same pattern*: Martin P. Paulus et al., "Subjecting Elite Athletes to Inspiratory Breathing Load Reveals Behavioral and Neural Signatures of Optimal Perform-

ers in Extreme Environments," *PLoS One* 7 (January 2012). See also Alan Simmons et al., "Altered Insula Activation in Anticipation of Changing Emotional States: Neural Mechanisms Underlying Cognitive Flexibility in Special Operations Forces Personnel," *NeuroReport* 24 (March 2012): 234–39.

a superior ability to sense: Martin P. Paulus et al., "A Neuroscience Approach to Optimizing Brain Resources for Human Performance in Extreme Environments," *Neuroscience & Biobehavioral Reviews* 33 (July 2009): 1080–88.

people with low resilience: Haase et al., "When the Brain Does Not Adequately Feel the Body."

"cognitive resilience": Amishi P. Jha et al., "Practice Is Protective: Mindfulness Training Promotes Cognitive Resilience in High-Stress Cohorts," *Mindfulness* (February 2017): 46–58.

"dig in, access deep wells": Elizabeth A. Stanley, *Widen the Window: Training Your Brain and Body to Thrive During Stress and Recover from Trauma* (New York: Avery, 2019), 5.

36 *created a program:* Elizabeth A. Stanley, "Mindfulness-Based Mind Fitness Training (MMFT): An Approach for Enhancing Performance and Building Resilience in High Stress Contexts," in *The Wiley Blackwell Handbook of Mindfulness,* ed. Amanda Ie, Christelle T. Ngnoumen, and Ellen J. Langer (Malden, MA: John Wiley & Sons, 2014), 964–85.

maintain their attentional focus: Amishi P. Jha et al., "Minds 'At Attention': Mindfulness Training Curbs Attentional Lapses in Military Cohorts," *PLoS One* 10 (February 2015).

preserve their working memory: Amishi P. Jha et al., "Short-Form Mindfulness Training Protects Against Working Memory Degradation over High-Demand Intervals," *Journal of Cognitive Enhancement* 1 (June 2017): 154–71.

a technique she calls "shuttling": Stanley, *Widen the Window,* 284–91.

37 *"to pay attention and notice what's happening":* Stanley, "Mindfulness-Based Mind Fitness Training (MMFT)."

feel their emotions more intensely: Lisa Feldman Barrett et al., "Interoceptive Sensitivity and Self-Reports of Emotional Experience," *Journal of Personality and Social Psychology* 87 (November 2004): 684–97. See also Beate M. Herbert, Olga Pollatos, and Rainer Schandry, "Interoceptive Sensitivity and Emotion Processing: An EEG Study," *International Journal of Psychophysiology* 65 (September 2007): 214–27.

managing their emotions more adeptly: Anne Kever et al., "Interoceptive Sensitivity Facilitates Both Antecedent- and Response-Focused Emotion Regulation Strategies," *Personality and Individual Differences* 87 (December 2015): 20–23.

"Common sense says": William James, "What Is an Emotion?," *Mind* 9 (April 1884): 188–205.

38 *constructed from more elemental parts:* Barrett, *How Emotions Are Made.* See also Lisa Feldman Barrett, "The Theory of Constructed Emotion: An Active Inference Account of Interoception and Categorization," *Social Cognitive and Affective Neuroscience* 12 (January 2017): 1–23.

In a series of three studies: Alison Wood Brooks, "Get Excited: Reappraising Pre-Performance Anxiety as Excitement," *Journal of Experimental Psychology: General* 143 (June 2014): 1144–58.

39 *reappraise debilitating "stress":* Alia J. Crum, Peter Salovey, and Shawn Achor, "Re-

thinking Stress: The Role of Mindsets in Determining the Stress Response," *Journal of Personality and Social Psychology* 104 (April 2013): 716–33.

the positive effects of stress: Jeremy P. Jamieson et al., "Turning the Knots in Your Stomach into Bows: Reappraising Arousal Improves Performance on the GRE," *Journal of Experimental Social Psychology* 46 (January 2010): 208–12.

neural effects of the reappraisal technique: Rachel Pizzie et al., "Neural Evidence for Cognitive Reappraisal as a Strategy to Alleviate the Effects of Math Anxiety," paper posted on PsyArXiv, March 2019.

40 *reappraisal works best:* Jürgen Füstös et al., "On the Embodiment of Emotion Regulation: Interoceptive Awareness Facilitates Reappraisal," *Social Cognitive and Affective Neuroscience* 8 (December 2013): 911–17.

the sensations we're actually feeling: Brooks, "Get Excited."

subtly and unconsciously mimic: Adrienne Wood et al., "Fashioning the Face: Sensorimotor Simulation Contributes to Facial Expression Recognition," *Trends in Cognitive Sciences* 20 (March 2016): 227–40.

can't engage in such mimicking: David T. Neal and Tanya L. Chartrand, "Embodied Emotion Perception: Amplifying and Dampening Facial Feedback Modulates Emotion Perception Accuracy," *Social Psychological and Personality Science* 2 (April 2011): 673–78.

41 *more likely to mimic:* Vivien Ainley, Marcel Brass, and Manos Tsakiris, "Heartfelt Imitation: High Interoceptive Awareness Is Linked to Greater Automatic Imitation," *Neuropsychologia* 60 (July 2014): 21–28.

more accurate in their interpretation: Yuri Terasawa et al., "Interoceptive Sensitivity Predicts Sensitivity to the Emotions of Others," *Cognition & Emotion* 28 (February 2014): 1435–48. See also Punit Shah, Caroline Catmur, and Geoffrey Bird, "From Heart to Mind: Linking Interoception, Emotion, and Theory of Mind," *Cortex* 93 (August 2017): 220–23.

tend to have more empathy: Delphine Grynberg and Olga Pollatos, "Perceiving One's Body Shapes Empathy," *Physiology & Behavior* 140 (March 2015): 54–60.

their own bodies' signals: Irene Messina et al., "Somatic Underpinnings of Perceived Empathy: The Importance of Psychotherapy Training," *Psychotherapy Research* 23 (March 2013): 169–77.

"For me it's like using": Psychotherapist, quoted in Robert Shaw, "The Embodied Psychotherapist: An Exploration of the Therapists' Somatic Phenomena Within the Therapeutic Encounter," *Psychotherapy Research* 14 (August 2006): 271–88.

"has helped me to realize": Susie Orbach, "The John Bowlby Memorial Lecture: The Body in Clinical Practice, Part One and Part Two," in *Touch: Attachment and the Body,* ed. Kate White (New York: Routledge, 2019), 17–48.

"social interoception": Andrew J. Arnold, Piotr Winkielman, and Karen Dobkins, "Interoception and Social Connection," *Frontiers in Psychology* 10 (November 2019).

directly in the eye: Tomoko Isomura and Katsumi Watanabe, "Direct Gaze Enhances Interoceptive Accuracy," *Cognition* 195 (February 2020). See also Matias Baltazar et al., "Eye Contact Elicits Bodily Self-Awareness in Human Adults," *Cognition* 133 (October 2014): 120–27.

a brief touch: Nesrine Hazem et al., "Social Contact Enhances Bodily Self-Awareness," *Scientific Reports* 8 (March 2018).

we tend to shift our focus: Arnold, Winkielman, and Dobkins, "Interoception and So-

cial Connection." See also Caroline Durlik and Manos Tsakiris, "Decreased Interoceptive Accuracy Following Social Exclusion," *International Journal of Psychophysiology* 96 (April 2015), 57–63.

a flexible back-and-forth movement: Arnold, Winkielman, and Dobkins, "Interoception and Social Connection."

42 *"sensitive parabolic reflectors":* Coates, *The Hour Between Dog and Wolf,* 120.

a similar device: Cassandra D. Gould van Praag et al., "HeartRater: Tools for the Evaluation of Interoceptive Dimensions of Experience," *Psychosomatic Medicine* 79 (May
2017).

deliberately distorted: "How doppel Works," post on feeldoppel.com, May 24, 2017,
https://feeldoppel.com/pages/science.

induces a sense of calm: Ruben T. Azevedo, "The Calming Effect of a New Wearable
Device During the Anticipation of Public Speech," *Scientific Reports* 7 (May 2017).

43 *more alert and more accurate:* Manos Tsakiris, "Investigating the Effect of *doppel* on
Alertness," working paper.

"to harvest our natural responses": Manos Tsakiris, quoted in "Wearing a Heart on
Your Sleeve: New Research Shows That a Tactile Heartbeat Significantly Reduces
Stress," post on the website of Royal Holloway, University of London, October 18, 2018,
https://www.royalholloway.ac.uk/research-and-teaching/departments-and-schools/
psychology/news/wearing-a-heart-on-your-sleeve/.

"What makes me the same person": Derek Parfit, *Reasons and Persons* (Oxford: Oxford
University Press, 1984), ix.

"I feel, therefore I am": A. D. (Bud) Craig, *How Do You Feel? An Interoceptive Moment
with Your Neurobiological Self* (Princeton: Princeton University Press, 2014), xvii, 182.

2. THINKING WITH MOVEMENT

44 *Fidler walks as he looks:* Jeff Fidler, interviewed by Sanjay Gupta, "Exercising at Work,"
CNN, July 7, 2007.

inspected a batch of images: Jeff L. Fidler et al., "Feasibility of Using a Walking Workstation During CT Image Interpretation," *Journal of the American College of Radiology*
5 (November 2008): 1130–36.

images of patients' lungs: Amee A. Patel et al., "Walking While Working: A Demonstration of a Treadmill-Based Radiologist Workstation," paper presented at the Radiological Society of North America Scientific Assembly and Annual Meeting, February
2008.

45 *used a treadmill workstation:* Cody R. Johnson et al., "Effect of Dynamic Workstation
Use on Radiologist Detection of Pulmonary Nodules on CT," *Journal of the American
College of Radiology* 16 (April 2019): 451–57.

visual sense is sharpened: Tom Bullock et al., "Acute Exercise Modulates Feature-
Selective Responses in Human Cortex," *Journal of Cognitive Neuroscience* 29 (April
2017): 605–18. See also Tom Bullock, Hubert Cecotti, and Barry Giesbrecht, "Multiple
Stages of Information Processing Are Modulated During Acute Bouts of Exercise,"
Neuroscience 307 (October 2015): 138–50.

the periphery of our gaze: Liyu Cao and Barbara Händel, "Walking Enhances Peripheral Visual Processing in Humans," *PLoS Biology* 17 (October 2019). See also Liyu Cao
and Barbara Händel, "Increased Influence of Periphery on Central Visual Processing
in Humans During Walking," bioRxiv (January 2018).

overall physical fitness: Fernando Gomez-Pinilla and Charles Hillman, "The Influence of Exercise on Cognitive Abilities," *Comprehensive Physiology* 3 (January 2013): 403–28.

larger than it "should" be: David A. Raichlen and John D. Polk, "Linking Brains and Brawn: Exercise and the Evolution of Human Neurobiology," *Proceedings of the Royal Society of London B: Biological Sciences* 280 (January 2013).

"At the same time": Raichlen and Polk, "Linking Brains and Brawn."

"Human ancestors transitioned": David A. Raichlen and Gene E. Alexander, "Adaptive Capacity: An Evolutionary Neuroscience Model Linking Exercise, Cognition, and Brain Health," *Trends in Neurosciences* 40 (July 2017): 408–21.

46 *vigorous, sustained physical activity:* Raichlen and Alexander, "Adaptive Capacity."

Hunting, too, poses: Ian J. Wallace, Clotilde Hainline, and Daniel E. Lieberman, "Sports and the Human Brain: An Evolutionary Perspective," in *Handbook of Clinical Neurology,* vol. 158, *Sports Neurology,* ed. Brian Hainline and Robert A. Stern (Amsterdam: Elsevier, 2018), 3–10.

135 minutes a day: David A. Raichlen et al., "Physical Activity Patterns and Biomarkers of Cardiovascular Disease Risk in Hunter-Gatherers," *American Journal of Human Biology* 29 (March–April 2017).

"I see that I can depend": Robert Feld, "Re: 'Walking Workstations: Ambulatory Medicine Redefined,'" *Journal of the American College of Radiology* 6 (March 2009): 213.

50 percent of the school day: Karl E. Minges et al., "Classroom Standing Desks and Sedentary Behavior: A Systematic Review," *Pediatrics* 137 (February 2016).

47 *two-thirds of the average workday:* Stacy A. Clemes et al., "Sitting Time and Step Counts in Office Workers," *Occupational Medicine* 64 (April 2014): 188–92.

"a mind on the hoof": Andy Clark, *Being There: Putting Brain, Body, and World Together Again* (Cambridge: MIT Press, 1998), 35.

students don't sit still: Mireya Villarreal, "California School Becomes First to Lose Chairs for Standing Desks," CBS News, October 12, 2015.

"I taught at sitting desks": Maureen Zink, quoted in Juliet Starrett, "Eight Reasons Why Kids Should Stand at School," *Medium,* June 3, 2019.

a limited resource: Roy F. Baumeister and Kathleen D. Vohs, "Willpower, Choice, and Self-Control," in *Time and Decision: Economic and Psychological Perspectives on Intertemporal Choice,* ed. George Loewenstein, Daniel Read, and Roy Baumeister (New York: Russell Sage Foundation, 2003), 201–16. See also Joel S. Warm, Raja Parasuraman, and Gerald Matthews, "Vigilance Requires Hard Mental Work and Is Stressful," *Human Factors* 50 (June 2008): 433–41.

solve a set of math problems: Christine Langhanns and Hermann Müller, "Effects of Trying 'Not to Move' Instruction on Cortical Load and Concurrent Cognitive Performance," *Psychological Research* 82 (January 2018): 167–76.

48 *simply by standing:* James A. Levine, Sara J. Schleusner, and Michael D. Jensen, "Energy Expenditure of Nonexercise Activity," *American Journal of Clinical Nutrition* 72 (December 2000): 1451–54.

enhancement in students' executive function: Ranjana K. Mehta, Ashley E. Shortz, and Mark E. Benden, "Standing Up for Learning: A Pilot Investigation on the Neurocognitive Benefits of Stand-Biased School Desks," *International Journal of Environmental Research and Public Health* 13 (December 2015).

"on-task engagement": Marianela Dornhecker et al., "The Effect of Stand-Biased

Desks on Academic Engagement: An Exploratory Study," *International Journal of Health Promotion and Education* 53 (September 2015): 271–80.

shown to boost productivity: Gregory Garrett et al., "Call Center Productivity over 6 Months Following a Standing Desk Intervention," *IIE Transactions on Occupational Ergonomics and Human Factors* 4 (July 2016): 188–95.

Such variable stimulation: Dustin E. Sarver et al., "Hyperactivity in Attention-Deficit/Hyperactivity Disorder (ADHD): Impairing Deficit or Compensatory Behavior?," *Journal of Abnormal Child Psychology* 43 (April 2015): 1219–32.

chronically under-*aroused:* Kerstin Mayer, Sarah Nicole Wyckoff, and Ute Strehl, "Underarousal in Adult ADHD: How Are Peripheral and Cortical Arousal Related?," *Clinical EEG and Neuroscience* 47 (July 2016): 171–79. See also Julia Geissler et al., "Hyperactivity and Sensation Seeking as Autoregulatory Attempts to Stabilize Brain Arousal in ADHD and Mania?," *ADHD Attention Deficit and Hyperactivity Disorders* 6 (July 2014): 159–73.

children aged ten to seventeen: Tadeus A. Hartanto et al., "A Trial by Trial Analysis Reveals More Intense Physical Activity Is Associated with Better Cognitive Control Performance in Attention-Deficit/Hyperactivity Disorder," *Child Neuropsychology* 22 (2016): 618–26.

49 *All of these activities:* Michael Karlesky and Katherine Isbister, "Understanding Fidget Widgets: Exploring the Design Space of Embodied Self-Regulation," paper presented at the 9th Nordic Conference on Human-Computer Interaction, October 2016.

"embodied self-regulation": Katherine Isbister and Michael Karlesky, "Embodied Self-Regulation with Tangibles," paper presented at the 8th International Conference on Tangible, Embedded and Embodied Interaction, February 2014.

"Changing what the body *does":* Katherine Isbister, quoted in Melissa Dahl, "Researchers Are Studying the Things People Fiddle With at Their Desks," *New York Magazine,* March 12, 2015.

a mildly positive mood state: Michael Karlesky and Katherine Isbister, "Designing for the Physical Margins of Digital Workspaces: Fidget Widgets in Support of Productivity and Creativity," paper presented at the 8th International Conference on Tangible, Embedded and Embodied Interaction, February 2014.

flexible and creative thinking: Alice M. Isen, "Positive Affect as a Source of Human Strength," in *A Psychology of Human Strengths: Fundamental Questions and Future Directions for a Positive Psychology,* ed. Lisa G. Aspinwall and Ursula M. Staudinger (Washington, DC: American Psychological Association, 2003), 179–95.

directed to doodle: Jackie Andrade, "What Does Doodling Do?," *Applied Cognitive Psychology* 24 (2010): 100–106.

"Today's digital devices": Karlesky and Isbister, "Understanding Fidget Widgets."

50 *"I usually keep track":* Kahneman, *Thinking, Fast and Slow,* 39–40.

51 *positive changes documented by scientists:* My discussion of the effects of physical activity on cognition draws on the following sources: Matthew B. Pontifex et al., "A Primer on Investigating the After Effects of Acute Bouts of Physical Activity on Cognition," *Psychology of Sport & Exercise* 40 (January 2019): 1–22; Annese Jaffery, Meghan K. Edwards, and Paul D. Loprinzi, "The Effects of Acute Exercise on Cognitive Function: Solomon Experimental Design," *Journal of Primary Prevention* 39 (February 2018): 37–46; Charles H. Hillman, Nicole E. Logan, and Tatsuya T. Shigeta, "A Review of Acute

Physical Activity Effects on Brain and Cognition in Children," *Translational Journal of the American College of Sports Medicine* 4 (September 2019): 132–36; Yu-Kai Chang et al., "The Effects of Acute Exercise on Cognitive Performance: A Meta-Analysis," *Brain Research* 1453 (May 2012): 87–101; Phillip D. Tomporowski, "Effects of Acute Bouts of Exercise on Cognition," *Acta Psychologica* 112 (March 2003): 297–324; Eveleen Sng, Emily Frith, and Paul D. Loprinzi, "Temporal Effects of Acute Walking Exercise on Learning and Memory Function," *American Journal of Health Promotion* 32 (September 2018): 1518–25. Paul D. Loprinzi et al., "The Temporal Effects of Acute Exercise on Episodic Memory Function: Systematic Review with Meta-Analysis," *Brain Sciences* 9 (April 2019); James T. Haynes 4th et al., "Experimental Effects of Acute Exercise on Episodic Memory Function: Considerations for the Timing of Exercise," *Psychological Reports* 122 (October 2019): 1744–54; Paul D. Loprinzi and Christy J. Kane, "Exercise and Cognitive Function: A Randomized Controlled Trial Examining Acute Exercise and Free-Living Physical Activity and Sedentary Effects," *Mayo Clinic Proceedings* 90 (April 2015): 450–60; Ai-Guo Chen et al., "Effects of Acute Aerobic Exercise on Multiple Aspects of Executive Function in Preadolescent Children," *Psychology of Sport and Exercise* 15 (November 2014): 627–36; and Julia C. Basso and Wendy A. Suzuki, "The Effects of Acute Exercise on Mood, Cognition, Neurophysiology, and Neurochemical Pathways: A Review," *Brain Plasticity* 2 (March 2017): 127–52.

focus their attention: Matthew T. Mahar, "Impact of Short Bouts of Physical Activity on Attention-to-Task in Elementary School Children," *Preventive Medicine* 52 (June 2011): S60–64.

executive function faculties: Anneke G. van der Niet et al., "Effects of a Cognitively Demanding Aerobic Intervention During Recess on Children's Physical Fitness and Executive Functioning," *Pediatric Exercise Science* 28 (February 2018): 64–70.

recess has been reduced: Jennifer McMurrer, "NCLB Year 5: Choices, Changes, and Challenges: Curriculum and Instruction in the NCLB Era," report produced by the Center on Education Policy, December 2007.

52 *"movement breaks":* Wendell C. Taylor, "Transforming Work Breaks to Promote Health," *American Journal of Preventive Medicine* 29 (December 2005): 461–65.

"inverted U-shaped curve": Pontifex et al., "A Primer on Investigating the After Effects of Acute Bouts of Physical Activity on Cognition."

conducive to creative thought: Yanyun Zhou et al., "The Impact of Bodily States on Divergent Thinking: Evidence for a Control-Depletion Account," *Frontiers in Psychology* 8 (September 2017).

more than two dozen marathons: Haruki Murakami, "The Running Novelist," *The New Yorker,* June 2, 2008.

fifty miles a week: Haruki Murakami, *What I Talk About When I Talk About Running: A Memoir* (New York: Alfred A. Knopf, 2008), 48.

"I'm often asked": Murakami, *What I Talk About When I Talk About Running,* 16–17.

53 *"transient hypofrontality":* Arne Dietrich, "Transient Hypofrontality as a Mechanism for the Psychological Effects of Exercise," *Psychiatry Research* 145 (November 2006): 79–83. See also Arne Dietrich and Michel Audiffren, "The Reticular-Activating Hypofrontality (RAH) Model of Acute Exercise," *Neuroscience & Biobehavioral Reviews* 35 (May 2011): 1305–25.

ideas and impressions mingle: Evangelia G. Chrysikou, "Creativity In and Out of (Cognitive) Control," *Current Opinion in Behavioral Sciences* 27 (June 2019): 94–99.

all kinds of altered states: Arne Dietrich and Laith Al-Shawaf, "The Transient Hypo-frontality Theory of Altered States of Consciousness," *Journal of Consciousness Studies* 25 (2018): 226–47. See also Arne Dietrich, "Functional Neuroanatomy of Altered States of Consciousness: The Transient Hypofrontality Hypothesis," *Consciousness and Cognition* 12 (June 2003): 231–56.

does not generate this disinhibiting effect: Chun-Chih Wang, "Executive Function During Acute Exercise: The Role of Exercise Intensity," *Journal of Sport and Exercise Psychology* 35 (August 2013): 358–67.

generally requires exercising: Jacqueline M. Del Giorno et al., "Cognitive Function During Acute Exercise: A Test of the Transient Hypofrontality Theory," *Journal of Sport and Exercise Psychology* 32 (June 2010): 312–23.

"provoke a kind of Cartesian collapse": Kathryn Schulz, "What We Think About When We Run," *The New Yorker,* November 3, 2015.

persuasive evidence: For a review, see Lawrence W. Barsalou, "Grounded Cognition," *Annual Review of Psychology* 59 (January 2008): 617–45.

54 *remarkably weak:* James Bigelow and Amy Poremba, "Achilles' Ear? Inferior Human Short-Term and Recognition Memory in the Auditory Modality," *PLoS One* 9 (2014).

much more robust: Ilaria Cutica, Francesco Iani, and Monica Bucciarelli, "Learning from Text Benefits from Enactment," *Memory & Cognition* 42 (May 2014): 1026–37. See also Christopher R. Madan and Anthony Singhal, "Using Actions to Enhance Memory: Effects of Enactment, Gestures, and Exercise on Human Memory," *Frontiers in Psychology* 3 (November 2012).

engage a process: Manuela Macedonia and Karsten Mueller, "Exploring the Neural Representation of Novel Words Learned Through Enactment in a Word Recognition Task," *Frontiers in Psychology* 7 (June 2016).

"enactment effect": Lars-Göran Nilsson, "Remembering Actions and Words," in *The Oxford Handbook of Memory,* ed. Endel Tulving and Fergus I. M. Craik (New York: Oxford University Press, 2000), 137–48.

can shed light: Helga Noice and Tony Noice, "What Studies of Actors and Acting Can Tell Us About Memory and Cognitive Functioning," *Current Directions in Psychological Science* 15 (February 2006): 14–18.

98 percent accuracy: Helga Noice and Tony Noice, "Long-Term Retention of Theatrical Roles," *Memory* 7 (May 1999): 357–82.

"you've got to have these two tracks": Actor, quoted in Helga Noice, "Elaborative Memory Strategies of Professional Actors," *Applied Cognitive Psychology* 6 (September–October 1992): 417–27.

six actors from a repertory company: Helga Noice, Tony Noice, and Cara Kennedy, "Effects of Enactment by Professional Actors at Encoding and Retrieval," *Memory* 8 (November 2000): 353–63.

55 *scene from the play:* A. R. Gurney, *The Dining Room* (New York: Dramatist Play Service, 1998), act 1.

series of studies: Helga Noice et al., "Improving Memory in Older Adults by Instructing Them in Professional Actors' Learning Strategies," *Applied Cognitive Psychology* 13 (August 1999): 315–28; Helga Noice, Tony Noice, and Graham Staines, "A Short-Term Intervention to Enhance Cognitive and Affective Functioning in Older Adults," *Journal of Aging and Health* 16 (August 2004): 562–85; Tony Noice and Helga Noice, "A Theatrical Intervention to Lower the Risk of Alzheimer's and Other Forms of Dementia," in *The Routledge Companion to Theatre, Performance and Cognitive Science*, ed. Rick Kemp and Bruce McConchie (New York: Routledge, 2019), 280–90.

56 *a study of undergraduates:* Helga Noice and Tony Noice, "Learning Dialogue With and Without Movement," *Memory & Cognition* 29 (September 2001): 820–27.

not a literal enactment: Helga Noice and Tony Noice, "The Non-Literal Enactment Effect: Filling in the Blanks," *Discourse Processes* 44 (August 2007): 73–89.

reproduce that same movement later: Christopher Kent and Koen Lamberts, "The Encoding-Retrieval Relationship: Retrieval as Mental Simulation," *Trends in Cognitive Sciences* 12 (March 2008): 92–98.

forming the intention *to move:* Elena Daprati, Angela Sirigu, and Daniele Nico, "Remembering Actions Without Proprioception," *Cortex* 113 (April 2019): 29–36.

57 *"One might paraphrase Descartes":* Noice and Noice, "The Non-Literal Enactment Effect."

designed a study: Sian L. Beilock et al., "Sports Experience Changes the Neural Processing of Action Language," *Proceedings of the National Academy of Sciences* 105 (September 2008): 13269–73.

58 *later demonstrate:* Tanja Link et al., "Walk the Number Line — An Embodied Training of Numerical Concepts," *Trends in Neuroscience and Education* 2 (June 2013): 74–84. See also Margina Ruiter, Sofie Loyens, and Fred Paas, "Watch Your Step Children! Learning Two-Digit Numbers Through Mirror-Based Observation of Self-Initiated Body Movements," *Educational Psychology Review* 27 (September 2015): 457–74.

"Moved by Reading" intervention: Arthur M. Glenberg, "How Reading Comprehension Is Embodied and Why That Matters," *International Electronic Journal of Elementary Education* 4 (2011): 5–18.

reading comprehension can actually double: Arthur M. Glenberg et al., "Activity and Imagined Activity Can Enhance Young Children's Reading Comprehension," *Journal of Educational Psychology* 96 (September 2004): 424–36.

read stories about life on a farm: Glenberg et al., "Activity and Imagined Activity Can Enhance Young Children's Reading Comprehension." See also Arthur Glenberg, "Thinking with the Body," *Scientific American,* March 3, 2008.

59 *help children with math:* Arthur M. Glenberg et al., "What Brains Are For: Action, Meaning, and Reading Comprehension," in *Reading Comprehension Strategies: Theories, Interventions, and Technologies,* ed. Danielle S. McNamara (Mahwah, NJ: Psychology Press, 2007), 221–40.

35 percent less likely: Arthur Glenberg et al., "Improving Reading to Improve Math," *Scientific Studies of Reading* 16 (July 2012): 316–40.

support the successful learning: Ayelet Segal, John Black, and Barbara Tversky, "Do Gestural Interfaces Promote Thinking? Congruent Gestures Promote Performance in Math," paper presented at the 51st Annual Meeting of the Psychonomic Society, November 2010. See also John B. Black et al., "Embodied Cognition and Learning Environment Design," in *Theoretical Foundations of Learning Environments,* ed. Susan Land and David Jonassen (New York: Routledge, 2012), 198–223.

number line estimation: Adam K. Dubé and Rhonda N. McEwen, "Do Gestures Matter? The Implications of Using Touchscreen Devices in Mathematics Instruction," *Learning and Instruction* 40 (December 2015): 89–98.

60 *research on physics education:* David Hestenes and Malcolm Wells, "Force Concept Inventory," *The Physics Teacher* 30 (March 1992): 141–58.

becomes less *accurate:* Carl Wieman, "The 'Curse of Knowledge,' or Why Intuition About Teaching Often Fails," *American Physical Society News* 16 (November 2017). See also Edward F. Redish, Jeffery M. Saul, and Richard N. Steinberg, "Student Expectations in Introductory Physics," *American Journal of Physics* 66 (October 1998): 212–24.

imagining themselves into a given scenario: Daniel L. Schwartz and Tamara Black, "Inferences Through Imagined Actions: Knowing by Simulated Doing," *Journal of Experimental Psychology: Learning, Memory, and Cognition* 25 (January 1999): 1–21.

a set of hands-on activities: Carly Kontra et al., "Physical Experience Enhances Science Learning," *Psychological Science* 26 (June 2015): 737–49.

61 *"Learning is moving":* Dor Abrahamson and Raúl Sánchez-García, "Learning Is Moving in New Ways: The Ecological Dynamics of Mathematics Education," *Journal of the Learning Sciences* 25 (2016): 203–39.

"empathy with entities": Elinor Ochs, Patrick Gonzales, and Sally Jacoby, "'When I Come Down I'm in the Domain State': Grammar and Graphic Representation in the Interpretive Activity of Physicists," in *Interaction and Grammar,* ed. Elinor Ochs, Emanuel A. Schegloff, and Sandra A. Thompson (Cambridge: Cambridge University Press, 1996), 328–69.

riding on a beam of light: Walter Isaacson, "The Light-Beam Rider," *New York Times,* October 30, 2015.

"No scientist thinks": Albert Einstein, quoted in Leopold Infeld, *Albert Einstein: His Work and Its Influence on Our World* (New York: Scribner, 1961), 312.

"visual" and even "muscular": Albert Einstein, quoted in Jacques Hadamard, *The Mathematician's Mind: The Psychology of Invention in the Mathematical Field* (Princeton: Princeton University Press, 1945), 143.

"When I was really working with them": Barbara McClintock, quoted in Evelyn Fox Keller, *A Feeling for the Organism: The Life and Work of Barbara McClintock* (New York: Henry Holt, 1983), 117.

"I would picture myself": Jonas Salk, *Anatomy of Reality: Merging of Intuition and Reason* (New York: Columbia University Press, 1983), 7.

62 *"embodied imagination":* Susan Gerofsky, "Approaches to Embodied Learning in Mathematics," in *Handbook of International Research in Mathematics Education,* ed. Lyn D. English and David Kirshner (New York: Routledge, 2016), 60–97.

"integrative glue": Jie Sui and Glyn W. Humphreys, "The Integrative Self: How Self-Reference Integrates Perception and Memory," *Trends in Cognitive Sciences* 19 (December 2015): 719–28.

the ability to alternate: Firat Soylu et al., "Embodied Perspective Taking in Learning About Complex Systems," *Journal of Interactive Learning Research* 28 (July 2017): 269–03. See also David J. DeLiema et al., "Learning Science by Being You, Being It, Being Both," in *Proceedings of the 10th International Conference of the Learning Sciences: Future of Learning,* vol. 2, ed. Peter Freebody et al. (Sydney: University of Sydney, 2012), 102–3.

"Energy Theater": Rachel E. Scherr et al., "Negotiating Energy Dynamics Through Embodied Action in a Materially Structured Environment," *Physical Review* 9 (July 2013).

"Students who use movement": Author interview with Rachel Scherr.

63 *"become" human chromosomes:* Joseph P. Chinnici, Joyce W. Yue, and Kieron M. Tor-

res, "Students as 'Human Chromosomes' in Role-Playing Mitosis & Meiosis," *The American Biology Teacher* 66 (2004): 35–39.

embody the solar system's planets: Ted Richards, "Using Kinesthetic Activities to Teach Ptolemaic and Copernican Retrograde Motion," *Science & Education* 21 (2012): 899–910.

embody carbon molecules: Pauline M. Ross, Deidre A. Tronson, and Raymond J. Ritchie, "Increasing Conceptual Understanding of Glycolysis & the Krebs Cycle Using Role-Play," *The American Biology Teacher* 70 (March 2008): 163–68.

embody amino acids: Diana Sturges, Trent W. Maurer, and Oladipo Cole, "Understanding Protein Synthesis: A Role-Play Approach in Large Undergraduate Human Anatomy and Physiology Classes," *Advances in Physiology Education* 33 (June 2009): 103–10.

64 "Being it": Carmen J. Petrick and H. Taylor Martin, "Learning Mathematics: You're It vs. It's It," in *Proceedings of the 10th International Conference of the Learning Sciences: Future of Learning,* vol. 2, ed. Peter Freebody et al. (Sydney: University of Sydney, 2012), 101–2.

form a triangle: Carmen Smith and Candace Walkington, "Four Principles for Designing Embodied Mathematics Activities," *Australian Mathematics Education Journal* 1 (2019): 16–20.

"body-based activities": Carmen Petrick Smith, "Body-Based Activities in Secondary Geometry: An Analysis of Learning and Viewpoint," *Social Science and Mathematics* 118 (May 2018): 179–89.

"Moving the body": Sian L. Beilock, *How the Body Knows Its Mind* (New York: Atria Books, 2015), 69–70.

"thinking outside the box": Angela K.-y. Leung et al., "Embodied Metaphors and Creative 'Acts,'" *Psychological Science* 23 (May 2012): 502–9.

65 *"on one hand":* Leung et al., "Embodied Metaphors and Creative 'Acts.'"

a loose kind of metaphor: Michael L. Slepian and Nalini Ambady, "Fluid Movement and Creativity," *Journal of Experimental Psychology: General* 141 (November 2012): 625–29. See also Shu Imaizumi, Ubuka Tagami, and Yi Yang, "Fluid Movements Enhance Creative Fluency: A Replication of Slepian and Ambady (2012)," paper posted on PsyArXiv, March 2020.

the effects of walking: Marily Oppezzo and Daniel L. Schwartz, "Give Your Ideas Some Legs: The Positive Effect of Walking on Creative Thinking," *Journal of Experimental Psychology: Learning, Memory, and Cognition* 40 (July 2014): 1142–52.

66 *a meandering, free-form route:* Chun-Yu Kuo and Yei-Yu Yeh, "Sensorimotor-Conceptual Integration in Free Walking Enhances Divergent Thinking for Young and Older Adults," *Frontiers in Psychology* 7 (October 2016).

"Only thoughts which come": Friedrich Nietzsche, *Twilight of the Idols* (Oxford: Oxford University Press, 1998), 9.

"I have walked myself": Søren Kierkegaard, *Søren Kierkegaard's Journals and Papers,* vol. 5, ed. and trans. Howard V. Hong and Edna H. Hong (Bloomington: Indiana University Press, 1978), 412.

"gymnastics for the mind": Ralph Waldo Emerson, *The Topical Notebooks of Ralph Waldo Emerson,* vol. 1, ed. Susan Sutton Smith (Columbia: University of Missouri Press, 1990), 260.

"I am unable to reflect": Jean-Jacques Rousseau, *The Confessions of Jean-Jacques Rousseau* (London: Reeves and Turner, 1861), 343.

"I have nothing": Michel de Montaigne, *The Essays: A Selection*, ed. and trans. M. A. Screech (New York: Penguin Books, 1987), 304.

"the mind at three miles per hour": Rebecca Solnit, *Wanderlust: A History of Walking* (New York: Penguin Books, 2000).

67   *"I think that I cannot preserve"*: Henry David Thoreau, "Walking," *Atlantic Monthly* 9 (June 1862): 657–74.

"How vain it is": Henry David Thoreau, *A Year in Thoreau's Journal: 1851* (New York: Penguin Books, 1993).

3. THINKING WITH GESTURE

68   *bounded onto the stage*: Gabriel Hercule, "HUDlog — Demo Day Pitch," video posted on YouTube, January 25, 2018, https://www.youtube.com/watch?v=dbBemSKGPbM.

69   *the motions of the hands*: Geoffrey Beattie and Heather Shovelton, "An Experimental Investigation of the Role of Different Types of Iconic Gesture in Communication: A Semantic Feature Approach," *Gesture* 1 (January 2001): 129–49.

impressionistic and holistic: Rowena A. E. Viney, Jean Clarke, and Joep Cornelissen, "Making Meaning from Multimodality: Embodied Communication in a Business Pitch Setting," in *The SAGE Handbook of Qualitative Business and Management Research Methods*, ed. Catherine Cassell, Ann L. Cunliffe, and Gina Grandy (Thousand Oaks, CA: Sage Publications, 2017), 193–214.

the effort to persuade: Nicole Torres, "When You Pitch an Idea, Gestures Matter More Than Words," *Harvard Business Review*, May–June 2019.

the center of the action: Joep P. Cornelissen, Jean S. Clarke, and Alan Cienki, "Sense-giving in Entrepreneurial Contexts: The Use of Metaphors in Speech and Gesture to Gain and Sustain Support for Novel Business," *International Small Business Journal: Researching Entrepreneurship* 30 (April 2012): 213–41.

human-scale, embodied terms: Alan Cienki, Joep P. Cornelissen, and Jean Clarke, "The Role of Human Scale, Embodied Metaphors/Blends in the Speech and Gestures of Entrepreneurs," paper presented at the 9th Conference on Conceptual Structure, Discourse, and Language, October 2008.

mentally simulate the gesturer's point of view: Francesco Ianì and Monica Bucciarelli, "Mechanisms Underlying the Beneficial Effect of a Speaker's Gestures on the Listener," *Journal of Memory and Language* 96 (October 2017): 110–21.

a palpable reality: Ruben van Werven, Onno Bouwmeester, and Joep P. Cornelissen, "Pitching a Business Idea to Investors: How New Venture Founders Use Micro-Level Rhetoric to Achieve Narrative Plausibility," *International Small Business Journal: Researching Entrepreneurship* 37 (May 2019): 193–214.

"entrepreneurs are operating": Van Werven, Bouwmeester, and Cornelissen, "Pitching a Business Idea to Investors."

70   *"the skilled use of gesture"*: Jean S. Clarke, Joep P. Cornelissen, and Mark P. Healey, "Actions Speak Louder Than Words: How Figurative Language and Gesturing in Entrepreneurial Pitches Influences Investment Judgments," *Academy of Management Journal* 62 (April 2019): 335–60.

"I talk with my hands": Author interview with Frederic Mishkin.

humankind's earliest language: Michael A. Arbib, Katja Liebal, and Simone Pika, "Primate Vocalization, Gesture, and the Evolution of Human Language," *Current Anthropology* 49 (December 2008): 1053–76. See also Erica A. Cartmill, Sian Beilock, and Susan Goldin-Meadow, "A Word in the Hand: Action, Gesture and Mental Representation in Humans and Non-Human Primates," *Philosophical Transactions of the Royal Society B: Biological Sciences* 367 (January 2012): 129–43.

71 *exert a powerful impact:* Spencer D. Kelly et al., "Offering a Hand to Pragmatic Understanding: The Role of Speech and Gesture in Comprehension and Memory," *Journal of Memory and Language* 40 (May 1999): 577–92.

"extra-verbal meaning": Jean A. Graham and Michael Argyle, "A Cross-Cultural Study of the Communication of Extra-Verbal Meaning by Gestures," *International Journal of Psychology* 10 (February 1975): 57–67.

more richly specify: Susan Goldin-Meadow, "Gesture and Cognitive Development," in *Handbook of Child Psychology and Developmental Science: Cognitive Processes,* ed. Lynn S. Liben, Ulrich Müller, and Richard M. Lerner (Hoboken, NJ: John Wiley & Sons, 2015), 339–80.

meaning that is not found anywhere: Susan Goldin-Meadow and Catherine Momeni Sandhofer, "Gestures Convey Substantive Information About a Child's Thoughts to Ordinary Listeners," *Developmental Science* 2 (December 1999): 67–74. See also Ruth Breckinridge Church and Susan Goldin-Meadow, "The Mismatch Between Gesture and Speech as an Index of Transitional Knowledge," *Cognition* 23 (June 1986): 43–71.

contradicts or departs entirely: David McNeill, Justine Cassell, and Karl-Erik McCullough, "Communicative Effects of Speech-Mismatched Gestures," *Research on Language and Social Interaction* 27 (1994): 223–37.

doctor-patient dialogue: Christian Heath, "Gesture's Discrete Tasks: Multiple Relevancies in Visual Conduct and in the Contextualization of Language," in *The Contextualization of Language,* ed. Peter Auer and Aldo Di Luzio (Philadelphia: John Benjamins, 1992), 101–28.

72 *"gestural foreshadowing":* Elaine M. Crowder, "Gestures at Work in Sense-Making Science Talk," *Journal of the Learning Sciences* 5 (1996), 173–208.

we've said something in error: Mandana Seyfeddinipur and Sotaro Kita, "Gesture as an Indicator of Early Error Detection in Self-Monitoring of Speech," paper presented at the Disfluency in Spontaneous Speech Workshop, August 2001.

mentally prime a word: Frances H. Rauscher, Robert M. Krauss, and Yihsiu Chen, "Gesture, Speech, and Lexical Access: The Role of Lexical Movements in Speech Production," *Psychological Science* 7 (July 1996): 226–31.

they talk less fluently: Karen J. Pine, Hannah Bird, and Elizabeth Kirk, "The Effects of Prohibiting Gestures on Children's Lexical Retrieval Ability," *Developmental Science* 10 (November 2007): 747–54. See also Sheena Finlayson et al., "Effects of the Restriction of Hand Gestures on Disfluency," paper presented at the Disfluency in Spontaneous Speech Workshop, September 2003.

remember less useful information: Donna Frick-Horbury and Robert E. Guttentag, "The Effects of Restricting Hand Gesture Production on Lexical Retrieval and Free Recall," *American Journal of Psychology* 111 (Spring 1998): 43–62.

solve problems less well: Monica Bucciarelli et al., "Children's Creation of Algorithms: Simulations and Gestures," *Journal of Cognitive Psychology* 28 (2016): 297–318.

less able to explain: Martha W. Alibali and Sotaro Kita, "Gesture Highlights Perceptually Present Information for Speakers," *Gesture* 10 (January 2010): 3–28.

before babies can talk: Núria Esteve-Gibert and Pilar Prietoba, "Infants Temporally Coordinate Gesture-Speech Combinations Before They Produce Their First Words," *Speech Communication* 57 (February 2014): 301–16.

one of children's first gestures: Elizabeth Bates et al., "Vocal and Gestural Symbols at 13 Months," *Merrill-Palmer Quarterly of Behavior and Development* 26 (October 1980): 407–23.

a request to point to their nose: Bari Walsh, "The Power of Babble," *Harvard Gazette,* July 2016.

elicit from their caretakers: Susan Goldin-Meadow et al., "Young Children Use Their Hands to Tell Their Mothers What to Say," *Developmental Science* 10 (November 2007): 778–85.

73 *lay the foundation:* Jana M. Iverson and Susan Goldin-Meadow, "Gesture Paves the Way for Language Development," *Psychological Science* 16 (May 2005): 367–71. See also Seyda Ozçalişkan and Susan Goldin-Meadow, "Gesture Is at the Cutting Edge of Early Language Development," *Cognition* 96 (July 2005): B101–13.

documented a link: Meredith L. Rowe and Susan Goldin-Meadow, "Early Gesture Selectively Predicts Later Language Learning," *Developmental Science* 12 (January 2009): 182–87.

a 30 million–"word gap": Betty Hart and Todd R. Risley, *Meaningful Differences in the Everyday Experience of Young American Children* (Baltimore: Paul H. Brookes Publishing, 1995).

other research has confirmed: Roberta Michnick Golinkoff et al., "Language Matters: Denying the Existence of the 30-Million-Word Gap Has Serious Consequences," *Child Development* 90 (May 2019): 985–92.

tend to talk more: Janellen Huttenlocher et al., "Sources of Variability in Children's Language Growth," *Cognitive Psychology* 61 (December 2010): 343–65.

employ a greater diversity of word types: Erika Hoff, "The Specificity of Environmental Influence: Socioeconomic Status Affects Early Vocabulary Development via Maternal Speech," *Child Development* 74 (October 2003): 1368–78.

more complex and more varied sentences: Huttenlocher et al., "Sources of Variability in Children's Language Growth"; Hoff, "The Specificity of Environmental Influence."

these differences are predictive: Huttenlocher et al., "Sources of Variability in Children's Language Growth"; Hoff, "The Specificity of Environmental Influence." See also Meredith L. Rowe, "Child-Directed Speech: Relation to Socioeconomic Status, Knowledge of Child Development and Child Vocabulary Skill," *Journal of Child Language* 35 (February 2008): 185–205.

high-income parents gesture more: Meredith L. Rowe and Susan Goldin-Meadow, "Differences in Early Gesture Explain SES Disparities in Child Vocabulary Size at School Entry," *Science* 323 (February 2009): 951–53.

74 *a mental word bank:* Betty Hart and Todd Risley, "The Early Catastrophe: The 30 Million Word Gap by Age 3," *American Educator,* Spring 2003.

a strong predictor: Dorthe Bleses et al., "Early Productive Vocabulary Predicts Academic Achievement 10 Years Later," *Applied Psycholinguistics* 37 (November 2016): 1461–76.

Engage in frequent pointing: Meredith L. Rowe and Kathryn A. Leech, "A Parent Intervention with a Growth Mindset Approach Improves Children's Early Gesture and Vocabulary Development," *Developmental Science* 22 (July 2019).

the reading of picture books: Eve Sauer LeBarton, Susan Goldin-Meadow, and Stephen Raudenbush, "Experimentally Induced Increases in Early Gesture Lead to Increases in Spoken Vocabulary," *Journal of Cognitive Development* 16 (2015): 199–220.

Come up with simple gestures: Susan W. Goodwyn, Linda P. Acredolo, and Catherine A. Brown, "Impact of Symbolic Gesturing on Early Language Development," *Journal of Nonverbal Behavior* 24 (June 2000): 81–103.

Rowe delivered this message: Rowe and Leech, "A Parent Intervention with a Growth Mindset Approach Improves Children's Early Gesture and Vocabulary Development."

"newest and most advanced ideas": Susan Goldin-Meadow, Martha W. Alibali, and Ruth Breckinridge Church, "Transitions in Concept Acquisition: Using the Hand to Read the Mind," *Psychological Review* 100 (April 1993): 279–97.

Such tasks were first employed: Jean Piaget, *The Child's Conception of Number* (New York: W. W. Norton and Company, 1965).

75 *The girl is shown:* Church and Goldin-Meadow, "The Mismatch Between Gesture and Speech as an Index of Transitional Knowledge."

40 percent of the time: Church and Goldin-Meadow, "The Mismatch Between Gesture and Speech as an Index of Transitional Knowledge."

when ten-year-olds solve math problems: Michelle Perry, Ruth Breckinridge Church, and Susan Goldin-Meadow, "Transitional Knowledge in the Acquisition of Concepts," *Cognitive Development* 3 (October 1988): 359–400.

fifteen-year-olds working on a problem-solving task: Addison Stone, Rebecca Webb, and Shahrzad Mahootian, "The Generality of Gesture-Speech Mismatch as an Index of Transitional Knowledge: Evidence from a Control-of-Variables Task," *Cognitive Development* 6 (July–September 1991): 301–13.

especially receptive to instruction: Martha W. Alibali and Susan Goldin-Meadow, "Gesture-Speech Mismatch and Mechanisms of Learning: What the Hands Reveal About a Child's State of Mind," *Cognitive Psychology* 25 (October 1993): 468–523.

a group of college students: Raedy Ping et al., "Gesture-Speech Mismatch Predicts Who Will Learn to Solve an Organic Chemistry Problem," paper presented at the Annual Meeting of the American Educational Research Association, April 2012.

76 *captures some aspect:* Martha W. Alibali et al., "Spontaneous Gestures Influence Strategy Choices in Problem Solving," *Psychological Science* 22 (September 2011): 1138–44.

"Gesture encourages experimentation": Susan Goldin-Meadow, quoted in Eric Jaffe, "Giving Students a Hand: William James Lecturer Goldin-Meadow Shows the Importance of Gesture in Teaching," *APS Observer,* July 2004.

actively trying to apprehend: Martha W. Alibali, Sotaro Kita, and Amanda J. Young, "Gesture and the Process of Speech Production: We Think, Therefore We Gesture," *Language and Cognitive Processes* 15 (December 2000): 593–613.

increases as a function of difficulty: Susan Goldin-Meadow, *Hearing Gesture: How Our Hands Help Us Think* (Cambridge: Belknap Press of Harvard University Press, 2003), 147. See also Autumn B. Hostetter, Martha W. Alibali, and Sotaro Kita, "I See It in My Hands' Eye: Representational Gestures Reflect Conceptual Demands," *Language and Cognitive Processes* 22 (April 2007): 313–36.

"muddled talk": Wolff-Michael Roth, "From Gesture to Scientific Language," *Journal of Pragmatics* 32 (October 2000): 1683–1714.

we divide the work: Susan Goldin-Meadow et al., "Explaining Math: Gesturing Lightens the Load," *Psychological Science* 12 (November 2001): 516–22.

the attainment of complex knowledge: Wolff-Michael Roth, "Gestures: Their Role in Teaching and Learning," *Review of Educational Research* 71 (Fall 2001): 365–92.

"virtual diagram": Barbara Tversky, *Mind in Motion: How Action Shapes Thought* (New York: Basic Books, 2019), 125.

77 *language becomes more precise:* Wolff-Michael Roth and Daniel Lawless, "Science, Culture, and the Emergence of Language," *Science Education* 86 (May 2002): 368–85.

Gestures are less frequent: Mingyuan Chu and Sotaro Kita, "The Nature of Gestures' Beneficial Role in Spatial Problem Solving," *Journal of Experimental Psychology: General* 140 (February 2011): 102–16.

more coordinated in meaning and timing: Crowder, "Gestures at Work in Sense-Making Science Talk."

scaffolding our own thinking: David DeLiema and Francis Steen, "Thinking with the Body: Conceptual Integration Through Gesture in Multiviewpoint Model Construction," in *Language and the Creative Mind,* ed. Michael Borkent, Barbara Dancygier, and Jennifer Hinnell (Stanford: CSLI Publications, 2013), 275–94.

asked to write on complex topics: Candace Walkington et al., "Being Mathematical Relations: Dynamic Gestures Support Mathematical Reasoning," paper presented at the 11th International Conference of the Learning Sciences, June 2014.

experts venturing into uncharted territory: Kim A. Kastens, Shruti Agrawal, and Lynn S. Liben, "The Role of Gestures in Geoscience Teaching and Learning," *Journal of Geoscience Education* 56 (September 2008): 362–68.

lab meetings of a biochemistry research group: L. Amaya Becvar, James D. Hollan, and Edwin Hutchins, "Hands as Molecules: Representational Gestures Used for Developing Theory in a Scientific Laboratory," *Semiotica* 156 (January 2005): 89–112.

78 *a range of specialized gestures:* Kinnari Atit et al., "Spatial Gestures Point the Way: A Broader Understanding of the Gestural Referent," paper presented at the 35th Annual Conference of the Cognitive Science Society, July–August 2013. See also Kinnari Atit, Thomas F. Shipley, and Basil Tipoff, "What Do a Geologist's Hands Tell You? A Framework for Classifying Spatial Gestures in Science Education," in *Space in Mind: Concepts for Spatial Learning and Education,* ed. Daniel R. Montello, Karl Grossner, and Donald G. Janelle (Cambridge: MIT Press, 2014), 173–94.

benefit from using gesture: Kastens, Agrawal, and Liben, "The Role of Gestures in Geoscience Teaching and Learning."

how they would use Play-Doh: Kinnari Atit, Kristin Gagnier, and Thomas F. Shipley, "Student Gestures Aid Penetrative Thinking," *Journal of Geoscience Education* 63 (February 2015): 66–72.

79 *"many college students":* Lynn S. Liben, Adam E. Christensen, and Kim A. Kastens, "Gestures in Geology: The Roles of Spatial Skills, Expertise, and Communicative Context," paper presented at the 7th International Conference on Spatial Cognition, July 2010, referring to another study by the same authors: Lynn S. Liben, Kim A. Kastens, and Adam E. Christensen, "Spatial Foundations of Science Education: The Illustrative Case of Instruction on Introductory Geological Concepts," *Cognition and Instruction* 29 (January–March 2011): 45–87.

Better performance on this task: Liben, Christensen, and Kastens, "Gestures in Geology."

"language can actually get in the way": Author interview with Michele Cooke.

an enhanced ability to process: Evguenia Malaia and Ronnie Wilbur, "Enhancement of Spatial Processing in Sign Language Users," in *Space in Mind: Concepts for Spatial Learning and Education,* ed. Daniel R. Montello, Karl Grossner, and Donald G. Janelle (Cambridge: MIT Press, 2014), 99–118. See also Karen Emmorey, Edward Klima, and Gregory Hickok, "Mental Rotation Within Linguistic and Non-Linguistic Domains in Users of American Sign Language," *Cognition* 68 (September 1998): 221–46.

exhibited by hearing people: Madeleine Keehner and Susan E. Gathercole, "Cognitive Adaptations Arising from Nonnative Experience of Sign Language in Hearing Adults," *Memory & Cognition* 35 (June 2007): 752–61.

80 *evoke a stronger reaction:* Anthony Steven Dick et al., "Co-Speech Gestures Influence Neural Activity in Brain Regions Associated with Processing Semantic Information," *Human Brain Mapping* 30 (April 2009): 3509–26.

captures listeners' attention: Spencer Kelly et al., "The Brain Distinguishes Between Gesture and Action in the Context of Processing Speech," paper presented at the 163rd Acoustical Society of America Meeting, May 2012.

roused to attention: Arne Nagels et al., "Hand Gestures Alert Auditory Cortices: Possible Impacts of Learning on Foreign Language Processing," in *Positive Learning in the Age of Information,* ed. Olga Zlatkin-Troitschanskaia, Gabriel Wittum, and Andreas Dengel (Wiesbaden: Springer VS, 2018), 53–66.

"Hand gestures appear to alert": Author interview with Spencer Kelly.

others better understand: Nicole Dargue and Naomi Sweller, "Two Hands and a Tale: When Gestures Benefit Adult Narrative Comprehension," *Learning and Instruction* 68 (August 2020).

33 percent more likely: Ruth Breckinridge Church, Philip Garber, and Kathryn Rogalski, "The Role of Gesture in Memory and Social Communication," *Gesture* 7 (July 2007): 137–58.

significantly more learning: Ji Y. Son et al., "Exploring the Practicing-Connections Hypothesis: Using Gesture to Support Coordination of Ideas in Understanding a Complex Statistical Concept," *Cognitive Research: Principles and Implications* 3 (December 2018).

direct their gaze more efficiently: Zhongling Pi et al., "Instructors' Pointing Gestures Improve Learning Regardless of Their Use of Directed Gaze in Video Lectures," *Computers & Education* 128 (January 2019): 345–52.

pay more attention: Pi et al., "Instructors' Pointing Gestures Improve Learning Regardless of Their Use of Directed Gaze in Video Lectures."

more readily transfer: Theodora Koumoutsakis, "Gesture in Instruction: Evidence from Live and Video Lessons," *Journal of Nonverbal Behavior* 40 (December 2016): 301–15.

seem to be especially helpful: Zhongling Pi et al., "All Roads Lead to Rome: Instructors' Pointing and Depictive Gestures in Video Lectures Promote Learning Through Different Patterns of Attention Allocation," *Journal of Nonverbal Behavior* 43 (July 2019): 549–59.

81 *appears to be even stronger:* Koumoutsakis et al., "Gesture in Instruction."

examined the top one hundred videos: Son et al., "Exploring the Practicing-Connections Hypothesis."

gesture as they teach on video: Jiumin Yang et al., "Instructors' Gestures Enhance Their Teaching Experience and Performance While Recording Video Lectures," *Journal of Computer Assisted Learning* 36 (April 2020): 189–98.

make more hand motions: Susan Wagner Cook and Susan Goldin-Meadow, "The Role of Gesture in Learning: Do Children Use Their Hands to Change Their Minds?," *Journal of Cognition and Development* 7 (April 2006): 211–32.

A simple request: Susan Goldin-Meadow, interviewed by Jaffe, "Giving Students a Hand."

encouraging them to gesture: Sara C. Broaders et al., "Making Children Gesture Brings Out Implicit Knowledge and Leads to Learning," *Journal of Experimental Psychology: General* 136 (November 2007): 539–50.

solving spatial problems: Chu and Kita, "The Nature of Gestures' Beneficial Role in Spatial Problem Solving."

82 *when asked to gesture more:* Alice Cravotta, M. Grazia Busà, and Pilar Prieto, "Effects of Encouraging the Use of Gestures on Speech," *Journal of Speech, Language, and Hearing Research* 62 (September 2019): 3204–19.

when teachers are told: Martha W. Alibali et al., "Students Learn More When Their Teacher Has Learned to Gesture Effectively," *Gesture* 13 (January 2013): 210–33.

the largest known cognitive gender difference: Jillian E. Lauer, Eukyung Yhang, and Stella F. Lourenco, "The Development of Gender Differences in Spatial Reasoning: A Meta-Analytic Review," *Psychological Bulletin* 145 (June 2019): 537–65.

five-year-old boys were already better: Stacy B. Ehrlich, Susan C. Levine, and Susan Goldin-Meadow, "The Importance of Gesture in Children's Spatial Reasoning," *Developmental Psychology* 42 (November 2006): 1259–68.

children who were encouraged to gesture: Raedy Ping et al., "Using Manual Rotation and Gesture to Improve Mental Rotation in Preschoolers," paper presented at the 33rd Annual Conference of the Cognitive Science Society, July 2011. See also Elizabeth M. Wakefield et al., "Breaking Down Gesture and Action in Mental Rotation: Understanding the Components of Movement That Promote Learning," *Developmental Psychology* 55 (May 2019): 981–93.

83 *in the face of its demands:* Carine Lewis, Peter Lovatt, and Elizabeth D. Kirk, "Many Hands Make Light Work: The Facilitative Role of Gesture in Verbal Improvisation," *Thinking Skills and Creativity* 17 (September 2015): 149–57.

"It is from the attempt": Wolff-Michael Roth, "Making Use of Gestures, the Leading Edge in Literacy Development," in *Crossing Borders in Literacy and Science Instruction: Perspectives on Theory and Practice,* ed. Wendy Saul (Newark, DE: International Reading Association; and Arlington, VA: National Science Teachers Association, 2004), 48–70.

research has confirmed: Andrew T. Stull et al., "Does Manipulating Molecular Models Promote Representation Translation of Diagrams in Chemistry?," paper presented at the International Conference on Theory and Application of Diagrams, August 2010. See also Lilian Pozzer-Ardenghi and Wolff-Michael Roth, "Photographs in Lectures: Gestures as Meaning-Making Resources," *Linguistics and Education* 15 (Summer 2004): 275–93.

"visual artifacts": Roth, "Gestures: Their Role in Teaching and Learning."

"more mature physics talk": Roth, "Making Use of Gestures, the Leading Edge in Literacy Development."

we may miss the clues: Susan Goldin-Meadow, Debra Wein, and Cecilia Chang, "Assessing Knowledge Through Gesture: Using Children's Hands to Read Their Minds," *Cognition and Instruction* 9 (1992): 201–19.

less than a third: Martha Wagner Alibali, Lucia M. Flevares, and Susan Goldin-Meadow, "Assessing Knowledge Conveyed in Gesture: Do Teachers Have the Upper Hand?," *Journal of Educational Psychology* 89 (March 1997): 183–93.

84 *recruited to watch video recordings:* Spencer D. Kelly et al., "A Helping Hand in Assessing Children's Knowledge: Instructing Adults to Attend to Gesture," *Cognition and Instruction* 20 (2002): 1–26.

"second" her gesture: Goldin-Meadow, *Hearing Gesture*, 128.

designed gestures: Shamin Padalkar and Jayashree Ramadas, "Designed and Spontaneous Gestures in Elementary Astronomy Education," *International Journal of Science Education* 33 (August 2011): 1703–39.

multiple mental "hooks": Beilock, *How the Body Knows Its Mind*, 92.

85 *hidden from our view:* Kensy Cooperrider, Elizabeth Wakefield, and Susan Goldin-Meadow, "More Than Meets the Eye: Gesture Changes Thought, Even Without Visual Feedback," paper presented at the 37th Annual Conference of the Cognitive Science Society, July 2015.

her students practice: Kerry Ann Dickson and Bruce Warren Stephens, "It's All in the Mime: Actions Speak Louder Than Words When Teaching the Cranial Nerves," *Anatomical Science Education* 8 (November–December 2015): 584–92.

act as an aid to memory: Manuela Macedonia, "Bringing Back the Body into the Mind: Gestures Enhance Word Learning in Foreign Language," *Frontiers in Psychology* 5 (December 2014).

That's not how anyone learns: Manuela Macedonia, "Learning a Second Language Naturally: The Voice Movement Icon Approach," *Journal of Educational and Developmental Psychology* 3 (September 2013): 102–16.

86 *use of gestures to enhance verbal memory:* Manuela Macedonia, "Sensorimotor Enhancing of Verbal Memory Through 'Voice Movement Icons' During Encoding of Foreign Language" (PhD diss., University of Salzburg, 2003).

areas in the brain: Manuela Macedonia et al., "Depth of Encoding Through Observed Gestures in Foreign Language Word Learning," *Frontiers in Psychology* 10 (January 2019).

found evidence that the motor cortex: Brian Mathias et al., "Motor Cortex Causally Contributes to Auditory Word Recognition Following Sensorimotor-Enriched Vocabulary Training," paper posted on ArXiv, May 2020.

achieve more lasting learning: Kirsten Bergmann and Manuela Macedonia, "A Virtual Agent as Vocabulary Trainer: Iconic Gestures Help to Improve Learners' Memory Performance," paper presented at the 13th International Conference on Intelligent Virtual Agents, August 2013.

those who observe the gesture: Manuela Macedonia, Kirsten Bergmann, and Friedrich Roithmayr, "Imitation of a Pedagogical Agent's Gestures Enhances Memory for Words in Second Language," *Science Journal of Education* 2 (2014): 162–69.

researchers have found that math students: Susan Wagner Cook et al., "Hand Gesture and Mathematics Learning: Lessons from an Avatar," *Cognitive Science* 41 (March 2017): 518–35.

offloads our cognitive burden: Goldin-Meadow et al., "Explaining Math." See also Raedy Ping and Susan Goldin-Meadow, "Gesturing Saves Cognitive Resources When Talking About Nonpresent Objects," *Cognitive Science* 34 (May 2010): 602–19.

87 *gesture more to compensate:* Angela M. Kessell and Barbara Tversky, "Using Diagrams

and Gestures to Think and Talk About Insight Problems," *Proceedings of the Annual Meeting of the Cognitive Science Society* 28 (2006).

"Academic language": Author interview with Brendan Jeffreys.

4. THINKING WITH NATURAL SPACES

91 *the artist Jackson Pollock:* My discussion of Jackson Pollock's move from New York City to the East End of Long Island draws on the following sources: Deborah Solomon, *Jackson Pollock: A Biography* (New York: Simon & Schuster, 1987); Jeffrey Potter, *To a Violent Grave: An Oral Biography of Jackson Pollock* (New York: G. P. Putnam, 1985); Lee F. Mindel, "Jackson Pollock and Lee Krasner's Long Island House and Studio," *Architectural Digest*, November 30, 2013; Ellen Maguire, "At Jackson Pollock's Hamptons House, a Life in Spatters," *New York Times*, July 14, 2006.

"It was a healing place": Audrey Flack, quoted in Maguire, "At Jackson Pollock's Hamptons House, a Life in Spatters."

92 *it was then that he realized:* Potter, *To a Violent Grave*, 81.

"a camping trip that lasts a lifetime": Gordon H. Orians and Judith H. Heerwagen, "Evolved Responses to Landscapes," in *The Adapted Mind: Evolutionary Psychology and the Generation of Culture*, ed. Jerome H. Barkow, Leda Cosmides, and John Tooby (New York: Oxford University Press, 1995), 555–79.

only about 7 percent of our time: Neil E. Klepeis et al., "The National Human Activity Pattern Survey (NHAPS): A Resource for Assessing Exposure to Environmental Pollutants," *Journal of Exposure Science & Environmental Epidemiology* 11 (July 2001): 231–52.

93 *meager even when compared:* F. Thomas Juster, Frank Stafford, and Hiromi Ono, "Changing Times of American Youth: 1981–2003," report produced by the Institute for Social Research, University of Michigan, January 2004.

five hours or less: Stephen R. Kellert et al., "The Nature of Americans: Disconnection and Recommendations for Reconnection," report produced April 2017.

only 26 percent of mothers: Rhonda Clements, "An Investigation of the Status of Outdoor Play," *Contemporary Issues in Early Childhood* 5 (March 2004): 68–80.

now live in cities: United Nations, "2018 Revision of World Urbanization Prospects," report produced by the Population Division of the UN Department of Economic and Social Affairs, May 2018.

Forty-two million people visit: Central Park Conservancy, "About Us," https://www.centralparknyc.org/about.

"Natural scenery": Frederick Law Olmsted, quoted in Witold Rybczynski, *A Clearing in the Distance: Frederick Law Olmsted and America in the 19th Century* (New York: Simon & Schuster, 1999), 258.

When Olmsted started his work: Rybczynski, *A Clearing in the Distance.* See also John G. Mitchell, "Frederick Law Olmsted's Passion for Parks," *National Geographic*, March 2005.

visits Olmsted made: Rybczynski, *A Clearing in the Distance.*

94 *preferences for certain kinds of natural spaces:* John D. Balling and John H. Falk, "Development of Visual Preference for Natural Environments," *Environment and Behavior* 14 (January 1982): 5–28. See also Mary Ann Fischer and Patrick E. Shrout, "Children's Liking of Landscape Paintings as a Function of Their Perceptions of Prospect, Refuge, and Hazard," *Environment and Behavior* 21 (May 2006): 373–93.

"prospect" and "refuge": Jay Appleton, *The Experience of Landscape* (New York: John Wiley & Sons, 1977).

we like a bit of mystery: Andrew M. Szolosi, Jason M. Watson, and Edward J. Ruddell, "The Benefits of Mystery in Nature on Attention: Assessing the Impacts of Presentation Duration," *Frontiers in Psychology* 5 (November 2014). See also Thomas R. Herzog and Anna G. Bryce, "Mystery and Preference in Within-Forest Settings," *Environment and Behavior* 39 (July 2007): 779–96.

intuited these preferences: Matt Ridley, "Why We Love a Bit of Africa in Our Parkland," *The Times* (London), December 28, 2015.

"ghosts of environments past": Gordon H. Orians, *Snakes, Sunrises, and Shakespeare: How Evolution Shapes Our Loves and Fears* (Chicago: University of Chicago Press, 2014), 20.

95 "environmental self-regulation": Kalevi Mikael Korpela, "Place-Identity as a Product of Environmental Self-Regulation," *Journal of Environmental Psychology* 9 (September 1989): 241–56.

reestablishing our mental equilibrium: Roger S. Ulrich et al., "Stress Recovery During Exposure to Natural and Urban Environments," *Journal of Environmental Psychology* 11 (September 1991): 201–30.

Drivers who travel: Russ Parsons et al., "The View from the Road: Implications for Stress Recovery and Immunization," *Journal of Environmental Psychology* 18 (June 1998): 113–40.

Laboratory studies: See, e.g., Bin Jiang et al., "A Dose-Response Curve Describing the Relationship Between Urban Tree Cover Density and Self-Reported Stress Recovery," *Environment and Behavior* 48 (September 2014): 607–29.

the more benefit they derive: Tytti Pasanen et al., "Can Nature Walks with Psychological Tasks Improve Mood, Self-Reported Restoration, and Sustained Attention? Results from Two Experimental Field Studies," *Frontiers in Psychology* 9 (October 2018).

a ninety-minute walk outside: Gregory N. Bratman et al., "Nature Experience Reduces Rumination and Subgenual Prefrontal Cortex Activation," *Proceedings of the National Academy of Sciences* 112 (July 2015): 8567–72.

a walk in nature lifts the mood: Marc G. Berman et al., "Interacting with Nature Improves Cognition and Affect for Individuals with Depression," *Journal of Affective Disorders* 140 (November 2012): 300–305.

96 *maintain our focus:* Rita Berto, "Exposure to Restorative Environments Helps Restore Attentional Capacity," *Journal of Environmental Psychology* 25 (September 2005): 249–59.

amid outdoor greenery: Gregory N. Bratman, J. Paul Hamilton, and Gretchen C. Daily, "The Impacts of Nature Experience on Human Cognitive Function and Mental Health," *Annals of the New York Academy of Sciences* 1249 (February 2012): 118–36.

benefits from time spent: Marc G. Berman, John Jonides, and Stephen Kaplan, "The Cognitive Benefits of Interacting with Nature," *Psychological Science* 19 (December 2008): 1207–12.

better able to focus: Andrea Faber Taylor and Frances E. Kuo, "Children with Attention Deficits Concentrate Better After Walk in the Park," *Journal of Attention Disorders* 12 (March 2009): 402–9. See also Frances E. Kuo and Andrea Faber Taylor, "A Potential Natural Treatment for Attention-Deficit/Hyperactivity Disorder: Evidence from a National Study," *American Journal of Public Health* 94 (September 2004): 1580–86.

"Doses of nature": Taylor and Kuo, "Children with Attention Deficits Concentrate Better After Walk in the Park."

"voluntary" and "passive": William James, *The Principles of Psychology*, vol. 1 (New York: Henry Holt and Company, 1890), 416–19.

97 *"soft fascination"*: Avik Basu, Jason Duvall, and Rachel Kaplan, "Attention Restoration Theory: Exploring the Role of Soft Fascination and Mental Bandwidth," *Environment and Behavior* 51 (November 2019): 1055–81.

"open monitoring": Freddie Lymeus, Per Lindberg, and Terry Hartig, "Building Mindfulness Bottom-Up: Meditation in Natural Settings Supports Open Monitoring and Attention Restoration," *Consciousness and Cognition* 59 (March 2018): 40–56.

"soft gazing": Patricia Morgan and Dor Abrahamson, "Contemplative Mathematics Pedagogy: Report from a Pioneering Workshop," paper presented at the Annual Meeting of the American Educational Research Association, April 2019.

easier to implement and maintain: Freddie Lymeus, Tobias Lundgren, and Terry Hartig, "Attentional Effort of Beginning Mindfulness Training Is Offset with Practice Directed Toward Images of Natural Scenery," *Environment and Behavior* 49 (June 2017): 536–59. See also Freddie Lymeus, Per Lindberg, and Terry Hartig, "A Natural Meditation Setting Improves Compliance with Mindfulness Training," *Journal of Environmental Psychology* 64 (August 2019): 98–106.

"substantially counteracts": Bin Jiang, Rose Schmillen, and William C. Sullivan, "How to Waste a Break: Using Portable Electronic Devices Substantially Counteracts Attention Enhancement Effects of Green Spaces," *Environment and Behavior* 51 (November 2019): 1133–60. See also Theresa S. S. Schilhab, Matt P. Stevenson, and Peter Bentsen, "Contrasting Screen-Time and Green-Time: A Case for Using Smart Technology and Nature to Optimize Learning Processes," *Frontiers in Psychology* 9 (June 2018).

98 *an app like ReTUNE*: Marc Berman and Kathryn Schertz, "ReTUNE (Restoring Through Urban Nature Experience)," post on the website of the University of Chicago, Summer 2017, https://appchallenge.uchicago.edu/retune/.

wants to walk: The first route described here was mapped out using the Waze app, downloaded from https://www.waze.com/. The second route was mapped out using the ReTUNE platform, found at https://retune-56d2e.firebaseapp.com/.

"Restoration Score": Kathryn E. Schertz, Omid Kardan, and Marc G. Berman, "ReTUNE: Restoring Through Urban Nature Experience," poster presented at the UChicago App Challenge, August 2017.

99 *81 percent accuracy*: Marc G. Berman et al., "The Perception of Naturalness Correlates with Low-Level Visual Features of Environmental Scenes," *PLoS One* 9 (December 2014).

contain more *visual information*: Hiroki P. Kotabe, Omid Kardan, and Marc G. Berman, "The Nature-Disorder Paradox: A Perceptual Study on How Nature Is Disorderly Yet Aesthetically Preferred," *Journal of Experimental Psychology: General* 146 (August 2017): 1126–42. See also Agnes van den Berg, Yannick Joye, and Sander L. Koole, "Why Viewing Nature Is More Fascinating and Restorative Than Viewing Buildings: A Closer Look at Perceived Complexity," *Urban Forestry & Urban Greening* 20 (October 2016): 397–401.

Roughly a third: Denise Grady, "The Vision Thing: Mainly in the Brain," *Discover*, June 1, 1993.

100 *"perceptual fluency"*: Yannick Joye and Agnes van den Berg, "Nature Is Easy on the

Mind: An Integrative Model for Restoration Based on Perceptual Fluency," paper presented at the 8th Biennial Conference on Environmental Psychology, September 2010.

makes us feel good: Rolf Reber, Norbert Schwarz, and Piotr Winkielman, "Processing Fluency and Aesthetic Pleasure: Is Beauty in the Perceiver's Processing Experience?," *Personality and Social Psychology Review* 8 (November 2004): 364–82. See also Piotr Winkelman and John T. Cacioppo, "Mind at Ease Puts a Smile on the Face: Psychophysiological Evidence That Processing Facilitation Elicits Positive Affect," *Journal of Personality and Social Psychology* 81(December 2001): 989–1000.

more redundant information: Yannick Joye et al., "When Complex Is Easy on the Mind: Internal Repetition of Visual Information in Complex Objects Is a Source of Perceptual Fluency," *Journal of Experimental Psychology: Human Perception and Performance* 42 (January 2016): 103–14.

"Natural environments are characterized": Yannick Joye and Agnes van den Berg, "Is Love for Green in Our Genes? A Critical Analysis of Evolutionary Assumptions in Restorative Environments Research," *Urban Forestry & Urban Greening* 10 (2011): 261–68.

101 *people prefer mid-range fractals*: Caroline M. Hägerhäll, Terry Purcell, and Richard Taylor, "Fractal Dimension of Landscape Silhouette Outlines as a Predictor of Landscape Preference," *Journal of Environmental Psychology* 24 (June 2004), 247–55.

a soothing effect: Richard P. Taylor et al., "Perceptual and Physiological Responses to the Visual Complexity of Fractal Patterns," *Nonlinear Dynamics, Psychology, and Life Sciences* 9 (January 2005): 89–114. See also Richard P. Taylor, "Reduction of Physiological Stress Using Fractal Art and Architecture," *Leonardo* 39 (June 2006): 245–51.

brain activity is being recorded: Caroline M. Hägerhäll et al., "Human Physiological Benefits of Viewing Nature: EEG Responses to Exact and Statistical Fractal Patterns," *Nonlinear Dynamics, Psychology, and Life Sciences* 19 (January 2015): 1–12; Caroline M. Hägerhäll et al., "Investigations of Human EEG Response to Viewing Fractal Patterns," *Perception* 37 (October 2008): 1488–94.

"fractal landscapes": Arthur W. Juliani et al., "Navigation Performance in Virtual Environments Varies with Fractal Dimension of Landscape," *Journal of Environmental Psychology* 47 (September 2016): 155–65.

complete challenging puzzles: Yannick Joye et al., "When Complex Is Easy on the Mind: Internal Repetition of Visual Information in Complex Objects Is a Source of Perceptual Fluency," *Journal of Experimental Psychology: Human Perception and Performance* 42 (January 2016): 103–14.

"tuned" our perceptual faculties: Richard P. Taylor and Branka Spehar, "Fractal Fluency: An Intimate Relationship Between the Brain and Processing of Fractal Stimuli," in *The Fractal Geometry of the Brain*, ed. Antonio Di Ieva (New York: Springer, 2016), 485–98.

exhibited a fractal pattern: Richard P. Taylor, Adam P. Micolich, and David Jonas, "Fractal Analysis of Pollock's Drip Paintings," *Nature* 399 (June 1999): E9–10. See also Jose Alvarez-Ramirez, Carlos Ibarra-Valdez, and Eduardo Rodriguez, "Fractal Analysis of Jackson Pollock's Painting Evolution," *Chaos, Solitons & Fractals* 83 (February 2016): 97–104.

102 *"If someone asked"*: Richard Taylor, quoted in Jennifer Ouellette, "Pollock's Fractals," *Discover*, October 2001.

"exploit nature like a drug": Roger Ulrich, quoted in Michael Waldholz, "The Leafy Green Road to Good Mental Health," *Wall Street Journal,* August 2003.

relieved pain and promoted healing: Roger S. Ulrich, "View Through a Window May Influence Recovery from Surgery," *Science* 224 (April 1984): 420–21.

"As a teenager": Roger Ulrich, quoted in Adam Alter, "Where We Are Shapes Who We Are," *New York Times,* June 14, 2013.

exposure to natural elements: Roger S. Ulrich et al., "A Review of the Research Literature on Evidence-Based Healthcare Design," *HERD: Health Environments Research & Design* 1 (April 2008): 61–125.

103 *our blood pressure drops*: Roger S. Ulrich, Robert F. Simons, and Mark Miles, "Effects of Environmental Simulations and Television on Blood Donor Stress," *Journal of Architectural and Planning Research* 20 (March 2003): 38–47.

brain activity becomes more relaxed: Roger S. Ulrich, "Natural Versus Urban Scenes: Some Psychophysiological Effects," *Environment and Behavior* 13 (September 1981): 523–56.

our eye movements change: Marek Franěk, Jan Petruzalek, and Denis Šefara, "Eye Movements in Viewing Urban Images and Natural Images in Diverse Vegetation Periods," *Urban Forestry & Urban Greening* 46 (December 2019). See also Deltcho Valtchanov and Colin Ellard, "Cognitive and Affective Responses to Natural Scenes: Effects of Low Level Visual Properties on Preference, Cognitive Load and Eye-Movements," *Journal of Environmental Psychology* 43 (September 2015): 184–95.

we remember details of natural scenes: Rita Berto et al., "An Exploratory Study of the Effect of High and Low Fascination Environments on Attentional Fatigue," *Journal of Environmental Psychology* 30 (December 2010): 494–500.

the brain's pleasure receptors: Irving Biederman and Edward A. Vessel, "Perceptual Pleasure and the Brain: A Novel Theory Explains Why the Brain Craves Information and Seeks It Through the Senses," *American Scientist* 94 (May–June 2006): 247–53.

the "biophilia hypothesis": Edward O. Wilson, *Biophilia: The Human Bond with Other Species* (Cambridge: Harvard University Press, 1984), 109–11.

"Everything we do at Second Home": Rohan Silva, quoted in Diana Budds, "There Are More Than 2,000 Plants in This Lush Coworking Space," *Fast Company,* February 12, 2017.

104 *presence of indoor plants*: Ruth K. Raanaas et al., "Benefits of Indoor Plants on Attention Capacity in an Office Setting," *Journal of Environmental Psychology* 31 (March 2011): 99–105. See also Tina Bringslimark, Terry Hartig, and Grete Grindal Patil, "Psychological Benefits of Indoor Plants in Workplaces: Putting Experimental Results into Context," *HortScience* 42 (June 2007): 581–87.

"green wall": Agnes E. van den Berg et al., "Green Walls for a Restorative Classroom Environment: A Controlled Evaluation Study," *Environment and Behavior* 49 (August 2017): 791–813.

Bank of America Tower: William Browning, "Constructing the Biophilic Community," in *Constructing Green: The Social Structures of Sustainability,* ed. Rebecca L. Henn and Andrew J. Hoffman (Cambridge: MIT Press, 2013), 341–50.

"We wanted the first thing": Author interview with Bill Browning.

an indoor environment: Jie Yin et al., "Physiological and Cognitive Performance of Exposure to Biophilic Indoor Environment," *Building and Environment* 132 (March 2018): 255–62.

105 *keep us alert:* Kevin Nute et al., "The Animation of the Weather as a Means of Sustaining Building Occupants and the Natural Environment," *International Journal of Environmental Sustainability* 8 (2012): 27–39.

forty-six minutes more: Mohamed Boubekri et al., "Impact of Windows and Daylight Exposure on Overall Health and Sleep Quality of Office Workers: A Case-Control Pilot Study," *Journal of Clinical Sleep Medicine* 10 (June 2014): 603–11.

Google has determined: Adele Peters, "Google Is Trying to Improve Its Workplaces with Offices Inspired by Nature," *Fast Company,* January 12, 2015.

the energy crisis of the 1970s: Lindsay Baker, *A History of School Design and Its Indoor Environmental Standards, 1900 to Today* (Washington, DC: National Institute of Building Sciences, 2012).

106 *"It changes the whole mood":* Lisa Sarnicola, quoted in Joann Gonchar, "Schools of the 21st Century: P.S. 62, the Kathleen Grimm School for Leadership and Sustainability at Sandy Ground," *Architectural Record,* January 2016.

gauging the "greenness": Wing Tuen Veronica Leung et al., "How Is Environmental Greenness Related to Students' Academic Performance in English and Mathematics?," *Landscape and Urban Planning* 181 (January 2019): 118–24.

assigned to a room: Dongying Li and William C. Sullivan, "Impact of Views to School Landscapes on Recovery from Stress and Mental Fatigue," *Landscape and Urban Planning* 148 (April 2016): 149–58.

effects of window access: Lisa Heschong, "Windows and Office Worker Performance: The SMUD Call Center and Desktop Studies," in *Creating the Productive Workplace,* ed. Derek Clements-Croome (New York: Taylor & Francis, 2006), 277–309.

107 *forty-second "micro-break":* Kate E. Lee et al., "40-Second Green Roof Views Sustain Attention: The Role of Micro-Breaks in Attention Restoration," *Journal of Environmental Psychology* 42 (June 2015): 182–89.

"microrestorative opportunities": Rachel Kaplan, "The Nature of the View from Home: Psychological Benefits," *Environment and Behavior* 33 (July 2001): 507–42.

"Oh, these vast calm": John Muir, *My First Summer in the Sierra* (Boston: Houghton Mifflin, 1916), 82.

"The first night we camped": Theodore Roosevelt, "John Muir: An Appreciation," in *The Outlook: A Weekly Newspaper,* January 16, 1915, 28.

108 *"I hope for the preservation":* Theodore Roosevelt, speech given at the Capitol Building in Sacramento, California, on May 19, 1903, in *A Compilation of the Messages and Speeches of Theodore Roosevelt, 1901–1905,* vol. 1, ed. Alfred Henry Lewis (Washington, DC: Bureau of National Literature and Art, 1906), 410.

delay immediate gratification: Meredith S. Berry et al., "Making Time for Nature: Visual Exposure to Natural Environments Lengthens Subjective Time Perception and Reduces Impulsivity," *PLoS One* 10 (November 2015).

restrain their impulse: Arianne J. van der Wal et al., "Do Natural Landscapes Reduce Future Discounting in Humans?," *Proceedings of the Royal Society B: Biological Sciences* 280 (December 2013).

capable of delaying gratification: Rebecca Jenkin et al., "The Relationship Between Exposure to Natural and Urban Environments and Children's Self-Regulation," *Landscape Research* 43 (April 2018): 315–28.

estimate how many seconds: Berry et al., "Making Time for Nature."

overestimate how long: Maria Davydenko and Johanna Peetz, "Time Grows on Trees:

The Effect of Nature Settings on Time Perception," *Journal of Environmental Psychology* 54 (December 2017): 20–26.

109 *Children's play is more imaginative:* Kellie Dowdell, Tonia Gray, and Karen Malone, "Nature and Its Influence on Children's Outdoor Play," *Journal of Outdoor and Environmental Education* 15 (December 2011): 24–35. See also Anne-Marie Morrissey, Caroline Scott, and Mark Rahimi, "A Comparison of Sociodramatic Play Processes of Preschoolers in a Naturalized and a Traditional Outdoor Space," *International Journal of Play* 6 (May 2017): 177–97.

expansively and open-mindedly: Ethan A. McMahan and David Estes, "The Effect of Contact with Natural Environments on Positive and Negative Affect: A Meta-Analysis," *Journal of Positive Psychology* 10 (November 2015): 507–19. See also George MacKerron and Susana Mourato, "Happiness Is Greater in Natural Environments," *Global Environmental Change* 23 (October 2013): 992–1000.

"three day effect": David Strayer, quoted in Florence Williams, "This Is Your Brain on Nature," *National Geographic,* January 2016. See also Frank M. Ferraro III, "Enhancement of Convergent Creativity Following a Multiday Wilderness Experience," *Ecopsychology* 7 (March 2015): 7–11.

a measure of creative thinking: Ruth Ann Atchley, David L. Strayer, and Paul Atchley, "Creativity in the Wild: Improving Creative Reasoning Through Immersion in Natural Settings," *PLoS One* 7 (December 2012).

110 *makes us feel tiny:* Yang Bai et al., "Awe, the Diminished Self, and Collective Engagement: Universals and Cultural Variations in the Small Self," *Journal of Personality and Social Psychology* 113 (August 2017): 185–209. See also Yannick Joye and Jan Willem Bolderdijk, "An Exploratory Study into the Effects of Extraordinary Nature on Emotions, Mood, and Prosociality," *Frontiers in Psychology* 5 (January 2015).

"in the upper reaches of pleasure": Dacher Keltner and Jonathan Haidt, "Approaching Awe, a Moral, Spiritual, and Aesthetic Emotion," *Cognition & Emotion* 17 (March 2003): 297–314.

we must work to accommodate: Michelle Shiota et al., "Transcending the Self: Awe, Elevation, and Inspiration," in *Handbook of Positive Emotions,* ed. Michele M. Tugade, Michelle N. Shiota, and Leslie D. Kirby (New York: Guilford Press, 2016), 362–77.

physical behavior that accompanies awe: Michelle N. Shiota, Belinda Campos, and Dacher Keltner, "The Faces of Positive Emotion: Prototype Displays of Awe, Amusement, and Pride," *Annals of the New York Academy of Science* 1000 (December 2003): 296–99. See also Belinda Campos et al., "What Is Shared, What Is Different? Core Relational Themes and Expressive Displays of Eight Positive Emotions," *Cognition & Emotion* 27 (January 2013): 37–52.

more curious and open-minded: Michelle N. Shiota, Dacher Keltner, and Amanda Mossman, "The Nature of Awe: Elicitors, Appraisals, and Effects on Self-Concept," *Cognition & Emotion* 21 (August 2007): 944–63. See also Laura L. Carstensen, Derek M. Isaacowitz, and Susan Turk Charles, "Taking Time Seriously: A Theory of Socioemotional Selectivity," *American Psychologist* 54 (March 1999): 165–81.

more willing to revise and update: Alexander F. Danvers and Michelle N. Shiota, "Going Off Script: Effects of Awe on Memory for Script-Typical and -Irrelevant Narrative Detail," *Emotion* 17 (September 2017): 938–52.

"a reset button": Jonathan Haidt, *The Righteous Mind: Why Good People Are Divided by Politics and Religion* (New York: Pantheon Books, 2012), 228.

111 *more prosocially:* Jia Wei Zhang et al., "An Occasion for Unselfing: Beautiful Nature Leads to Prosociality," *Journal of Environmental Psychology* 37 (March 2014): 61–72. See also John M. Zelenski, Raelyne L. Dopko, and Colin A. Capaldi, "Cooperation Is in Our Nature: Nature Exposure May Promote Cooperative and Environmentally Sustainable Behavior," *Journal of Environmental Psychology* 42 (June 2015): 24–31.

more altruistically: Nicolas Guéguen and Jordy Stefan, "'Green Altruism': Short Immersion in Natural Green Environments and Helping Behavior," *Environment and Behavior* 48 (February 2016): 324–42. See also Claire Prade and Vassilis Saroglou, "Awe's Effects on Generosity and Helping," *Journal of Positive Psychology* 11 (September 2016): 522–30.

The "functional" account of awe: Dacher Keltner and James J. Gross, "Functional Accounts of Emotions," *Cognition and Emotion* 13 (September 1999): 467–80.

put aside their individual interests: Bai et al., "Awe, the Diminished Self, and Collective Engagement." See also Jennifer E. Stellar et al., "Self-Transcendent Emotions and Their Social Functions: Compassion, Gratitude, and Awe Bind Us to Others Through Prosociality," *Emotion* 9 (July 2017): 200–207.

"the overview effect": Frank White, *The Overview Effect: Space Exploration and Human Evolution* (Reston, VA: American Institute of Aeronautics and Astronautics, 1998), 3–4. See also David B. Yaden et al., "The Overview Effect: Awe and Self-Transcendent Experience in Space Flight," *Psychology of Consciousness: Theory, Research, and Practice* 3 (March 2016): 1–11.

"If somebody'd said before the flight": Alan Shepard, quoted in Don Nardo, *The Blue Marble: How a Photograph Revealed Earth's Fragile Beauty* (North Mankato, MN: Compass Point Books, 2014), 46.

"you identify with Houston": Rusty Schweikart, quoted in White, *The Overview Effect*, 11.

112 *"explosion of awareness":* Edgar Mitchell, quoted in White, *The Overview Effect*, 38.

"overwhelming sense of oneness": Edgar Mitchell, quoted in Yaden et al., *"The Overview Effect."*

preserve the psychological well-being: Clay Morgan, "Long-Duration Psychology: The Real Final Frontier?," in *Shuttle Mir: The United States and Russia Share History's Highest Stage* (Houston: Lyndon B. Johnson Space Center), 50–51.

"space broccoli": Monica Edwards and Laurie Abadie, "Zinnias from Space! NASA Studies the Multiple Benefits of Gardening," post on the website of the NASA Human Research Program, January 19, 2016.

"Farmer Foale": Morgan, "Long-Duration Psychology."

"I loved the greenhouse experiment": Michael Foale, quoted in "Mir-24 Mission Interviews," October 29, 1997, posted on the website of NASA History, https://history.nasa.gov/SP-4225/documentation/mir-summaries/mir24/interviews.htm.

113 *"homesick":* David Jagneaux, "Virtual Reality Could Provide Healthy Escape for Homesick Astronauts," *Vice,* January 10, 2016.

"we feel 'at home'": Harry Francis Mallgrave, *Architecture and Embodiment: The Implications of the New Sciences and Humanities for Design* (New York: Routledge, 2013), 74.

5. THINKING WITH BUILT SPACES

114 *take leave of the lab:* My discussion about Jonas Salk's visit to the monastery in Assisi draws on the following sources: John Paul Eberhard, "Architecture and Neuroscience:

A Double Helix," in *Mind in Architecture: Neuroscience, Embodiment, and the Future of Design,* ed. Sarah Robinson and Juhani Pallasmaa (Cambridge: MIT Press, 2015), 123–36; Nathaniel Coleman, *Utopias and Architecture* (New York: Routledge, 2007); Norman L. Koonce, "Stewardship: An Architect's Perspective," in *Historic Cities and Sacred Sites: Cultural Roots for Urban Futures,* ed. Ismail Serageldin, Ephim Shluger, and Joan Martin-Brown (Washington, DC: World Bank, 2001), 30–32.

"The spirituality of the architecture": Jonas Salk, quoted in Coleman, *Utopias and Architecture,* 185.

build an intellectual community: My discussion of Jonas Salk's collaboration with Louis Kahn draws on the following sources: Wendy Lesser, *You Say to Brick: The Life of Louis Kahn* (New York: Farrar, Straus and Giroux, 2017); Charlotte DeCroes Jacobs, *Jonas Salk: A Life* (New York: Oxford University Press, 2015); Jonathan Salk, "Reflections on the Relationship Between Lou Kahn and Jonas Salk," talk given at the DFC Technology and Innovation Summit, 2015, posted on the DesignIntelligence website, https://www.di.net/articles/reflections-on-the-relationship-between-lou-kahn-and -jonas-salk/; Stuart W. Leslie, "A Different Kind of Beauty: Scientific and Architectural Style in I. M. Pei's Mesa Laboratory and Louis Kahn's Salk Institute," *Historical Studies in the Natural Sciences* 38 (Spring 2008): 173–221; Carter Wiseman, *Louis I. Kahn: Beyond Time and Style; A Life in Architecture* (New York: W. W. Norton, 2007); David B. Brownlee and David G. De Long, *Louis I. Kahn: In the Realm of Architecture* (New York: Rizzoli, 1991).

115 *an optimal place to think:* Carolina A. Miranda, "Louis Kahn's Salk Institute, the Building That Guesses Tomorrow, Is Aging — Very, Very Gracefully," *Los Angeles Times,* November 22, 2016.

"I would say the building": Jonas Salk, quoted in Wiseman, *Louis I. Kahn,* 135.

tradition of folk architecture: Christopher Alexander, Sara Ishikawa, and Murray Silverstein, *A Pattern Language* (New York: Oxford University Press, 1977).

"neuroarchitecture": Alex Coburn, Oshin Vartanian and Anjan Chatterjee, "Buildings, Beauty, and the Brain: A Neuroscience of Architectural Experience," *Journal of Cognitive Neuroscience* 29 (September 2017): 1521–31.

116 *it was a disaster:* Roger N. Goldstein, "Architectural Design and the Collaborative Research Environment," *Cell* 127 (October 2006): 243–46.

Roger Barker had set out: Roger G. Barker, *Habitats, Environments, and Human Behavior: Studies in Ecological Psychology and Eco-Behavioral Science from the Midwest Psychological Field Station, 1947–1972* (San Francisco: Jossey-Bass, 1978). See also Harry Heft, *Ecological Psychology in Context: James Gibson, Roger Barker, and the Legacy of William James* (Mahwah, NJ: Lawrence Erlbaum Associates, 2001).

"Barker and his colleagues": Benjamin R. Meagher, "Ecologizing Social Psychology: The Physical Environment as a Necessary Constituent of Social Processes," *Personality and Social Psychology Review* 24 (February 2020): 3–23.

"The characteristics of the behavior": Roger Barker, *Ecological Psychology: Concepts and Methods for Studying the Environment of Human Behavior* (Palo Alto: Stanford University Press, 1968), 152.

"the arrogance of the belief": Christopher Alexander, *The Timeless Way of Building* (New York: Oxford University Press, 1979), 106.

118 *"The wall was designed":* Colin Ellard, *Places of the Heart: The Psychogeography of Everyday Life* (New York: Bellevue Literary Press, 2015), 24–25.

"Our distant ancestors": John L. Locke, *Eavesdropping: An Intimate History* (Oxford: Oxford University Press, 2010), 5.

the studiolo *of Federico da Montefeltro:* Joscelyn Godwin, *The Pagan Dream of the Renaissance* (Boston: Weiser Books, 2005).

119 *such "thinking rooms" served a need:* Philippe Ariès and Georges Duby, eds., *A History of Private Life,* vol. 2, *Revelations of the Medieval World* (Cambridge: Belknap Press of Harvard University Press, 1988).

"we must reserve a back room": Michel de Montaigne, "Of Solitude," in *The Essays of Michael Seigneur De Montaigne,* trans. Peter Coste (London: S. and E. Ballard, 1759), 277.

70 to 80 percent of American office workers: Matthew C. Davis, Desmond J. Leach, and Chris W. Clegg, "The Physical Environment of the Office: Contemporary and Emerging Issues," *International Review of Industrial and Organizational Psychology* 26 (2011): 193–237.

120 *"fertilized countless Enlightenment-era innovations":* Steven Johnson, *Where Good Ideas Come From: A Natural History of Invention* (New York: Riverhead Books, 2010), 162–63.

many hours at a coffeehouse: Steven Johnson, *The Invention of Air: A Story of Science, Faith, Revolution, and the Birth of America* (New York: Riverhead, 2008).

"There should be a plaque": Steven Johnson, interviewed by Guy Raz, "How Cafe Culture Helped Make Good Ideas Happen," *All Things Considered,* NPR, October 17, 2010.

"Allen curve": Thomas J. Allen, *Managing the Flow of Technology* (Cambridge: MIT Press, 1977). See also Peter Dizikes, "The Office Next Door," *MIT Technology Review,* October 2011.

121 *MIThenge:* "Infinite Corridor," post on the *Atlas Obscura* website, https://www .atlasobscura.com/places/infinite-corridor.

the Allen curve still applies: Thomas J. Allen and Gunter W. Henn, *The Organization and Architecture of Innovation* (New York: Routledge, 2007). See also Matthew Claudel et al., "An Exploration of Collaborative Scientific Production at MIT Through Spatial Organization and Institutional Affiliation," *PLoS One* 12 (June 2017).

"ineluctably break through": Fabrice B. R. Parmentier, "Deviant Sounds Yield Distraction Irrespective of the Sounds' Informational Value," *Journal of Experimental Psychology: Human Perception and Performance* 42 (June 2016): 837–46.

"involuntary capture": Fabrice B. R. Parmentier, Jacqueline Turner, and Laura Perez, "A Dual Contribution to the Involuntary Semantic Processing of Unexpected Spoken Words," *Journal of Experimental Psychology: General* 143 (February 2014): 38–45.

122 *intelligible speech is particularly distracting:* Niina Venetjoki et al., "The Effect of Speech and Speech Intelligibility on Task Performance," *Ergonomics* 49 (September 2006): 1068–91. See also Valtteri Hongisto, "A Model Predicting the Effect of Speech of Varying Intelligibility on Work Performance," *Indoor Air* 15 (December 2005): 458–68.

its semantic meaning is processed: Fabrice B. R. Parmentier, Jacqueline Turner, and Laura Perez, "A Dual Contribution to the Involuntary Semantic Processing of Unexpected Spoken Words," *Journal of Experimental Psychology: General* 143 (February 2014): 38–45. See also Helena Jahncke, "Open-Plan Office Noise: The Susceptibility and Suitability of Different Cognitive Tasks for Work in the Presence of Irrelevant Speech," *Noise Health* 14 (November–December 2012), 315–20.

drawing on the same limited resource: Marijke Keus van de Poll and Patrik Sörqvist, "Effects of Task Interruption and Background Speech on Word Processed Writing," *Applied Cognitive Psychology* 30 (May–June 2016): 430–39.

participants were asked to write: Marijke Keus van de Poll et al., "Disruption of Writing by Background Speech: The Role of Speech Transmission Index," *Applied Acoustics* 81 (July 2014): 15–18.

tune out one-sided conversations: John E. Marsh et al., Why Are Background Telephone Conversations Distracting?," *Journal of Experimental Psychology: Applied* 24 (June 2018): 222–35. See also Veronica V. Galván, Rosa S. Vessal, and Matthew T. Golley, "The Effects of Cell Phone Conversations on the Attention and Memory of Bystanders," *PLoS One* 8 (2013).

"halfalogue": Lauren L. Emberson et al., "Overheard Cell-Phone Conversations: When Less Speech Is More Distracting," *Psychological Science* 21 (October 2010): 1383–88.

competes for mental resources: Nick Perham and Harriet Currie, "Does Listening to Preferred Music Improve Reading Comprehension Performance?," *Applied Cognitive Psychology* 28 (March–April 2014): 279–84.

123 *tasks that are difficult or complex:* Manuel F. Gonzalez and John R. Aiello, "More Than Meets the Ear: Investigating How Music Affects Cognitive Task Performance," *Journal of Experimental Psychology: Applied* 25 (September 2019): 431–44.

those requiring creativity: Emma Threadgold et al., "Background Music Stints Creativity: Evidence from Compound Remote Associate Tasks," *Applied Cognitive Psychology* 33 (September–October 2019): 873–88.

high intensity, fast tempos: Peter Tze-Ming Chou, "Attention Drainage Effect: How Background Music Effects [*sic*] Concentration in Taiwanese College Students," *Journal of the Scholarship of Teaching and Learning* (January 2010): 36–46.

frequent variation: Nick Perham and Martinne Sykora, "Disliked Music Can Be Better for Performance Than Liked Music," *Applied Cognitive Psychology* 26 (June 2012): 550–55.

"attention drainage effect": Chou, "Attention Drainage Effect."

disrupts cognition for young people: Martin R. Vasilev, Julie A. Kirkby, and Bernhard Angele, "Auditory Distraction During Reading: A Bayesian Meta-Analysis of a Continuing Controversy," *Perspectives on Psychological Science* 12 (September 2018): 567–97.

while listening to music they prefer: Perham and Sykora, "Disliked Music Can Be Better for Performance Than Liked Music."

prioritizes the processing of faces: Romina Palermo and Gillian Rhodes, "Are You Always on My Mind? A Review of How Face Perception and Attention Interact," *Neuropsychologia* 45 (2007): 75–92.

the gaze of other people: Takemasa Yokoyama et al., "Attentional Capture by Change in Direct Gaze," *Perception* 40 (July 2011): 785–97. See also Atsushi Senju and Mark H. Johnson, "The Eye Contact Effect: Mechanisms and Development," *Trends in Cognitive Sciences* 13 (March 2009): 127–34.

the processing of eye contact: J. Jessica Wang and Ian A. Apperly, "Just One Look: Direct Gaze Briefly Disrupts Visual Working Memory," *Psychonomic Bulletin & Review* 24 (2017): 393–99. See also Laurence Conty et al., "The Cost of Being Watched: Stroop Interference Increases Under Concomitant Eye Contact," *Cognition* 115 (April 2010): 133–39.

increases our physiological arousal: Laurence Conty et al., "The Mere Perception of

Eye Contact Increases Arousal During a Word-Spelling Task," *Social Neuroscience* 5 (2010): 171–86. See also Terhi M. Helminen, Suvi M. Kaasinen, and Jari K. Hietanen, "Eye Contact and Arousal: The Effects of Stimulus Duration," *Biological Psychology* 88 (September 2011): 124–30.

when we close our eyes: Annelies Vredeveldt and Timothy J. Perfect, "Reduction of Environmental Distraction to Facilitate Cognitive Performance," *Frontiers in Psychology* 5 (August 2014).

"helps people to disengage": Arthur M. Glenberg, Jennifer L. Schroeder, and David A. Robertson, "Averting the Gaze Disengages the Environment and Facilitates Remembering," *Memory & Cognition* 26 (July 1998): 651–58.

experience less cognitive load: Annelies Vredeveldt, Graham J. Hitch, and Alan D. Baddeley, "Eyeclosure Helps Memory by Reducing Cognitive Load and Enhancing Visualisation," *Memory & Cognition* 39 (April 2011): 1253–63.

retrieve elusive information: Rémi Radel and Marion Fournier, "The Influence of External Stimulation in Missing Knowledge Retrieval," *Memory* 25 (October 2017): 1217–24.

124 *both visual and auditory:* Timothy J. Perfect et al., "How Can We Help Witnesses to Remember More? It's an (Eyes) Open and Shut Case," *Law and Human Behavior* 32 (August 2008): 314–24.

a 23 percent increase: Robert A. Nash et al., "Does Rapport-Building Boost the Eyewitness Eyeclosure Effect in Closed Questioning?," *Legal and Criminological Psychology* 21 (September 2016): 305–18.

"sensory reduction": Radel and Fournier, "The Influence of External Stimulation in Missing Knowledge Retrieval."

"Good fences make good neighbors": Robert Frost, "Mending Wall," in *Robert Frost: Collected Poems, Prose, and Plays* (New York: Library of America, 1995), 39–40.

putting oneself on display: Kathleen D. Vohs, Roy F. Baumeister, and Natalie J. Ciarocco, "Self-Regulation and Self-Presentation: Regulatory Resource Depletion Impairs Impression Management and Effortful Self-Presentation Depletes Regulatory Resources," *Journal of Personality and Social Psychology* 88 (April 2005): 632–57.

particularly draining for members: Alison Hirst and Christina Schwabenland, "Doing Gender in the 'New Office,'" *Gender, Work & Organization* 25 (March 2018): 159–76.

relieved of the cognitive load: Shira Baror and Moshe Bar, "Associative Activation and Its Relation to Exploration and Exploitation in the Brain," *Psychological Science* 27 (June 2016): 776–89.

"statistically common": Moshe Bar, "Think Less, Think Better," *New York Times*, June 17, 2016.

125 *less likely to try new approaches:* Ethan Bernstein, "The Transparency Trap," *Harvard Business Review*, October 2014. See also Julian Birkinshaw and Dan Cable, "The Dark Side of Transparency," *McKinsey Quarterly*, February 2017.

a mobile phone factory in China: Ethan S. Bernstein, "The Transparency Paradox: A Role for Privacy in Organizational Learning and Operational Control," *Administrative Science Quarterly* 57 (June 2012): 181–216.

privacy leads to feelings of empowerment: Darhl M. Pedersen, "Psychological Functions of Privacy," *Journal of Environmental Psychology* 17 (June 1997): 147–56.

fewer and more superficial work-related conversations: Anne-Laure Fayard and John Weeks, "Who Moved My Cube?," *Harvard Business Review*, July–August 2011.

interactions among employees actually decrease: Ben Waber, "Do Open Offices Really Increase Collaboration?," *Quartz*, April 13, 2018.

trust and cooperativeness among co-workers: Rachel L. Morrison and Keith A. Macky, "The Demands and Resources Arising from Shared Office Spaces," *Applied Ergonomics* 60 (April 2017): 103–15.

exactly the kind of behavior: Rachel Morrison, "Open-Plan Offices Might Be Making Us Less Social and Productive, Not More," *Quartz*, September 19, 2016.

126 *"hot-desking" or "hoteling":* Alison Hirst, "Settlers, Vagrants and Mutual Indifference: Unintended Consequences of Hot-Desking," *Journal of Organizational Change* 24 (2011): 767–88. See also Alison Hirst, "How Hot-Deskers Are Made to Feel Like the Homeless People of the Office World," *The Conversation*, February 13, 2017.

"home advantage": Jeremy P. Jamieson, "The Home Field Advantage in Athletics: A Meta-Analysis," *Journal of Applied Social Psychology* 40 (July 2010): 1819–48. See also Mark S. Allen and Marc V. Jones, "The 'Home Advantage' in Athletic Competitions," *Current Directions in Psychological Science* 23 (February 2014): 48–53.

play more aggressively: Kerry S. Courneya and Albert V. Carron, "The Home Advantage in Sport Competitions: A Literature Review," *Journal of Sport and Exercise Psychology* 14 (March 1992): 13–27.

higher levels of testosterone: Nick Neave and Sandy Wolfson, "Testosterone, Territoriality, and the 'Home Advantage,'" *Physiology & Behavior* 78 (February 2003): 269–75.

more confident and capable: Lorraine E. Maxwell and Emily J. Chmielewski, "Environmental Personalization and Elementary School Children's Self-Esteem," *Journal of Environmental Psychology* 28 (June 2008): 143–53. See also Clare Ulrich, "A Place of Their Own: Children and the Physical Environment," *Human Ecology* 32 (October 2004): 11–14.

more efficient and productive: David C. Glass and Jerome E. Singer, "Experimental Studies of Uncontrollable and Unpredictable Noise," *Representative Research in Social Psychology* 4 (1973): 165–83. See also David C. Glass, Jerome E. Singer, and Lucy N. Friedman, "Psychic Cost of Adaptation to an Environmental Stressor," *Journal of Personality and Social Psychology* 12 (1969): 200–210.

more focused and less distractible: So Young Lee and Jay L. Brand, "Can Personal Control over the Physical Environment Ease Distractions in Office Workplaces?," *Ergonomics* 53 (March 2010): 324–35. See also So Young Lee and Jay L. Brand, "Effects of Control over Office Workspace on Perceptions of the Work Environment and Work Outcomes," *Journal of Environmental Psychology* 25 (September 2005): 323–33.

advance their own interests: Graham Brown, "Setting (and Choosing) the Table: The Influence of the Physical Environment in Negotiation," in *Negotiation Excellence: Successful Deal Making*, ed. Michael Benoliel (Singapore: World Scientific Publishing, 2015), 23–37.

60 and 160 percent more value: Graham Brown and Markus Baer, "Location in Negotiation: Is There a Home Field Advantage?," *Organizational Behavior and Human Decision Processes* 114 (March 2011): 190–200.

operate more efficiently: Meagher, "Ecologizing Social Psychology."

"our cognition is distributed": Benjamin R. Meagher, "The Emergence of Home Advantage from Differential Perceptual Activity" (PhD diss., University of Connecticut, July 2014).

127 *volunteers were given a set of tasks:* Craig Knight and S. Alexander Haslam, "The Rela-

tive Merits of Lean, Enriched, and Empowered Offices: An Experimental Examination of the Impact of Workspace Management Strategies on Well-Being and Productivity," *Journal of Experimental Psychology: Applied* 16 (2010): 158–72.

"I wanted to hit you": Study participant, quoted in Tim Harford, *Messy: The Power of Disorder to Transform Our Lives* (New York: Riverhead Books 2016), 66.

"intermittent collaboration": Ethan Bernstein, Jesse Shore, and David Lazer, "How Intermittent Breaks in Interaction Improve Collective Intelligence," *Proceedings of the National Academy of Sciences* 115 (August 2018): 8734–39. See also Ethan Bernstein, Jesse Shore, and David Lazer, "Improving the Rhythm of Your Collaboration," *MIT Sloan Management Review,* September 2019.

128 *"serves as the key element of connection":* Richard D. G. Irvine, "The Architecture of Stability: Monasteries and the Importance of Place in a World of Non-Places," *Etnofoor* 23 (2011): 29–49.

129 *a swaggering motto:* Una Roman D'Elia, *Raphael's Ostrich* (University Park: Pennsylvania State University Press, 2015).

130 *people perform better:* Katharine H. Greenaway et al., "Spaces That Signal Identity Improve Workplace Productivity," *Journal of Personnel Psychology* 15 (2016): 35–43.

one-third of the personal tokens: Kris Byron and Gregory A. Laurence, "Diplomas, Photos, and *Tchotchkes* as Symbolic Self-Representations: Understanding Employees' Individual Use of Symbols," *Academy of Management Journal* 58 (February 2015): 298–323.

it is in fact quite fluid: Daphna Oysterman, Kristen Elmore, and George Smith, "Self, Self-Concept, and Identity," in *Handbook of Self and Identity,* ed. Mark R. Leary and June Price Tangney (New York: Guilford Press, 2012), 69–104.

dependent on external structure: Brandi Pearce et al., "What Happened to My Office? The Role of Place Identity at Work," *Academy of Management Annual Meeting Proceedings* (November 2017). See also Russell W. Belk, "Possessions and the Extended Self," *Journal of Consumer Research* 15 (September 1988): 139–68.

"they tell us things about ourselves": Mihaly Csikszentmihalyi, "Why We Need Things," in *History from Things: Essays on Material Culture,* ed. Steven Lubar and W. David Kingery (Washington, DC: Smithsonian Institution, 1993), 20–29. See also Mihaly Csikszentmihalyi, "The Symbolic Function of Possessions: Towards a Psychology of Materialism," paper presented at the 90th Annual Convention of the American Psychological Association, August 1982.

131 *"Which identity is salient in the moment":* Daphna Oyserman and Mesmin Destin, "Identity-Based Motivation: Implications for Intervention," *The Counseling Psychologist* 38 (October 2010): 1001–43.

cues that remind Asian-American girls: Nalini Ambady et al., "Stereotype Susceptibility in Children: Effects of Identity Activation on Quantitative Performance," *Psychological Science* 12 (September 2001): 385–90.

"environmental self-regulation": Korpela, "Place-Identity as a Product of Environmental Self-Regulation."

"emotional exhaustion": Gregory A. Laurence, Yitzhak Fried, and Linda H. Slowik, "'My Space': A Moderated Mediation Model of the Effect of Architectural and Experienced Privacy," *Journal of Environmental Psychology* 36 (December 2013): 144–52.

an ethnographic study: Ryoko Imai and Masahide Ban, "Disrupting Workspace: Designing an Office That Inspires Collaboration and Innovation," *Ethnographic Praxis in*

Industry Conference, Case Studies 1: Pathmaking with Ethnographic Approaches in and for Organizations (November 2016): 444–64.

132 *discourage or even ban*: Krystal D'Costa, "Resisting the Depersonalization of the Work Space," *Scientific American*, July 18, 2018. See also Alex Haslam and Craig Knight, "Your Place or Mine?," BBC News, November 17, 2006.

bring a different figure to mind: Knight and Haslam, "The Relative Merits of Lean, Enriched, and Empowered Offices."

"Making people feel": Alexander Haslam, quoted in Richard Webb, "Sit and Arrange," *Ahmedabad Mirror*, February 5, 2019.

see themselves reflected in it: Craig Knight and S. Alexander Haslam, "Your Place or Mine? Organizational Identification and Comfort as Mediators of Relationships Between the Managerial Control of Workspace and Employees' Satisfaction and Well-Being," *British Journal of Management* 21 (September 2010): 717–35.

especially attuned to the signals: Valerie Purdie-Vaughns et al., "Social Identity Contingencies: How Diversity Cues Signal Threat or Safety for African Americans in Mainstream Institutions," *Journal of Personality and Social Psychology* 94 (April 2008): 615–30.

"When we think of prejudice": Mary C. Murphy, Kathryn M. Kroeper, and Elise M. Ozier, "Prejudiced Places: How Contexts Shape Inequality and How Policy Can Change Them," *Policy Insights from the Behavioral and Brain Sciences* 5 (March 2018): 66–74.

133 *"prejudiced places"*: Mary C. Murphy and Gregory M. Walton, "From Prejudiced People to Prejudiced Places: A Social-Contextual Approach to Prejudice," in *Stereotyping and Prejudice (Frontiers of Social Psychology)*, ed. Charles Stangor and Christian S. Crandall (New York: Psychology Press, 2013), 181–203.

unimpressive results: Frank Dobbin and Alexandra Kalev, "Why Doesn't Diversity Training Work? The Challenge for Industry and Academia," *Anthropology Now* 10 (September 2018): 48–55.

the environments Sapna Cheryan encountered: My discussion of Sapna Cheryan's personal experiences with a sense of belonging draw on the following sources: Sapna Cheryan, "Redesigning Environments Increases Girls' Interest in Computer Science," video presented as part of the STEM for All Video Showcase: Transforming the Educational Landscape, May 2018, https://stemforall2018.videohall.com/presentations/1198; Sapna Cheryan, "Stereotypes as Gatekeepers," talk given at TEDxSeattle, Apr 27, 2010, https://www.youtube.com/watch?v=TYwI-qM2oX4; Sapna Cheryan, interviewed by Manola Secaira, "This Gender and Race Researcher Explains Why Techies Don't Have to Be Trekkies," ;May 3, 2019, https://crosscut.com/2019/05/gender-and-race-researcher-explains-why-techies-dont-have-be-trekkies. Lisa Grossman, "Of Geeks and Girls," *Science Notes* (2009).

"the action figures": Cheryan, "Stereotypes as Gatekeepers."

134 *Cheryan commandeered space*: Sapna Cheryan et al., "Ambient Belonging: How Stereotypical Cues Impact Gender Participation in Computer Science," *Journal of Personality and Social Psychology* 97 (December 2009): 1045–60.

more likely to predict: Sapna Cheryan, Andrew N. Meltzoff, and Saenam Kim, "Classrooms Matter: The Design of Virtual Classrooms Influences Gender Disparities in Computer Science Classes," *Computers & Education* 57 (September 2011): 1825–35.

"we know from past work": Cheryan, "Stereotypes as Gatekeepers."

not to eliminate *stereotypes:* Sapna Cheryan, Allison Master, and Andrew N. Melt-zoff, "Cultural Stereotypes as Gatekeepers: Increasing Girls' Interest in Computer Science and Engineering by Diversifying Stereotypes," *Frontiers in Psychology* 6 (February 2015).

135 *reached 32 percent:* Sapna Cheryan, "A New Study Shows How Star Trek Jokes and Geek Culture Make Women Feel Unwelcome in Computer Science," *Quartz,* October 31, 2016.

especially attuned to cues of exclusion: Rene F. Kizilcec et al., "Welcome to the Course: Early Social Cues Influence Women's Persistence in Computer Science," paper presented at the ACM Conference on Human Factors in Computing Systems, April 2020.

signals of non-belonging: Cheryan, Meltzoff, and Kim, "Classrooms Matter." See also Danaë Metaxa-Kakavouli et al., "Gender-Inclusive Design: Sense of Belonging and Bias in Web Interfaces," paper presented at the CHI Conference on Human Factors in Computing Systems, April 2018.

how actively they engage: Christopher Brooks, Josh Gardner, and Kaifeng Chen, "How Gender Cues in Educational Video Impact Participation and Retention," paper presented at the 13th International Conference of the Learning Sciences, June 2018.

"gender-inclusive": René F. Kizilcec and Andrew J. Saltarelli, "Psychologically Inclusive Design: Cues Impact Women's Participation in STEM Education," paper presented at the CHI Conference on Human Factors in Computing Systems, May 2019.

lower socioeconomic status: René F. Kizilcec and Andrew J. Saltarelli, "Can a Diversity Statement Increase Diversity in MOOCs?," paper presented at the 6th ACM Conference on Learning at Scale, June 2019.

"strategically placing content": Kizilcec and Saltarelli, "Psychologically Inclusive Design."

136 *"You say to brick":* Louis Kahn, quoted in Lesser, *You Say to Brick,* 5.

"the vision of an artist": Jonas Salk, quoted in Lesser, *You Say to Brick,* 33.

the way our brains respond: Amber Dance, "Science and Culture: The Brain Within Buildings," *Proceedings of the National Academy of Sciences* 114 (January 2017): 785–87. See also Heeyoung Choo et al., "Neural Codes of Seeing Architectural Styles," *Scientific Reports* 7 (January 2017).

high-ceilinged places: Joan Meyers-Levy and Rui (Juliet) Zhu, "The Influence of Ceiling Height: The Effect of Priming on the Type of Processing That People Use," *Journal of Consumer Research* 34 (August 2007): 174–86.

symmetrical shapes: Sarah Williams Goldhagen, *Welcome to Your World: How the Built Environment Shapes Our Lives* (New York: HarperCollins, 2017). See also Ann Sussman and Justin B. Hollander, *Cognitive Architecture: Designing for How We Respond to the Built Environment* (New York: Routledge, 2015).

architectural designs that resemble faces: Sussman and Hollander, *Cognitive Architecture.* See also Ann Sussman, "Why Brain Architecture Matters for Built Architecture," *Metropolis,* August 19, 2015.

Curved shapes: Oshin Vartanian et al., "Impact of Contour on Aesthetic Judgments and Approach-Avoidance Decisions in Architecture," *Proceedings of the National Academy of Sciences* 110 (June 2013): 10446–53. See also Letizia Palumbo, Nicole Ruta, and Marco Bertamini, "Comparing Angular and Curved Shapes in Terms of Implicit Associations and Approach/Avoidance Responses," *PLoS One* 10 (October 2015).

babies as young as one week old: Robert L. Fantz and Simon B. Miranda, "Newborn Infant Attention to Form of Contour," *Child Development* 46 (1975): 224–28.

even across species: Enric Munar et al., "Common Visual Preference for Curved Contours in Humans and Great Apes," *PLoS One* 10 (November 2015).

the brains of rhesus monkeys: Research cited in Michael Anft, "This Is Your Brain on Art," *Johns Hopkins Magazine,* March 6, 2010.

137 *Philip Johnson first saw it:* Matt Tyrnauer, "Architecture in the Age of Gehry," *Vanity Fair,* August 2010.

"cognitive deficiency of non-places": Richard Coyne, "Thinking Through Virtual Reality: Place, Non-Place and Situated Cognition," *Technè: Research in Philosophy and Technology* 10 (Spring 2007): 26–38. See also Marc Augé, *Non-Places: Introduction to an Anthropology of Supermodernity,* trans. John Howe (London: Verso, 1995).

"rich layering of custom": Coyne, "Thinking Through Virtual Reality."

138 *"If you look at the Baths of Caracalla":* Louis Kahn, quoted in "Marin City Redevelopment," *Progressive Architecture* 41 (November 1960).

6. THINKING WITH THE SPACE OF IDEAS

139 *He was prominently featured:* Joshua Foer, *Moonwalking with Einstein: The Art and Science of Remembering Everything* (New York: Penguin Press, 2011).

his "lucky hat": Sarah Chalmers, "Mindboggling — Meet the Forgetful Memory Champ Who's Battling for Britain," *Daily Mail,* August 30, 2007.

"I am famously bad": Ben Pridmore, quoted in Adam Lusher, "World Memory Champion Reveals His Secrets," *Sunday Telegraph,* August 5, 2007.

"method of loci": Stephen Robb, "How a Memory Champ's Brain Works," BBC News, April 7, 2009.

a venerable technique: Kurt Danziger, *Marking the Mind: A History of Memory* (New York: Cambridge University Press, 2008).

140 *"Using neuropsychological measures":* Eleanor A. Maguire et al., "Routes to Remembering: The Brains Behind Superior Memory," *Nature Neuroscience* 6 (January 2003): 90–95.

the brain's built-in navigational system: Russell A. Epstein et al., "The Cognitive Map in Humans: Spatial Navigation and Beyond," *Nature Neuroscience* 20 (October 2017): 1504–13. See also Alexandra O. Constantinescu, Jill X. O'Reilly, and Timothy E. J. Behrens, "Organizing Conceptual Knowledge in Humans with a Gridlike Code," *Science* 352 (June 2016): 1464–68.

reflected in the language: George Lakoff and Mark Johnson, "Conceptual Metaphor in Everyday Language," *Journal of Philosophy* 77 (August 1980): 453–86. See also George Lakoff and Mark Johnson, *Metaphors We Live By* (Chicago: University of Chicago Press, 1980).

141 *"We are far better":* Barbara Tversky, *Mind in Motion: How Action Shapes Thought* (New York: Basic Books, 2019), 57.

maps abstract spaces: Arne D. Ekstrom and Charan Ranganath, "Space, Time, and Episodic Memory: The Hippocampus Is All Over the Cognitive Map," *Hippocampus* 28 (September 2018): 680–87. See also Mona M. Garvert, Raymond J. Dolan, and Timothy E. J. Behrens, "A Map of Abstract Relational Knowledge in the Human Hippocampal-Entorhinal Cortex," *eLife* 6 (April 2017).

scanned the brains: Branka Milivojevic et al., "Coding of Event Nodes and Narrative Context in the Hippocampus," *Journal of Neuroscience* 36 (December 2016): 12412–24. See also Vishnu Sreekumar, "Hippocampal Activity Patterns Reflect the Topology of Spaces: Evidence from Narrative Coding," *Journal of Neuroscience* 37 (June 2017): 5975–77.

"infantile amnesia": Arthur M. Glenberg and Justin Hayes, "Contribution of Embodiment to Solving the Riddle of Infantile Amnesia," *Frontiers in Psychology* 7 (January 2016).

142 *continue to be tagged:* Jessica Robin, Jordana Wynn, and Morris Moscovitch, "The Spatial Scaffold: The Effects of Spatial Context on Memory for Events," *Journal of Experimental Psychology: Learning, Memory, and Cognition* 42 (February 2016): 308–15. See also Jonathan F. Miller et al., "Neural Activity in Human Hippocampal Formation Reveals the Spatial Context of Retrieved Memories," *Science* 342 (November 2013): 1111–14.

The automatic place log: Timothy P. McNamara and Christine M. Valiquette, "Remembering Where Things Are," in *Human Spatial Memory: Remembering Where,* ed. Gary L. Allen (Mahwah, NJ: Lawrence Erlbaum Associates, 2004), 3–24. See also Andrea N. Suarez et al., "Gut Vagal Sensory Signaling Regulates Hippocampus Function Through Multi-Order Pathways," *Nature Communications* 9 (June 2018): 1–15.

the world's top memory competitors: Martin Dresler et al., "Mnemonic Training Reshapes Brain Networks to Support Superior Memory," *Neuron* 93 (March 2017): 1227–35.

course on civil liberties: Charles (Trey) Wilson, "The Chow Hall as Mind Palace," post on Teaching Academic: A CTLL Blog, August 29, 2018, https://blog.ung.edu/ctll/the-chow-hall-as-mind-palace-2/.

144 *the work of historian Robert Caro:* My discussion of Robert Caro and his working methods draws on the following sources: James Santel, "Robert Caro, the Art of Biography," *Paris Review,* Spring 2016; Scott Porch, "'The Power Broker' Turns 40: How Robert Caro Wrote a Masterpiece," *Daily Beast,* September 16, 2014; Charles McGrath, "Robert Caro's Big Dig," *New York Times Magazine,* April 12, 2012; Chris Jones, "The Big Book," *Esquire,* April 12, 2012; Scott Sherman, "Caro's Way," *Columbia Journalism Review,* May–June 2002; Stephen Harrigan, "The Man Who Never Stops," *Texas Monthly,* April 1990; William Goldstein, "Robert Caro Talks About His Art, His Methods and LBJ," *Publishers Weekly,* November 25, 1983.

"It was so big": Robert Caro, quoted in Jones, "The Big Book."

"I can't start writing a book": Robert Caro, quoted in Santel, "Robert Caro, the Art of Biography."

145 *"I don't want to stop":* Robert Caro, quoted in Goldstein, "Robert Caro Talks About His Art, His Methods and LBJ."

"When thought overwhelms the mind": Barbara Tversky and Angela Kessell, "Thinking in Action," *Pragmatics & Cognition* 22 (January 2014): 206–33.

"the cognitive congeniality of a space": David Kirsh, "Adapting the Environment Instead of Oneself," *Adaptive Behavior* 4 (1996): 415–52.

offload facts and ideas: Evan F. Risko and Sam J. Gilbert, "Cognitive Offloading," *Trends in Cognitive Sciences* 20 (September 2016): P676–88. See also Andy Clark, "Minds in Space," in *The Spatial Foundations of Language and Cognition,* ed. Kelly S. Mix, Linda B. Smith, and Michael Gasser (New York: Oxford University Press, 2010), 7–15.

146 *investigating the way children learn science:* Joseph D. Novak, "A Search to Create a Science of Education: The Life of an Ivy League Professor, Business Consultant, and Research Scientist," self-published autobiography, https://www.ihmc.us/files/JNovak-ASearchToCreateAScienceOfEducation.pdf.

"the knowledge structure necessary to understand": Joseph D. Novak, *Learning, Creating, and Using Knowledge: Concept Maps as Facilitative Tools in Schools and Corporations* (Mahwah, NJ: Lawrence Erlbaum Associates, 1998), 177.

147 *"The most successful cinematic technologies"*: David A. Kirby, *Lab Coats in Hollywood: Science, Scientists, and Cinema* (Cambridge: MIT Press, 2010), 200.

Spielberg invited computer scientists: John Underkoffler, "Pointing to the Future of UI," TED Talk, February 2010, https://www.ted.com/talks/john_underkoffler_pointing _to_the_future_of_ui?language=en.

"Is that real?": John Underkoffler, quoted in Kirby, *Lab Coats in Hollywood,* 201.

148 *increase by more than tenfold:* Robert Ball and Chris North, "Realizing Embodied Interaction for Visual Analytics Through Large Displays," *Computers & Graphics* 31 (June 2007): 380–400.

When using a large display: Robert Ball and Chris North, "The Effects of Peripheral Vision and Physical Navigation on Large Scale Visualization," paper presented at the 34th Annual Graphics Interface Conference, May 2008. See also Khairi Reda et al., "Effects of Display Size and Resolution on User Behavior and Insight Acquisition in Visual Exploration," *Proceedings of the 33rd Annual ACM Conference on Human Factors in Computing Systems* (April 2015): 2759–68.

everyone *who engages with the larger display:* Robert Ball, "Three Ways Larger Monitors Can Improve Productivity," *Graziadio Business Review* 13 (January 2010): 1–5.

"physical embodied resources": Ball and North, "Realizing Embodied Interaction for Visual Analytics Through Large Displays." See also Alex Endert et al., "Visual Encodings That Support Physical Navigation on Large Displays," *Proceedings of Graphics Interface 2011* (May 2011): 103–10.

enables us to gather more knowledge: Ball and North, "The Effects of Peripheral Vision and Physical Navigation on Large Scale Visualization." See also Reda et al., "Effects of Display Size and Resolution on User Behavior and Insight Acquisition in Visual Exploration."

allows us to be more efficient: Ball and North, "Realizing Embodied Interaction for Visual Analytics Through Large Displays."

encourage a narrower visual focus: Ball and North, "The Effects of Peripheral Vision and Physical Navigation on Large Scale Visualization."

our spatial memory: Yvonne Jansen, Jonas Schjerlund, and Kasper Hornbæk, "Effects of Locomotion and Visual Overview on Spatial Memory When Interacting with Wall Displays," *Proceedings of the 2019 CHI Conference on Human Factors in Computing Systems* (May 2019): 1–12. See also Joey Scarr, Andy Cockburn, and Carl Gutwin, "Supporting and Exploiting Spatial Memory in User Interfaces," *Foundations and Trends in Human-Computer Interaction,* December 2013.

149 *56 percent more information:* Desney S. Tan et al., "The Infocockpit: Providing Location and Place to Aid Human Memory," *Proceedings of the 2001 Workshop on Perceptive User Interfaces* (November 2001): 1–4.

Other embodied resources: Ball and North, "Realizing Embodied Interaction for Visual Analytics Through Large Displays." See also Endert et al., "Visual Encodings That Support Physical Navigation on Large Displays."

the mental version of our map: Carl Gutwin and Andy Cockburn, "A Field Experiment of Spatially-Stable Overviews for Document Navigation," *Proceedings of the 2017 CHI Conference on Human Factors in Computing Systems* (May 2017): 5905–16.

more intuitive physical *navigation*: Mikkel R. Jakobsen and Kasper Hornbæk, "Is Moving Improving? Some Effects of Locomotion in Wall-Display Interaction," *Proceedings of the 33rd Annual ACM Conference on Human Factors in Computing Systems* (April 2015): 4169–78. See also Roman Rädle et al., "The Effect of Egocentric Body Movements on Users' Navigation Performance and Spatial Memory in Zoomable User Interfaces," *Proceedings of the 2013 ACM International Conference on Interactive Tabletops and Surfaces* (October 2013): 23–32.

as display size increases: Ball and North, "Realizing Embodied Interaction for Visual Analytics Through Large Displays."

150   *90 percent less "window management"*: Ball, "Three Ways Larger Monitors Can Improve Productivity."

"will most likely be more productive": Ball, "Three Ways Larger Monitors Can Improve Productivity."

expedition aboard the HMS Beagle: My discussion of Charles Darwin and his field notebooks draws on the following sources: Charles Darwin, *The Beagle Record: Selections from the Original Pictorial Records and Written Accounts of the Voyage of HMS Beagle*, ed. Richard Darwin Keynes (New York: Cambridge University Press, 1979); Charles Darwin, *The Complete Work of Charles Darwin Online*, ed. John van Wyhe, http://darwin-online.org.uk; E. Janet Browne, *Charles Darwin: Voyaging* (Princeton: Princeton University Press, 1995); John Gribbin and Mary Gribbin, *FitzRoy: The Remarkable Story of Darwin's Captain and the Invention of the Weather Forecast* (New Haven: Yale University Press, 2004); National Archives, *Tales from the Captain's Log* (London: Bloomsbury, 2017).

151   *"In keeping such copious records"*: Browne, *Charles Darwin*, 194.

152   *"I must confess however"*: Charles Darwin, quoted in Gordon Chancellor, "Darwin's Geological Diary from the Voyage of the Beagle," post on Darwin Online, http://darwin-online.org.uk/EditorialIntroductions/Chancellor_GeologicalDiary.html.

"acquire the habit of writing": Chancellor, "Darwin's Geological Diary from the Voyage of the Beagle."

throughout his long career: Erick Greene, "Why Keep a Field Notebook?," in *Field Notes on Science and Nature*, ed. Michael Canfield (Cambridge: Harvard University Press, 2011), 251–76.

153   *what psychologists call their "affordances"*: James J. Gibson, *The Senses Considered as Perceptual Systems* (New York: Houghton Mifflin, 1966), 285.

a classic thought experiment: Daniel C. Dennett, *Content and Consciousness* (London: Routledge, 1969), 154.

becomes maddeningly slippery: Daniel Reisberg, "External Representations and the Advantages of Externalizing One's Thoughts," paper presented at the 9th Annual Conference of the Cognitive Science Society, July 1987.

154   *"detachment gain"*: Daniel Reisberg, "The Detachment Gain: The Advantage of Thinking Out Loud," in *Perception, Cognition, and Language: Essays in Honor of Henry and Lila Gleitman*, ed. Barbara Landau et al. (Cambridge: MIT Press, 2000), 139–56.

experimenters asked eighth-grade students: Eliza Bobek and Barbara Tversky, "Creating Visual Explanations Improves Learning," *Cognitive Research: Principles and Implications* 1 (December 2016).

"check for completeness": Bobek and Tversky, "Creating Visual Explanations Improves

Learning." See also Barbara Tversky, "Some Ways of Thinking," in *Model-Based Reasoning in Science and Technology: Theoretical and Cognitive Issues,* ed. Lorenzo Magnani (New York: Springer, 2014), 3–8.

"there can be neither doubt": Reisberg, "External Representations and the Advantages of Externalizing One's Thoughts."

"discover" elements in their own work: Masaki Suwa, John Gero, and Terry Purcell, "Unexpected Discoveries and S-Invention of Design Requirements: Important Vehicles for a Design Process," *Design Studies* 21 (November 2000): 539–67. See also Masaki Suwa et al., "Seeing into Sketches: Regrouping Parts Encourages New Interpretations," in *Visual and Spatial Reasoning in Design II,* ed. John S. Gero, Barbara Tversky, and Terry Purcell (Sydney: Key Centre of Design Computing and Cognition, 2001), 207–19.

155 *"One reads off the sketch"*: Gabriela Goldschmidt, "On Visual Design Thinking: The Vis Kids of Architecture," *Design Studies* 15 (April 1994): 158–74.

a "conversation" carried on: Barbara Tversky, "Multiple Models: In the Mind and in the World," *Historical Social Research* 31 (2018): 59–65.

"the backtalk of self-generated sketches": Gabriela Goldschmidt, "The Backtalk of Self-Generated Sketches," *Design Issues* 19 (Winter 2003): 72–88.

expert architects are far more adept: Barbara Tversky and Masaki Suwa, "Thinking with Sketches," in *Tools for Innovation: The Science Behind the Practical Methods That Drive New Ideas,* ed. Arthur B. Markman and Kristin L. Wood (New York: Oxford University Press, 2009), 75–84. See also Barbara Tversky, "The Cognitive Design of Tools of Thought," *Review of Philosophy and Psychology* 6 (2015): 99–116.

some promising prescriptions: Andrea Kantrowitz, Michelle Fava, and Angela Brew, "Drawing Together Research and Pedagogy," *Art Education* 70 (May 2017): 50–60.

their skillful use of externalization: David Kirsh and Paul Maglio, "On Distinguishing Epistemic from Pragmatic Action," *Cognitive Science* 18 (October 1994): 513–49. See also Paul P. Maglio, Michael J. Wenger, and Angelina M. Copeland, "Evidence for the Role of Self-Priming in Epistemic Action: Expertise and the Effective Use of Memory," *Acta Psychologica* 127 (January 2008): 72–88.

"Better players use the world better": Paul P. Maglio and David Kirsh, "Epistemic Action Increases with Skill," paper presented at the 18th Annual Conference of the Cognitive Science Society, July 1996.

156 *"they are literally thinking"*: Daniel Smithwick and David Kirsh, "Let's Get Physical: Thinking with Things in Architectural Design," paper presented at the 37th Annual Conference of the Cognitive Science Society, July 2015. See also Daniel Smithwick, Larry Sass, and David Kirsh, "Creative Interaction with Blocks and Robots," paper presented at the 38th Annual Conference of the Cognitive Science Society, August 2016.

"enable forms of thought": Smithwick and Kirsh, "Let's Get Physical."

James Watson in low spirits: My discussion of James Watson and his discovery of the structure of DNA draws from the following sources: James D. Watson, *The Double Helix: A Personal Account of the Discovery of the Structure of DNA* (New York: Scribner, 1996); James Watson, "Discovering the Double Helix Structure of DNA," interview transcript posted on the website of Cold Spring Harbor Laboratory, https://www.cshl.edu/dnalcmedia/discovering-the-double-helix-structure-of-dna-james-watson

-video-with-3d-animation-and-narration/#transcript; Cavendish Laboratory Educational Outreach, "The Structure of DNA: Crick and Watson, 1953," post on CambridgePhysics.org, http://www.cambridgephysics.org/dna/dna_index.htm; US National Library of Medicine, "The Discovery of the Double Helix, 1951–1953," post on the website of the US National Library of Medicine, https://profiles.nlm.nih.gov/spotlight/sc/feature/doublehelix.

"not to waste any more time": Watson, *The Double Helix*, 192.

"in desperation": Watson, "Discovering the Double Helix Structure of DNA."

"When I got to our still empty office": Watson, *The Double Helix*, 194–96.

157 *"implies that simulating a situation"*: Gaëlle Vallée-Tourangeau and Frédéric Vallée-Tourangeau, "Why the Best Problem-Solvers Think with Their Hands, as Well as Their Heads," *The Conversation*, November 10, 2016.

follow a similar pattern: Frédéric Vallée-Tourangeau and Gaëlle Vallée-Tourangeau, "Diagrams, Jars, and Matchsticks: A Systemicist's Toolkit," in *Diagrammatic Reasoning*, ed. Riccardo Fusaroli and Kristian Tylén (Philadelphia: John Benjamins Publishing, 2014), 187–205.

"inevitably benefits performance": Vallée-Tourangeau and Vallée-Tourangeau, "Why the Best Problem-Solvers Think with Their Hands, as Well as Their Heads."

basic arithmetic: Lisa G. Guthrie and Frédéric Vallée-Tourangeau, "Interactivity and Mental Arithmetic: Coupling Mind and World Transforms and Enhances Performance," *Studies in Logic, Grammar and Rhetoric* 41 (2015): 41–59. See also Wendy Ross, Frédéric Vallée-Tourangeau, and Jo Van Herwegen, "Mental Arithmetic and Interactivity: The Effect of Manipulating External Number Representations on Older Children's Mental Arithmetic Success," *International Journal of Science and Mathematics Education* (June 2019).

complex reasoning: Gaëlle Vallée-Tourangeau, Marlène Abadie, and Frédéric Vallée-Tourangeau, "Interactivity Fosters Bayesian Reasoning Without Instruction," *Journal of Experimental Psychology: General* 144 (June 2015): 581–603.

planning for future events: Emma Henderson, Gaëlle Vallée-Tourangeau, and Frédéric Vallée-Tourangeau, "Planning in Action: Interactivity Improves Planning Performance," paper presented at the 39th Annual Conference of the Cognitive Science Society, July 2017.

solving creative "insight" problems: Anna Weller, Gaëlle Villejoubert, and Frédéric Vallée-Tourangeau, "Interactive Insight Problem Solving," *Thinking & Reasoning* 17 (2011): 424–39. See also Lisa G. Guthrie et al., "Learning and Interactivity in Solving a Transformation Problem," *Memory & Cognition* 43 (July 2015): 723–35.

158 *enjoy increased working memory:* Frédéric Vallée-Tourangeau, Miroslav Sirota, and Gaëlle Vallée-Tourangeau, "Interactivity Mitigates the Impact of Working Memory Depletion on Mental Arithmetic Performance," *Cognitive Research: Principles and Implications* 1 (December 2016).

they learn more: Mariana Lozada and Natalia Carro, "Embodied Action Improves Cognition in Children: Evidence from a Study Based on Piagetian Conservation Tasks," *Frontiers in Psychology* 7 (March 2016).

better able to transfer their learning: Guthrie et al., "Learning and Interactivity in Solving a Transformation Problem."

less likely to engage in "symbol pushing": Daniel L. Schwartz, Jessica M. Tsang, and

Kristen P. Blair, *The ABCs of How We Learn: 26 Scientifically Proven Approaches, How They Work, and When to Use Them* (New York: W. W. Norton, 2016), 86–101.

more motivated and engaged: Guthrie and Vallée-Tourangeau, "Interactivity and Mental Arithmetic: Coupling Mind and World Transforms and Enhances Performance."

experience less anxiety: Michael Allen and Frédéric Vallée-Tourangeau, "Interactivity Defuses the Impact of Mathematics Anxiety in Primary School Children," *International Journal of Science and Mathematics Education* 14 (July 2015): 1553–66.

"Moves in the World": Frédéric Vallée-Tourangeau, Lisa Guthrie, and Gaëlle Villejoubert, "Moves in the World Are Faster Than Moves in the Head: Interactivity in the River Crossing Problem," paper presented at the 35th Annual Conference of the Cognitive Science Society, July–August 2013.

"a record of the day-to-day work": My discussion of Richard Feynman's exchange with Charles Weiner is drawn from the account in James Gleick, *Genius: The Life and Science of Richard Feynman* (New York: Pantheon Books, 1992), 409.

"Feynman was actually thinking": Clark, *Supersizing the Mind,* xxv.

159 *"It is because we are so prone to think":* Clark, *Natural-Born Cyborgs,* 5.

"put brain, body, and world": Clark, *Supersizing the Mind,* 23.

7. THINKING WITH EXPERTS

163 *half a million young Germans:* John Ydstie, "Robust Apprenticeship Program Key to Germany's Manufacturing Might," *Morning Edition,* NPR, January 4, 2018.

course on theoretical computer science: Maria Knobelsdorf, Christoph Kreitz, and Sebastian Böhne, "Teaching Theoretical Computer Science Using a Cognitive Apprenticeship Approach," paper presented at the 45th ACM Technical Symposium on Computer Science Education, March 2014.

164 *cognitive apprenticeship:* Allan Collins, John Seely Brown, and Ann Holum, "Cognitive Apprenticeship: Making Thinking Visible," *American Educator,* Winter 1991.

165 *Inside the Hôpital Universitaire Pitié-Salpêtrière:* "Les Étudiants Miment les Maladies Neurologiques pour Mieux les Appréhender [Students Mimic Neurological Diseases to Better Understand Them]," video posted on the website of Assistance Hôpitaux Publique de Paris, https://www.aphp.fr/contenu/avec-allodocteurs-les-etudiants-miment-les-maladies-neurologiques-pour-mieux-les-apprehender.

"mime-based role-play training program": Emmanuel Roze et al., "Miming Neurological Syndromes Improves Medical Students' Long-Term Retention and Delayed Recall of Neurology," *Journal of the Neurological Sciences* 391 (August 2018): 143–48. See also Emmanuel Roze et al., "'The Move,' an Innovative Simulation-Based Medical Education Program Using Roleplay to Teach Neurological Semiology: Students' and Teachers' Perceptions," *Revue Neurologique* 172 (April–May 2016): 289–94.

had become concerned: Emmanuel Roze, interviewed in "'The Move' Program Presented in Dublin," video posted on the website of the Institut du Cerveau, December 19, 2016, https://icm-institute.org/en/actualite/the-move-program-presented-in-dublin/.

166 *recalled neurological signs and symptoms:* Roze et al., "Miming Neurological Syndromes Improves Medical Students' Long-Term Retention and Delayed Recall of Neurology."

deepened their understanding: Joshua A. Cuoco, "Medical Student Neurophobia: A

Review of the Current Pandemic and Proposed Educational Solutions," *European Journal of Educational Sciences* 3 (September 2016): 41–46.

intentionally imitating someone's accent: Patti Adank, Peter Hagoort, and Harold Bekkering, "Imitation Improves Language Comprehension," *Psychological Science* 21 (December 2010): 1903–09.

come to feel more positive: Patti Adank et al., "Accent Imitation Positively Affects Language Attitudes," *Frontiers in Psychology* 4 (May 2013).

167 *an art in its own right:* Edward P. J. Corbett, "The Theory and Practice of Imitation in Classical Rhetoric," *College Composition and Communication* 22 (October 1971): 243–50. See also James J. Murphy, "Roman Writing as Described by Quintilian," in *A Short History of Writing Instruction: From Ancient Greece to Contemporary America,* ed. James J. Murphy (New York: Routledge, 2012), 36–76.

students would begin: James J. Murphy, "The Modern Value of Ancient Roman Methods of Teaching Writing, with Answers to Twelve Current Fallacies," *Writing on the Edge* 1 (Fall 1989): 28–37. See also J. Scott Shields, "The Art of Imitation," *The English Journal* 96 (July 2007): 56–60.

"worthy of our study": The *Institutio Oratoria* of Quintilian, trans. H. E. Butler (New York: G. P. Putnam's Sons, 1922), 75.

168 *"Nature has fashioned man":* Juan Luis Vives, cited in Don Paul Abbott, "Reading, Writing, and Rhetoric in the Renaissance," in *A Short History of Writing Instruction: From Ancient Greece to Contemporary America,* ed. James J. Murphy (New York: Routledge, 2012), 148–71.

the Romantics arrived: Jessica Millen, "Romantic Creativity and the Ideal of Originality: A Contextual Analysis," *Cross-Sections* 6 (2010): 91–104. See also Elaine K. Gazda, *The Ancient Art of Emulation: Studies in Artistic Originality and Tradition from the Present to Classical Antiquity* (Ann Arbor: University of Michigan Press, 2002).

"the burden of the past": Walter Jackson Bate, *The Burden of the Past and the English Poet* (Cambridge: Harvard University Press, 2013).

a technique he had invented himself: Robert N. Essick and Joseph Viscomi, "An Inquiry into William Blake's Method of Color Printing," *Blake: An Illustrated Quarterly* 35 (Winter 2002): 74–103.

169 *"I must create a system":* William Blake, *Jerusalem: The Emanation of the Giant Albion* (Princeton: Princeton University Press, 1991), 144.

children, women, and "savages": Bennett Galef, "Breathing New Life into the Study of Imitation by Animals: What and When Do Chimpanzees Imitate?," in *Perspectives on Imitation: Mechanisms of Imitation and Imitation in Animals,* ed. Susan L. Hurley and Nick Chater (Cambridge: MIT Press, 2005), 295–96.

"Here's to the crazy ones": The commercial from Apple can be viewed at https://www.youtube.com/watch?v=tjgtLSHhTPg.

the "proudest moment": Kevin N. Laland, *Darwin's Unfinished Symphony: How Culture Made the Human Mind* (Princeton: Princeton University Press, 2017), 50.

170 *"We were expecting someone":* Luke Rendell, quoted in Sarah Boesveld, "Post-Grad Copycats Prove That Innovation Is Highly Overrated," *Globe and Mail,* April 16, 2010.

how valuable imitation can be: Luke Rendell et al., "Cognitive Culture: Theoretical and Empirical Insights into Social Learning Strategies," *Trends in Cognitive Sciences* 15 (February 2011): 68–76. See also R. Alexander Bentley, Mark Earls, and Michael J. O'Brien, *I'll Have What She's Having: Mapping Social Behavior* (Cambridge: MIT Press, 2011).

most efficient and effective route: Peter Duersch, Jörn Oechssler, and Burkhard C. Schipper, "Unbeatable Imitation," *Games and Economic Behavior* 76 (September 2012): 88–96.

did nothing but copy: Gerald S. Martin and John Puthenpurackal, "Imitation Is the Sincerest Form of Flattery: Warren Buffett and Berkshire Hathaway," paper posted on the Social Science Research Network, September 2005.

171 *"Zara is engaged in a permanent quest":* Enrique Badía, *Zara and Her Sisters: The Story of the World's Largest Clothing Retailer* (New York: Palgrave Macmillan, 2009), 23.

copies its own customers: Patricia Kowsmann, "Fast Fashion: How a Zara Coat Went from Design to Fifth Avenue in 25 Days," *Wall Street Journal,* December 6, 2016.

"depends on a constant exchange": Kasra Ferdows, Michael A. Lewis, and José A. D. Machuca, "Rapid-Fire Fulfillment," *Harvard Business Review,* November 2004.

172 *a historical analysis:* Gerard J. Tellis and Peter N. Golder, "First to Market, First to Fail? Real Causes of Enduring Market Leadership," *MIT Sloan Management Review,* January 15, 1996.

"honest signals": Alex (Sandy) Pentland, *Honest Signals: How They Shape Our World* (Cambridge: MIT Press, 2008).

competitive interactions among yachts: Jan-Michael Ross and Dmitry Sharapov, "When the Leader Follows: Avoiding Dethronement Through Imitation," *Academy of Management Journal* 58 (2015): 658–79.

sailors often engaged in "covering": Jan-Michael Ross, "The Highest Form of Flattery? What Imitation in the America's Cup Can Teach Business," posted on the website of Imperial College Business School, May 25, 2017, https://www.imperial.ac.uk/business-school/blogs/ib-knowledge/highest-form-flattery-what-imitation-americas-cup-can-teach-business/.

173 *the imitator's costs:* Oded Shenkar, *Copycats: How Smart Companies Use Imitation to Gain a Strategic Edge* (Cambridge: Harvard Business Press, 2010), 9.

the "correspondence problem": Christopher L. Nehaniv and Kerstin Dautenhahn, "The Correspondence Problem," in *Imitation in Animals and Artifacts,* ed. Kerstin Dautenhahn and Chrystopher L. Nehaniv (Cambridge: MIT Press, 2002), 41–62.

a landmark report: Institute of Medicine, *To Err Is Human: Building a Safer Health System,* ed. Linda T. Kohn, Janet M. Corrigan, and Molla S. Donaldson (Washington, DC: National Academies Press, 2000).

174 *the moments of highest risk:* Valerie E. Barnes and William P. Monan, "Cockpit Distractions: Precursors to Emergencies," *Proceedings of the Human Factors and Ergonomics Society Annual Meeting* 34 (October 1990): 1142–44.

the riskiest moments happen: Magdalena Z. Raban and Johanna I. Westbrook, "Are Interventions to Reduce Interruptions and Errors During Medication Administration Effective? A Systematic Review," *BMJ Quality & Safety* 23 (May 2014): 414–21.

distractions and interruptions of the pilot: William P. Monan, "Distraction — A Human Factor in Air Carrier Hazard Events," Technical Memorandum 78608 (Washington, DC: National Aeronautics and Space Administration, 1979).

interruptions of health care professionals: Kyle Anthony et al., "No Interruptions Please: Impact of a No Interruption Zone on Medication Safety in Intensive Care Units," *Critical Care Nurse* 30 (June 2010): 21–29.

one nurse, dispensing one medication: Lew McCreary, "Kaiser Permanente's Innovation on the Front Lines," *Harvard Business Review,* September 2010.

the *"sterile cockpit rule"*: Robert L. Sumwalt, "Accident and Incident Reports Show Importance of 'Sterile Cockpit' Compliance," *Flight Safety Digest* 13 (July 1994): 1–8.

her 2002 dissertation: Theresa M. Pape, "The Effect of Nurses' Use of a Focused Protocol to Reduce Distractions During Medication Administration" (PhD diss., Texas Woman's University, 2002).

"The key to preventing medication errors": Theresa M. Pape, "Applying Airline Safety Practices to Medication Administration," *MEDSURG Nursing* 12 (April 2003): 77–93.

175 *"virtual elimination of nurse distractions":* Agency for Healthcare Research and Quality, "AHRQ Health Care Innovations Exchange: Checklists with Medication Vest or Sash Reduce Distractions During Medication Administration" (2009).

a golden "age of imitation": Shenkar, *Copycats*, 41.

Shenkar would like to see: Oded Shenkar, "The Challenge of Imovation," *Ivey Business Journal*, March–April 2011.

led hospitals to imitate: Paul F. Nunes, Narendra P. Mulani, and Trevor J. Gruzin, "Leading by Imitation," *Accenture Outlook*, 2007, 1–9.

the onboard "checklist": Ellen Barlow, "A Simple Checklist That Saves Lives," *Harvard Public Health Magazine*, Fall 2008.

a nineteen-item checklist: Alex B. Haynes et al., "A Surgical Safety Checklist to Reduce Morbidity and Mortality in a Global Population," *New England Journal of Medicine* 360 (January 2009): 491–99.

"peer-to-peer assessment technique": Elizabeth Mort et al., "Improving Health Care Quality and Patient Safety Through Peer-to-Peer Assessment: Demonstration Project in Two Academic Medical Centers," *American Journal of Medical Quality* 32 (September 2017): 472–79. See also Peter J. Pronovost and Daniel W. Hudson, "Improving Healthcare Quality Through Organisational Peer-to-Peer Assessment: Lessons from the Nuclear Power Industry," *BMJ Quality & Safety* 21 (October 2012): 872–75.

176 *on a visit to Xerox PARC:* My discussion of Steve Jobs's imitation of innovations he first saw at Xerox PARC is drawn from Walter Isaacson, *Steve Jobs* (New York: Simon & Schuster, 2011), 92–101.

three steps: Oded Shenkar, "Just Imitate It! A Copycat Path to Strategic Agility," *Ivey Business Journal*, May–June 2012.

177 *our success as a* species: Andrew N. Meltzoff and Peter J. Marshall, "Human Infant Imitation as a Social Survival Circuit," *Current Opinion in Behavioral Sciences* 24 (December 2018): 130–36.

"one-shot imitation learning": Hyacinth Mascarenhas, "Elon Musk's OpenAI Is Teaching Robots How to Imitate Humans After Seeing Them Do a Task Once," *International Business Times*, May 17, 2017.

"are still far from being able": Alison Gopnik, "AIs Versus Four-Year-Olds," in *Possible Minds: Twenty-Five Ways of Looking at AI*, ed. John Brockman (New York: Penguin Press, 2019), 219–30.

their mimicry differs: Zanna Clay and Claudio Tenie, "Is Overimitation a Uniquely Human Phenomenon? Insights from Human Children as Compared to Bonobos," *Child Development* 89 (September–October 2018): 1535–44.

children are quite selective: Emily R. R. Burdett et al., "Do Children Copy an Expert or a Majority? Examining Selective Learning in Instrumental and Normative Contexts," *PLoS One* 11 (October 2016). See also Diane Poulin-Dubois, Ivy Brooker, and Alexandra Polonia, "Infants Prefer to Imitate a Reliable Person," *Infant Behavior and Development* 34 (April 2011): 303–9.

178 *copy their mothers:* Amanda J. Lucas, Emily R. R. Burdett et al., "The Development of Selective Copying: Children's Learning from an Expert Versus Their Mother," *Child Development* 88 (November 2017): 2026–42.

strikingly unselective: Francys Subiaul, "What's Special About Human Imitation? A Comparison with Enculturated Apes," *Behavioral Sciences* 6 (September 2016). See also Mark Nielsen and Cornelia Blank, "Imitation in Young Children: When Who Gets Copied Is More Important Than What Gets Copied," *Developmental Psychology* 47 (July 2011): 1050–53.

"high-fidelity" copiers: Francys Subiaul et al., "Becoming a High-Fidelity-Super-Imitator: What Are the Contributions of Social and Individual Learning?," *Developmental Science* 18 (November 2015): 1025–35.

faithfully imitate every step: Andrew Whiten et al., "Emulation, Imitation, Over-Imitation and the Scope of Culture for Child and Chimpanzee," *Philosophical Transactions of the Royal Society B: Biological Sciences* 364 (August 2009): 2417–28.

Humans' tendency to "overimitate": Clay and Tenie, "Is Overimitation a Uniquely Human Phenomenon?"

"cognitively opaque": Derek E. Lyons et al., "The Scope and Limits of Overimitation in the Transmission of Artefact Culture," *Philosophical Transactions of the Royal Society B: Biological Sciences* 366 (Aparil 2011): 1158–67.

the customs of one's culture: Joseph Henrich, "A Cultural Species: How Culture Drove Human Evolution," *Psychological Science Agenda,* November 2011. See also Mark Nielsen and Keyan Tomaselli, "Overimitation in Kalahari Bushman Children and the Origins of Human Cultural Cognition," *Psychological Science* 21 (May 2010): 729–36.

more likely to overimitate: Yue Yu and Tamar Kushnir, "Social Context Effects in 2- and 4-Year-Olds' Selective Versus Faithful Imitation," *Developmental Psychology* 50 (March 2014): 922–33.

continues to increase across development: Nicola McGuigan, Jenny Makinson, and Andrew Whiten, "From Over-Imitation to Super-Copying: Adults Imitate Causally Irrelevant Aspects of Tool Use with Higher Fidelity Than Young Children," *British Journal of Psychology* 102 (February 2011): 1–18. See also Andrew Whiten et al., "Social Learning in the Real-World: 'Over-Imitation' Occurs in Both Children and Adults Unaware of Participation in an Experiment and Independently of Social Interaction," *PLoS One* 11 (July 2016).

babies who were days or even hours old: Andrew N. Meltzoff and M. Keith Moore, "Newborn Infants Imitate Adult Facial Gestures," *Child Development* 54 (June 1983): 702–09.

179 *Mayan children from Guatemala:* Maricela Correa-Chávez and Barbara Rogoff, "Children's Attention to Interactions Directed to Others: Guatemalan Mayan and European American Patterns," *Developmental Psychology* 45 (May 2009): 630–41.

American children aren't granted: Alison Gopnik, "What's Wrong with the Teenage Mind?," *Wall Street Journal,* January 28, 2012.

towed around a rolling suitcase: Ron Berger, "Deeper Learning: Highlighting Student Work," *Edutopia,* January 3, 2013. See also Ron Berger, "Models of Excellence: What Do Standards Really Look Like?," post on the website of *Education Week,*

April 20, 2015, http://blogs.edweek.org/edweek/learning_deeply/2015/04/models_of
_excellence_what_do_standards_really_look_like.html.

a picture he calls "Austin's Butterfly": Ron Berger, featured in the video "Austin's Butterfly: Building Excellence in Student Work," posted on the website of EL Education, https://modelsofexcellence.eleducation.org/resources/austins-butterfly.

teachers of composition and rhetoric: Paul Butler, "Imitation as Freedom: (Re)Forming Student Writing," *The Quarterly* 24 (Spring 2002): 25–32. See also Donna Gorrell, "Freedom to Write — Through Imitation," *Journal of Basic Writing* 6 (Fall 1987): 53–59.

180 *who never relinquished:* Tom Pace, "Style and the Renaissance of Composition Studies," in *Refiguring Prose Style: Possibilities for Writing Pedagogy,* ed. T. R. Johnson and Tom Pace (Logan: Utah State University Press, 2005), 3–22.

"Imitate, that you may be different": Corbett, "The Theory and Practice of Imitation in Classical Rhetoric."

write within particular academic genres: Davida H. Charney and Richard A. Carlson, "Learning to Write in a Genre: What Student Writers Take from Model Texts," *Research in the Teaching of English* 29 (February 1995): 88–125.

"disciplinary writing": Fredricka L. Stoller et al., "Demystifying Disciplinary Writing: A Case Study in the Writing of Chemistry," *Across the Disciplines* 2 (2005).

"authentic texts": Stoller et al., "Demystifying Disciplinary Writing."

teaching legal writing: Terrill Pollman, "The Sincerest Form of Flattery: Examples and Model-Based Learning in the Classroom," *Journal of Legal Education* 64 (November 2014): 298–333.

181 *It was too much:* Terri L. Enns and Monte Smith, "Take a (Cognitive) Load Off: Creating Space to Allow First-Year Legal Writing Students to Focus on Analytical and Writing Processes," *Legal Writing: The Journal of the Legal Writing Institute* 20 (2015): 109–40.

skill has become "automatized": David F. Feldon, "The Implications of Research on Expertise for Curriculum and Pedagogy," *Educational Psychology Review* 19 (June 2006): 91–110.

expert trauma surgeons: Richard E. Clark et al., "The Use of Cognitive Task Analysis to Improve Instructional Descriptions of Procedures," *Journal of Surgical Research* 173 (March 2012): e37–42.

expert experimental psychologists: David F. Feldon, "Do Psychology Researchers Tell It Like It Is? A Microgenetic Analysis of Research Strategies and Self-Report Accuracy," *Instructional Science* 38 (July 2010): 395–415.

expert computer programmers: Chin-Jung Chao and Gavriel Salvendy, "Percentage of Procedural Knowledge Acquired as a Function of the Number of Experts from Whom Knowledge Is Acquired for Diagnosis, Debugging, and Interpretation Tasks," *International Journal of Human-Computer Interaction* 6 (July 1994): 221–33.

182 *"re-enactive empathy":* Karsten Stueber, *Rediscovering Empathy: Agency, Folk Psychology, and the Human Sciences* (Cambridge: MIT Press, 2006), 21.

stage such a reenactment: Ting Zhang, "Back to the Beginning: Rediscovering Inexperience Helps Experts Give Advice," *Academy of Management Proceedings* 2015 (November 2017). See also Carmen Nobel, "How to Break the Expert's Curse," *Harvard Business School Working Knowledge,* February 23, 2015.

experts habitually engage in "chunking": Fernand Gobet, "Chunking Models of Exper tise: Implications for Education," *Applied Cognitive Psychology* 19 (March 2005): 183–204.

math education expert John Mighton: My discussion of John Mighton and JUMP draws on the following sources: John Mighton, "If You Want to Make Math Appealing to Children, Think Like a Child," *New York Times,* December 10, 2013; John Mighton, interviewed by Ingrid Wickelgren, "Kids JUMP for Math," *Scientific American* podcast, August 7, 2013; John Mighton, *The End of Ignorance: Multiplying Our Human Potential* (Toronto: Alfred A. Knopf Canada, 2007); John Mighton, *The Myth of Ability: Nurturing Mathematical Talent in Every Child* (New York: Walker & Company, 2003); Jenny Anderson, "A Mathematician Has Created a Teaching Method That's Proving There's No Such Thing as a Bad Math Student," *Quartz,* February 15, 2017; David Bornstein, "A Better Way to Teach Math," *New York Times,* April 18, 2011; Sue Ferguson, "The Math Motivator," *Maclean's,* September 22, 2003.

183 *evaluation of the JUMP program:* Tracy Solomon et al., "A Cluster-Randomized Controlled Trial of the Effectiveness of the JUMP Math Program of Math Instruction for Improving Elementary Math Achievement," *PLoS One* 14 (October 2019).

digitally morphing the outlines: Itiel E. Dror, Sarah V. Stevenage, and Alan R. S. Ashworth, "Helping the Cognitive System Learn: Exaggerating Distinctiveness and Uniqueness," *Applied Cognitive Psychology* 22 (May 2008): 573–84.

"the caricature advantage": Gillian Rhodes, Susan Brennan, and Susan Carey, "Identification and Ratings of Caricatures: Implications for Mental Representations of Faces," *Cognitive Psychology* 19 (October 1987): 473–97.

184 *A classic experiment:* Michelene T. H. Chi, Paul J. Feltovich, and Robert Glaser, "Categorization and Representation of Physics Problems by Experts and Novices," *Cognitive Science* 5 (April 1981): 121–52.

"I heard the same questions": Joshua Wesson, quoted by Adam Fiore in "Wine Talk: The Taste and Flavor Guru," *Jerusalem Post,* May 9, 2012.

185 *in ways different from novices:* Andreas Gegenfurtner, Erno Lehtinen, and Roger Säljö, "Expertise Differences in the Comprehension of Visualizations: A Meta-Analysis of Eye-Tracking Research in Professional Domains," *Educational Psychology Review* 23 (December 2011): 523–52.

even high school teachers: Charlotte E. Wolff et al., "Teacher Vision: Expert and Novice Teachers' Perception of Problematic Classroom Management Scenes," *Instructional Science* 44 (June 2016): 243–65.

how they engage in looking: Ellen M. Kok and Halszka Jarodzka, "Before Your Very Eyes: The Value and Limitations of Eye Tracking in Medical Education," *Medical Education* 51 (January 2017): 114–22.

guiding their gaze with unobtrusive cues: Reynold J. Bailey et al., "Subtle Gaze Direction," *ACM Transactions on Graphics* 28 (September 2009): 1–22. See also Brett Roads, Michael C. Mozer, and Thomas A. Busey, "Using Highlighting to Train Attentional Expertise," *PLoS One* 11 (January 2016).

a way of "cheating experience": Samuel J. Vine et al., "Cheating Experience: Guiding Novices to Adopt the Gaze Strategies of Experts Expedites the Learning of Technical Laparoscopic Skills," *Surgery* 152 (July 2012): 32–40.

reduce cognitive load and improve performance: Janet van der Linden et al., "Buzzing to Play: Lessons Learned from an In the Wild Study of Real-Time Vibrotactile Feedback," paper presented at the SIGCHI Conference on Human Factors in Computing Systems, May 2011. See also Jeff Lieberman and Cynthia Breazeal, "TIKL: Development of a Wearable Vibrotactile Feedback Suit for Improved Human Motor Learning," *IEEE Transactions on Robotics* 23 (October 2007): 919–26.

186 *"the trade, art, and mystery"*: United States Bureau of Apprenticeship and Training, *Apprenticeship: Past and Present* (Washington, DC: United States Department of Labor, 1969).

8. THINKING WITH PEERS

187 *did not extend to the classroom:* My discussion of Wieman's efforts to improve physics teaching, including his own, are drawn from the following sources: Wieman, "The 'Curse of Knowledge'"; Carl E. Wieman, "Expertise in University Teaching & the Implications for Teaching Effectiveness, Evaluation & Training," *Dædalus* 148 (Fall 2019): 47–78; Carl Wieman, *Improving How Universities Teach Science: Lessons from the Science Education Initiative* (Cambridge: Harvard University Press, 2017); Susanne Dambeck, "Carl Wieman: Teaching Science More Effectively," post on the website of the Lindau Nobel Laureate Meetings, July 7, 2016, https://www.lindau-nobel.org/carl-wieman-teaching-science-more-effectively/; Carl E. Wieman, "Ideas for Improving Science Education," *New York Times,* September 2, 2013; Carl Wieman, "Why Not Try a Scientific Approach to Science Education?," *Change: The Magazine of Higher Learning* 39 (September–October 2007): 9–15.

more the rule than the exception: M. Mitchell Waldrop, "Why We Are Teaching Science Wrong, and How to Make It Right," *Nature* 523 (July 2015).

work carried out in his lab: Carl E. Wieman, *Collected Papers of Carl Wieman* (Singapore: World Scientific, 2008).

"sing in unison": The Royal Swedish Academy of Sciences, "New State of Matter Revealed: Bose-Einstein Condensate," post on website of the Nobel Prize, October 9, 2001, https://www.nobelprize.org/prizes/physics/2001/press-release/.

188 *"It was clear to me"*: Wieman, *Collected Papers,* 4.

A 2019 study: David F. Feldon et al., "Postdocs' Lab Engagement Predicts Trajectories of PhD Students' Skill Development," *Proceedings of the National Academy of Sciences* 116 (October 2019): 20910–16.

189 *"multiple brief small-group discussions"*: Wieman, *Improving How Universities Teach Science.*

"active learning" approach: Scott Freeman et al., "Active Learning Increases Student Performance in Science, Engineering, and Mathematics," *Proceedings of the National Academy of Sciences* 111 (June 2014): 8410–15.

students who engage in active learning: Louis Deslauriers, Ellen Schelew, and Carl Wieman, "Improved Learning in a Large-Enrollment Physics Class," *Science* 332 (May 2011): 862–64.

"this big pot of money": Carl Wieman, "Why I Donated to PhET," video posted on the website of PhET Interactive Simulations/University of Colorado Boulder, https://phet.colorado.edu/en/about.

a kind of internalized conversation: Charles Fernyhough, "Dialogic Thinking," in *Private Speech, Executive Functioning, and the Development of Verbal Self-Regulation,* ed. Adam Winsler, Charles Fernyhough, and Ignacio Montego (New York: Cambridge University Press, 2009), 42–52.

the brain stores social information: Jason P. Mitchell et al., "Thinking About Others: The Neural Substrates of Social Cognition," in *Social Neuroscience: People Thinking About Thinking People,* ed. Karen T. Litfin (Cambridge: MIT Press, 2006), 63–82.

remember social information more accurately: Jason P. Mitchell, C. Neil Macrae, and Mahzarin R. Banaji, "Encoding-Specific Effects of Social Cognition on the Neural Correlates of Subsequent Memory," *Journal of Neuroscience* 24 (May 2004): 4912–17.

"social encoding advantage": Matthew D. Lieberman, *Social: Why Our Brains Are Wired to Connect* (New York: Crown, 2013), 284.

190 *served to reinforce this notion:* Thalia Wheatley et al., "Beyond the Isolated Brain: The Promise and Challenge of Interacting Minds," *Neuron* 103 (July 2019): 186–88.

hampered by technical constraints: James McPartland and Joy Hirsch, "Imaging of Social Brain Enters Real World," *Spectrum,* January 31, 2017.

the "interactive brain hypothesis": Ezequiel Di Paolo and Hanne De Jaegher, "The Interactive Brain Hypothesis," *Frontiers of Human Neuroscience* 6 (June 2012).

191 *experiments that track brain activity:* Joy Hirsch et al., "Frontal Temporal and Parietal Systems Synchronize Within and Across Brains During Live Eye-To-Eye Contact," *Neuroimage* 157 (August 2017): 314–30.

people playing poker: Matthew Piva et al., "Distributed Neural Activity Patterns During Human-to-Human Competition," *Frontiers in Human Neuroscience* 11 (November 2017).

Other studies have found: For example, Sören Krach et al., "Are Women Better Mindreaders? Sex Differences in Neural Correlates of Mentalizing Detected with Functional MRI," *BMC Neuroscience* 10 (February 2009).

brain regions associated with reward: Jari Kätsyri et al., "The Opponent Matters: Elevated fMRI Reward Responses to Winning Against a Human Versus a Computer Opponent During Interactive Video Game Playing," *Cerebral Cortex* 23 (December 2013): 2829–39.

192 *it was applied in a study:* Barbara T. Conboy et al., "Social Interaction in Infants' Learning of Second-Language Phonetics: An Exploration of Brain-Behavior Relations," *Developmental Neuropsychology* 40 (April 2015): 216–29.

the Wason Selection Task: Peter C. Wason, "Reasoning," in *New Horizons in Psychology,* ed. Brian M. Foss (Harmondsworth: Penguin Books, 1966), 135–51.

193 *"Take a look at the cards":* Raymond S. Nickerson, *Conditional Reasoning: The Unruly Syntactics, Semantics, Thematics, and Pragmatics of "If"* (New York: Oxford University Press, 2015), 33.

People's performance on this task: Philip N. Johnson-Laird and Peter C. Wason, "Insight into a Logical Relation," *Quarterly Journal of Experimental Psychology* 22 (1970): 49–61.

only about 10 percent of subjects: Dan Sperber and Hugo Mercier, "Reasoning as a Social Competence," in *Collective Wisdom: Principles and Mechanisms,* ed. Hélène Landemore and Jon Elster (New York: Cambridge University Press, 2012), 368–92.

the language of the task is rephrased: Leda Cosmides, "The Logic of Social Exchange: Has Natural Selection Shaped How Humans Reason? Studies with the Wason Selection Task," *Cognition* 31 (April 1989): 187–276.

one particular aspect of the task: Leda Cosmides and John Tooby, "Cognitive Adaptations for Social Exchange," in *The Adapted Mind: Evolutionary Psychology and the Generation of Culture,* ed. Jerome H. Barkow, Leda Cosmides, and John Tooby (New York: Oxford University Press, 1995), 163–228.

Make it social: Richard A. Griggs and James R. Cox, "The Elusive Thematic-Materials Effect in Wason's Selection Task," *British Journal of Social Psychology* 73 (August 1982): 407–20. See also Ken I. Manktelow and David E. Over, "Social Roles and Utilities in Reasoning with Deontic Conditionals," *Cognition* 39 (May 1991): 85–105.

the social version of the task: Christopher Badcock, "Making Sense of Wason," post on the website of *Psychology Today,* May 5, 2012, https://www.psychologytoday.com/us/blog/the-imprinted-brain/201205/making-sense-wason.

evolutionary psychologists have speculated: Cosmides and Tooby, "Cognitive Adaptations for Social Exchange."

simply because it is social in nature: Nicola Canessa et al., "The Effect of Social Content on Deductive Reasoning: An fMRI Study," *Human Brain Mapping* 26 (September 2005): 30–43.

developed our oversized brains: Robin I. M. Dunbar, "The Social Brain Hypothesis," *Evolutionary Anthropology* 6 (December 1998): 178–90.

a specialized "social brain": Lieberman, *Social,* 31.

194 *The social brain with its "superpowers":* Matthew D. Lieberman, "The Social Brain and Its Superpowers," post on the website of *Psychology Today,* October 8, 2013, https://www.psychologytoday.com/us/blog/social-brain-social-mind/201310/the-social-brain-and-its-superpowers.

powerfully driven to form bonds: Sarah-Jayne Blakemore and Kathryn L. Mills, "Is Adolescence a Sensitive Period for Sociocultural Processing?," *Annual Review of Psychology* 65 (2014): 187–207.

teens' brains become more sensitive: This development in teenagers has been found in a number of studies, including Maya L. Rosen et al., "Salience Network Response to Changes in Emotional Expressions of Others Is Heightened During Early Adolescence: Relevance for Social Functioning," *Developmental Science* 21 (May 2018); William E. Moore III et al., "Facing Puberty: Associations Between Pubertal Development and Neural Responses to Affective Facial Displays," *Social Cognitive and Affective Neuroscience* 7 (January 2012): 35–43; Jennifer H. Pfeifer et al., "Entering Adolescence: Resistance to Peer Influence, Risky Behavior, and Neural Changes in Emotion Reactivity," *Neuron* 69 (March 2011): 1029–36.

become more attuned to reward: Linda Van Leijenhorst et al., "What Motivates the Adolescent? Brain Regions Mediating Reward Sensitivity Across Adolescence," *Cerebral Cortex* 20 (January 2010): 61–69. See also Monique Ernst et al., "Amygdala and Nucleus Accumbens in Responses to Receipt and Omission of Gains in Adults and Adolescents," *NeuroImage* 25 (May 2005): 1279–91.

the sweetest reward for a teen: Jason Chein et al., "Peers Increase Adolescent Risk Taking by Enhancing Activity in the Brain's Reward Circuitry," *Developmental Science* 14 (March 2011): F1–10.

195 *appears to be "turned on":* Iroise Dumontheil et al., "Developmental Differences in the Control of Action Selection by Social Information," *Journal of Cognitive Neuroscience* 24 (October 2012): 2080–95.

"What the brain really wants": Matthew D. Lieberman, "Education and the Social Brain," *Trends in Neuroscience and Education* 1 (December 2012): 3–9.

leverage their burgeoning sociability: Mary Helen Immordino-Yang and Antonio Damasio, "We Feel, Therefore We Learn: The Relevance of Affective and Social Neuroscience to Education," *Mind, Brain and Education* 1 (March 2007): 3–10.

did not evolve to care: David C. Geary and Daniel B. Berch, "Evolution and Children's Cognitive and Academic Development," in *Evolutionary Perspectives on Child Development and Education,* ed. David C. Geary and Daniel B. Berch (New York: Springer, 2016), 217–49. See also David C. Geary, "An Evolutionarily Informed Education Science," *Educational Psychologist* 43 (October 2008): 179–95.

Evidence of teaching has been found: Jamshid J. Tehrani and Felix Riede, "Towards an Archaeology of Pedagogy: Learning, Teaching and the Generation of Material Culture Traditions," *World Archaeology* 40 (September 2008): 316–31.

the act of teaching has been observed: Gergely Csibra and György Gergely, "Natural Pedagogy as Evolutionary Adaptation," *Philosophical Transactions of the Royal Society B: Biological Sciences* 366 (April 2011): 1149–57. See also Barry S. Hewlett and Casey J. Roulette, "Teaching in Hunter-Gatherer Infancy," *Royal Society Open Science* 3 (January 2016).

"teaching instinct": Cecilia I. Calero, A. P. Goldin, and M. Sigman, "The Teaching Instinct," *Review of Philosophy and Psychology* 9 (December 2018): 819–30.

we unconsciously offer cues: György Gergely and Gergely Csibra, "Natural Pedagogy," in *Navigating the Social World: What Infants, Children, and Other Species Can Teach Us,* ed. Mahzarin R. Banaji and Susan A. Gelman (New York: Oxford University Press, 2013), 127–32. See also György Gergely, Katalin Egyed, and Ildikó Király, "On Pedagogy," *Developmental Science* 10 (January 2007): 139–46.

start speaking to their infants: Anne Fernald and Thomas Simon, "Expanded Intonation Contours in Mothers' Speech to Newborns," *Developmental Psychology* 20 (January 1984): 104–13.

learn new words more readily: Nairán Ramírez-Esparza, Adrián García-Sierra, and Patricia K. Kuhl, "Look Who's Talking: Speech Style and Social Context in Language Input to Infants Are Linked to Concurrent and Future Speech Development," *Developmental Science* 17 (November 2014): 880–91.

196 *teaching behavior has been observed:* Sidney Strauss, Margalit Ziv, and Adi Stein, "Teaching as a Natural Cognition and Its Relations to Preschoolers' Developing Theory of Mind," *Cognitive Development* 17 (September–December 2002): 1473–87.

an area of the social brain was activated: Joy Hirsch et al., "A Two-Person Neural Mechanism for Sharing Social Cues During Real Eye Contact," paper presented at the 49th Annual Meeting of the Society for Neuroscience, October 2019.

"Eye contact opens the gate": Joy Hirsch, quoted in Sarah Deweerdt, "Looking Directly in the Eyes Engages Region of the Social Brain," *Spectrum,* October 20, 2019.

Another factor that seems to "gate": Patricia K. Kuhl, "Is Speech Learning 'Gated' by the Social Brain?," *Developmental Science* 10 (January 2007): 110–20.

pick up almost nothing: Judy S. DeLoache et al., "Do Babies Learn from Baby Media?," *Psychological Science* 21 (November 2010): 1570–74. See also Patricia K. Kuhl, Feng-Ming Tsao, and Huei-Mei Liu, "Foreign-Language Experience in Infancy: Effects of Short-Term Exposure and Social Interaction on Phonetic Learning," *Proceedings of the National Academy of Sciences* 100 (July 2003): 9096–9101.

the "video deficit": Daniel R. Anderson and Tiffany A. Pempek, "Television and Very Young Children," *American Behavioral Scientist* 48 (January 2005): 505–22.

firstborn children have an IQ: Petter Kristensen and Tor Bjerkedal, "Explaining the Relation Between Birth Order and Intelligence," *Science* 316 (June 2007): 1717.

benefits for all involved: Cynthia A. Rohrbeck et al., "Peer-Assisted Learning Interventions with Elementary School Students: A Meta-Analytic Review," *Journal of Educational Psychology* 95 (June 2003): 240–57. See also Peter A. Cohen, James A. Kulik, and Chen-Lin C. Kulik, "Educational Outcomes of Tutoring: A Meta-Analysis of Findings," *American Educational Research Journal* 19 (January 1982): 237–48.

One such process kicks in: John F. Nestojko et al., "Expecting to Teach Enhances Learning and Organization of Knowledge in Free Recall of Text Passages," *Memory & Cognition* 42 (October 2014): 1038–48. See also John A. Bargh and Yaacov Schul, "On the Cognitive Benefits of Teaching," *Journal of Educational Psychology* 72 (1980): 593–604.

197 *More learning happens:* Logan Fiorella and Richard E. Mayer, "The Relative Benefits of Learning by Teaching and Teaching Expectancy," *Contemporary Educational Psychology* 38 (October 2013): 281–88. See also Stewart Ehly, Timothy Z. Keith, and Barry Bratton, "The Benefits of Tutoring: An Exploration of Expectancy and Outcomes," *Contemporary Educational Psychology* 12 (April 1987): 131–34.

When explaining academic content: Jonathan Galbraith and Mark Winterbottom, "Peer-Tutoring: What's in It for the Tutor?," *Educational Studies* 37 (July 2011): 321–32.

to the most important aspects: Rod D. Roscoe and Michelene T. H. Chi, "Understanding Tutor Learning: Knowledge-Building and Knowledge-Telling in Peer Tutors' Explanations and Questions," *Review of Educational Research* 77 (December 2007): 534–74.

fielding the pupil's questions: Rod D. Roscoe and Michelene T. H. Chi, "Tutor Learning: The Role of Explaining and Responding to Questions," *Instructional Science* 36 (July 2008): 321–50.

explain academic content on camera: Vincent Hoogerheide, Sofie M. M. Loyens, and Tamara van Gog, "Effects of Creating Video-Based Modeling Examples on Learning and Transfer," *Learning and Instruction* 33 (October 2014): 108–19. See also Vincent Hoogerheide et al., "Generating an Instructional Video as Homework Activity Is Both Effective and Enjoyable," *Learning and Instruction* 64 (December 2019).

198 *does not generate the same gains:* Vincent Hoogerheide et al., "Gaining from Explaining: Learning Improves from Explaining to Fictitious Others on Video, Not from Writing to Them," *Contemporary Educational Psychology* 44–45 (January–April 2016): 95–106.

increases the explainers' physiological arousal: Vincent Hoogerheide et al., "Enhancing Example-Based Learning: Teaching on Video Increases Arousal and Improves Problem-Solving Performance," *Journal of Educational Psychology* 111 (January 2019): 45–56.

positively affect students' identity: Marika D. Ginsburg-Block, Cynthia A. Rohrbeck, and John W. Fantuzzo, "A Meta-Analytic Review of Social, Self-Concept, and Behavioral Outcomes of Peer-Assisted Learning," *Journal of Educational Psychology* 98 (November 2006): 732–49.

Valued Youth Partnership: Formerly named the Coca-Cola Valued Youth Program; see https://www.idra.org/valued-youth/.

does just the opposite: Nurit Bar-Eli and Amiram Raviv, "Underachievers as Tutors," *Journal of Educational Research* 75 (1982): 139–43. See also Vernon L. Allen and Robert S. Feldman, "Learning Through Tutoring: Low-Achieving Children as Tutors," *Journal of Experimental Education* 42 (1973): 1–5.

Evaluations of the program: Robert E. Slavin, "Evidence-Based Reform: Advancing the Education of Students at Risk," report prepared by Renewing Our Schools, Securing Our Future: A National Task Force on Public Education, March 2005. See also Olatokunbo S. Fashola and Robert E. Slavin, "Effective Dropout Prevention and College

Attendance Programs for Students Placed at Risk," *Journal of Education for Students Placed at Risk* 3 (1998): 159–83.

"productive agency": Sandra Y. Okita and Daniel L. Schwartz, "Learning by Teaching Human Pupils and Teachable Agents: The Importance of Recursive Feedback," *Journal of the Learning Sciences* 22 (2013): 375–412. See also Daniel L. Schwartz and Sandra Okita, "The Productive Agency in Learning by Teaching," working paper, available at http://citeseerx.ist.psu.edu/viewdoc/download?doi=10.1.1.90.5549&rep=rep1&type=pdf.

the program has been shown: Behnoosh Afghani et al., "A Novel Enrichment Program Using Cascading Mentorship to Increase Diversity in the Health Care Professions," *Academic Medicine* 88 (September 2013): 1232–38.

199 *sharing job-related knowledge:* Yu-Qian Zhu, Holly Chiu, and Eduardo Jorge Infante Holguin-Veras, "It Is More Blessed to Give Than to Receive: Examining the Impact of Knowledge Sharing on Sharers and Recipients," *Journal of Knowledge Management* 22 (2018): 76–91.

Family Playlists: PowerMyLearning, "Engage Families Using Family Playlists," post on the website of PowerMyLearning, https://powermylearning.org/learn/connect/family-playlists/. See also David Bornstein, "When Parents Teach Children (and Vice Versa)," *New York Times*, March 13, 2018.

200 *devilish in its design:* Emmanuel Trouche et al., "The Selective Laziness of Reasoning," *Cognitive Science* 40 (November 2016): 2122–36.

Thinking Skills Assessment: Fabio Paglieri, "A Plea for Ecological Argument Technologies," *Philosophy & Technology* 30 (June 2017): 209–38.

Cognitive Reflection Test: Shane Frederick, "Cognitive Reflection and Decision Making," *Journal of Economic Perspectives* 19 (Fall 2005): 25–42.

201 *confirmation bias:* Peter C. Wason, "On the Failure to Eliminate Hypotheses in a Conceptual Task," *Quarterly Journal of Experimental Psychology* 12 (July 1960): 129–40.

"Contrary to the rules": Kahneman, *Thinking, Fast and Slow*, 81.

a "flawed superpower": Hugo Mercier and Dan Sperber, *The Enigma of Reason* (Cambridge: Harvard University Press, 2017), 1.

We did not evolve to solve: Hugo Mercier, "Why So Smart? Why So Dumb?," post on the website of *Psychology Today*, July 28, 2011, https://www.psychologytoday.com/intl/blog/social-design/201107/why-so-smart-why-so-dumb?

"argumentative theory of reasoning": Hugo Mercier and Dan Sperber, "Why Do Humans Reason? Arguments for an Argumentative Theory," *Behavioral and Brain Sciences* 34 (April 2011): 57–74.

202 *makes specific predictions:* Hugo Mercier, "The Argumentative Theory: Predictions and Empirical Evidence," *Trends in Cognitive Sciences* 20 (September 2016): 689–700.

how thinking is usually done: Hugo Mercier et al., "Experts and Laymen Grossly Underestimate the Benefits of Argumentation for Reasoning," *Thinking & Reasoning* 21 (July 2015): 341–55.

urge a different approach: Emmanuel Trouche, Emmanuel Sander, and Hugo Mercier, "Arguments, More Than Confidence, Explain the Good Performance of Reasoning Groups," *Journal of Experimental Psychology: General* 143 (October 2014): 1958–71. See also Hugo Mercier, "Making Science Education More Natural—Some Ideas from the Argumentative Theory of Reasoning," *Zeitschrift für Pädagogische Psychologie* 30 (2016): 151–53.

"famous for fighting openly": Brad Bird, quoted in Hayagreeva Rao, Robert Sutton, and Allen P. Webb, "Innovation Lessons from Pixar: An Interview with Oscar-Winning Director Brad Bird," *McKinsey Quarterly*, April 2008.

"a vigorous practitioner": Robert I. Sutton, *Good Boss, Bad Boss: How to Be the Best . . . and Learn from the Worst* (New York: Hachette Book Group, 2010), 85, 83. See also Robert I. Sutton, "It's Up to You to Start a Good Fight," *Harvard Business Review*, August 3, 2010.

203   *"a general rule of teaching"*: David W. Johnson, Roger T. Johnson, and Karl A. Smith, "Constructive Controversy: The Educative Power of Intellectual Conflict," *Change: The Magazine of Higher Learning* 32 (January–February 2000): 28–37. See also David W. Johnson and Roger T. Johnson, "Energizing Learning: The Instructional Power of Conflict," *Educational Researcher* 38 (January 2009): 37–51.

"the accountability effect": Philip E. Tetlock, "Accountability and Complexity of Thought," *Journal of Personality and Social Psychology* 45 (July 1983): 74–83. See also Jennifer S. Lerner and Philip E. Tetlock, "Accounting for the Effects of Accountability," *Psychological Bulletin*, 125 (March 1999): 255–75.

204   *relieves cognitive load*: Baruch B. Schwarz, "Argumentation and Learning," in *Argumentation and Education: Theoretical Foundations and Practices*, ed. Nathalie Muller Mirza and Anne-Nelly Perret-Clermont (New York: Springer, 2009), 91–126.

Children as young as two or three: Nancy L. Stein and Elizabeth R. Albro, "The Origins and Nature of Arguments: Studies in Conflict Understanding, Emotion, and Negotiation," *Discourse Processes* 32 (2001): 113–33. See also Nancy Stein and Ronan S. Bernas, "The Early Emergence of Argumentative Knowledge and Skill," in *Foundations of Argumentative Text Processing*, ed. Jerry Andriessen and Pierre Courier (Amsterdam: Amsterdam University Press, 1999), 97–116.

critically evaluate others' arguments: Thomas Castelain, Stéphane Bernard, and Hugo Mercier, "Evidence That Two-Year-Old Children Are Sensitive to Information Presented in Arguments," *Infancy* 23 (January 2018): 124–35. See also Hugo Mercier, Stéphane Bernard, and Fabrice Clément, "Early Sensitivity to Arguments: How Preschoolers Weight Circular Arguments," *Journal of Experimental Child Psychology* 125 (September 2014): 102–9.

"natural-born arguers": Hugo Mercier et al., "Natural-Born Arguers: Teaching How to Make the Best of Our Reasoning Abilities," *Educational Psychologist* 52 (2017): 1–16.

"strong opinions, weakly held": Sutton, *Good Boss, Bad Boss*, 273, 98.

study of educational methods: Diana J. Arya and Andrew Maul, "The Role of the Scientific Discovery Narrative in Middle School Science Education: An Experimental Study," *Journal of Educational Psychology*, 104 (November 2012): 1022–32. See also Diana Jaleh Arya, "Discovery Stories in the Science Classroom" (PhD diss., University of California, Berkeley, 2010).

206   *this "depersonalized" approach*: Janet Ahn et al., "Motivating Students' STEM Learning Using Biographical Information," *International Journal of Designs for Learning* 7 (February 2016): 71–85.

wielded by narrative: Karen D. Larison, "Taking the Scientist's Perspective: The Nonfiction Narrative Engages Episodic Memory to Enhance Students' Understanding of Scientists and Their Practices," *Science & Education* 27 (March 2018): 133–57.

"psychologically privileged": Daniel T. Willingham, *Why Don't Students Like School? A*

Cognitive Scientist Answers Questions About How the Mind Works and What It Means for the Classroom (San Francisco: Jossey-Bass, 2009), 51–58.

we attend to stories more closely: Daniel T. Willingham, "The Privileged Status of Story," *American Educator,* Summer 2004.

evidence of causal relationships: Willingham, "The Privileged Status of Story."

207 *hear about characters emoting:* Paul J. Zak, "Why Inspiring Stories Make Us React: The Neuroscience of Narrative," *Cerebrum* 2 (January–February 2015).

hear about characters moving vigorously: Nicole K. Speer et al., "Reading Stories Activates Neural Representations of Visual and Motor Experiences," *Psychological Science* 20 (August 2009): 989–99.

We even tend to remember: Danielle N. Gunraj et al., "Simulating a Story Character's Thoughts: Evidence from the Directed Forgetting Task," *Journal of Memory and Language* 96 (October 2017): 1–8.

running a simulation: Diana I. Tamir et al., "Reading Fiction and Reading Minds: The Role of Simulation in the Default Network," *Social Cognitive and Affective Neuroscience* 11 (February 2016): 215–24.

watching medical transport teams: Christopher G. Myers, "The Stories We Tell: Organizing for Vicarious Learning in Medical Transport Teams," working paper, November 2015. See also Christopher G. Myers, "That Others May Learn: Three Views on Vicarious Learning in Organizations" (PhD diss., University of Michigan, 2015).

"I don't want to read": Transport nurse, quoted in Myers, "That Others May Learn."

208 *"I had never actually seen":* Transport nurse, quoted in Myers, "The Stories We Tell."

vicarious learning is increasingly necessary: Christopher G. Myers, "Performance Benefits of Reciprocal Vicarious Learning in Teams," *Academy of Management Journal* (in press), published online April 2020.

"fourteen hundred experiences a year": Transport nurse, quoted in Myers, "That Others May Learn."

"There's too much going on": Transport nurse, quoted in Myers, "That Others May Learn."

209 *1 percent reduction:* Ben Waber, "How Tracking Worker Productivity Could Actually Make Amazon Warehouses Less Efficient," *Quartz,* April 26, 2018.

"But what is gossip?": Sandy Pentland, quoted in Larry Hardesty, "Social Studies," *MIT News,* November 1, 2010.

a ten-by-fifteen-foot area: Christopher G. Myers, "Is Your Company Encouraging Employees to Share What They Know?," *Harvard Business Review,* November 6, 2015.

"cleaner," he reports: Myers, "That Others May Learn."

"tacit knowledge": Philippe Baumard, *Tacit Knowledge in Organizations,* trans. Samantha Wauchope (Thousand Oaks, CA: Sage Publications, 1999).

"Much of the knowledge needed": Christopher Myers, "Try Asking the Person at the Next Desk," *Medium,* January 16, 2017.

210 *"I use the knowledge management system":* Tech employee, interviewed by Christopher Myers and quoted in Michael Blanding, "Knowledge Transfer: You Can't Learn Surgery by Watching," *Harvard Business School Working Knowledge,* September 8, 2015.

9. THINKING WITH GROUPS

211 *After several days conducting military drills:* Details of this incident are drawn from Edwin Hutchins, *Cognition in the Wild* (Cambridge: MIT Press, 1995). The

name of the ship, and the names of the crew members, are pseudonyms devised by Hutchins.

"Bridge, Main Control": Hutchins, *Cognition in the Wild*, 1.

"Shutting throttles, aye": Hutchins, *Cognition in the Wild*.

"Captain, the engineer is losing steam": Hutchins, *Cognition in the Wild*.

212 *"Sir, I have no helm"*: Hutchins, *Cognition in the Wild*, 3.

"Come on, damn it": Hutchins, *Cognition in the Wild*.

a *"casualty"*: Hutchins, *Cognition in the Wild*, 2.

"anything but routine": Hutchins, *Cognition in the Wild*.

"socially distributed cognition": Hutchins, *Cognition in the Wild*, xiii.

"move the boundaries": Hutchins, *Cognition in the Wild*, xiv.

"may have interesting cognitive properties": Hutchins, *Cognition in the Wild*, xiii.

213 *"subvocally rehearsed"*: Hutchins, *Cognition in the Wild*, 233.

"external memory": Hutchins, *Cognition in the Wild*, 328.

"Normally the Palau*"*: Hutchins, *Cognition in the Wild*, 4.

"basically a bicycle pump": Hutchins, *Cognition in the Wild*.

"Sailboat crossing Palau's bow": Hutchins, *Cognition in the Wild*, 5.

214 *"a consistent pattern of action"*: Hutchins, *Cognition in the Wild*, 322.

"they perform what will be": Hutchins, *Cognition in the Wild*, 339.

"Twenty-five minutes after the engineering casualty": Hutchins, *Cognition in the Wild*, 5.

215 *"cognitive individualism"*: Eviatar Zerubavel and Eliot R. Smith, "Transcending Cognitive Individualism," *Social Psychology Quarterly* 73 (December 2010): 321–25. See also Stephen M. Downes, "Socializing Naturalized Philosophy of Science," *Philosophy of Science* 60 (September 1993): 452–68.

"only accessible to crowds": Gustave Le Bon, *The Crowd: A Study of the Popular Mind* (New York: Macmillan, 1896), 51.

"Not only mobs": William McDougall, *The Group Mind: A Sketch of the Principles of Collective Psychology, with Some Attempt to Apply Them to the Interpretation of National Life and Character* (New York: G. P. Putnam's Sons, 1920), 58.

This conception of the group mind: My discussion of early notions about the "group mind" draws on the following sources: Georg Theiner, "A Beginner's Guide to Group Minds," in *New Waves in Philosophy of Mind*, ed. Mark Sprevak and Jesper Kallestrup (New York: Palgrave Macmillan, 2014), 301–22; Georg Theiner, Colin Allen, and Robert L. Goldstone, "Recognizing Group Cognition," *Cognitive Systems Research* 11 (December 2010): 378–95; Georg Theiner and Timothy O'Connor, "The Emergence of Group Cognition," in *Emergence in Science and Philosophy*, ed. Antonella Corradini and Timothy O'Connor (New York: Routledge, 2010): 78–120; Daniel M. Wegner, "Transactive Memory: A Contemporary Analysis of the Group Mind," in *Theories of Group Behavior*, ed. Brian Mullen and George R. Goethals (Berlin: Springer Verlag, 1987), 185–208; Daniel M. Wegner, Toni Giuliano, and Paula T. Hertel, "Cognitive Interdependence in Close Relationships," in *Compatible and Incompatible Relationships*, ed. William Ickes (New York: Springer, 1985), 253–76.

216 *"magnetic influence"*: Le Bon, *The Crowd*, 10–11.

"telepathic communication": McDougall, *The Group Mind*, 41.

"genetic ectoplasm": Carl Jung, cited in Theiner and O'Connor, "The Emergence of Group Cognition."

"slipped ignominiously into the history": Wegner, Giuliano, and Hertel, "Cognitive Interdependence in Close Relationships."

"banished from the realm": Theiner, "A Beginner's Guide to Group Minds."

staging a surprising comeback: Jay J. Van Bavel, Leor M. Hackel, and Y. Jenny Xiao, "The Group Mind: The Pervasive Influence of Social Identity on Cognition," in *New Frontiers in Social Neuroscience,* ed. Jean Decety and Yves Christen (New York: Springer, 2014), 41–56.

the program starts up: For a sense of what the routine looks like, see "Radio Taiso Exercise Routine," video posted on YouTube, January 7, 2017, https://www.youtube.com/watch?v=I6ZRH9Mraqw.

broadcast daily in Japan: Justin McCurry, "Listen, Bend and Stretch: How Japan Fell in Love with Exercise on the Radio," *The Guardian,* July 20, 2019. See also Agence France-Presse, "Japan Limbers Up with Monkey Bars, Radio Drills for Company Employees," *Straits Times,* June 15, 2017.

217 *pairs of four-year-old children:* Tal-Chen Rabinowitch and Andrew N. Meltzoff, "Synchronized Movement Experience Enhances Peer Cooperation in Preschool Children," *Journal of Experimental Child Psychology* 160 (August 2017): 21–32.

Comparable results were found: Tal-Chen Rabinowitch and Ariel Knafo-Noam, "Synchronous Rhythmic Interaction Enhances Children's Perceived Similarity and Closeness Towards Each Other," *PLoS One* 10 (April 2015).

makes us better collaborators: Scott S. Wiltermuth and Chip Heath, "Synchrony and Cooperation," *Psychological Science* 20 (January 2009): 1–5.

synchrony sends a tangible signal: Günther Knoblich and Natalie Sebanz, "Evolving Intentions for Social Interaction: From Entrainment to Joint Action," *Philosophical Transactions of the Royal Society B: Biological Sciences* 363 (June 2008): 2021–31.

The recognition that we're moving: Edward H. Hagen and Gregory A. Bryant, "Music and Dance as a Coalition Signaling System," *Human Nature* 14 (February 2003): 21–51.

others are making motions: Thalia Wheatley et al., "From Mind Perception to Mental Connection: Synchrony as a Mechanism for Social Understanding," *Social and Personality Psychology Compass* 6 (August 2012): 589–606.

more apt to "mentalize": Adam Baimel, Susan A. J. Birch, and Ara Norenzayan, "Coordinating Bodies and Minds: Behavioral Synchrony Fosters Mentalizing," *Journal of Experimental Social Psychology* 74 (January 2018): 281–90. See also Adam Baimel et al., "Enhancing 'Theory of Mind' Through Behavioral Synchrony," *Frontiers in Psychology* 6 (June 2015).

alters the nature of our perception: Piercarlo Valdesolo, Jennifer Ouyang, and David DeSteno, "The Rhythm of Joint Action: Synchrony Promotes Cooperative Ability," *Journal of Experimental Social Psychology* 46 (July 2010): 693–95.

218 *we form more accurate memories:* Lynden K. Miles et al., "Moving Memories: Behavioral Synchrony and Memory for Self and Others," *Journal of Experimental Social Psychology* 46 (March 2010): 457–60. See also C. Neil Macrae et al., "A Case of Hand Waving: Action Synchrony and Person Perception," *Cognition* 109 (October 2008): 152–56.

We learn from them: Patricia A. Herrmann et al., "Stick to the Script: The Effect of Witnessing Multiple Actors on Children's Imitation," *Cognition* 129 (December 2013): 536–43.

We communicate with them: Rabinowitch and Meltzoff, "Synchronized Movement Experience Enhances Peer Cooperation in Preschool Children."

we pursue shared goals: Valdesolo, Ouyang, and DeSteno, "The Rhythm of Joint Action." See also Martin Lang et al., "Endorphin-Mediated Synchrony Effects on Cooperation," *Biological Psychology* 127 (July 2017): 191–97.

seem a bit like friends: Bahar Tunçgenç and Emma Cohen, "Movement Synchrony Forges Social Bonds Across Group Divides," *Frontiers in Psychology* 7 (May 2016). See also Piercarlo Valdesolo and David DeSteno, "Synchrony and the Social Tuning of Compassion," *Emotion* 11 (April 2011): 262–66.

We feel more warmly: Tanya Vacharkulksemsuk and Barbara L. Fredrickson, "Strangers in Sync: Achieving Embodied Rapport Through Shared Movements," *Journal of Experimental Social Psychology* 48 (January 2012): 399–402. See also Michael J. Hove and Jane L. Risen, "It's All in the Timing: Interpersonal Synchrony Increases Affiliation," *Social Cognition* 27 (December 2009): 949–61.

more willing to help them out: Bahar Tunçgenç and Emma Cohen, "Interpersonal Movement Synchrony Facilitates Pro-Social Behavior in Children's Peer-Play," *Developmental Science,* 21 (January 2018): 1–9. See also Laura K. Cirelli, "How Interpersonal Synchrony Facilitates Early Prosocial Behavior," *Current Opinion in Psychology* 20 (April 2018): 35–39.

a blurring of the boundaries: Maria-Paola Paladino et al., "Synchronous Multisensory Stimulation Blurs Self-Other Boundaries," *Psychological Science* 21 (September 2010): 1202–7. See also Elisabeth Pacherie, "How Does It Feel to Act Together?," *Phenomenology and the Cognitive Sciences* 13 (2014): 25–46.

the perception of physical pain: Bronwyn Tarr et al., "Synchrony and Exertion During Dance Independently Raise Pain Threshold and Encourage Social Bonding," *Biology Letters* 11 (October 2015). See also Emma E. A. Cohen et al., "Rowers' High: Behavioural Synchrony Is Correlated with Elevated Pain Thresholds," *Biology Letters* 6 (February 2010): 106–8.

a "social eddy": Kerry L. Marsh, "Coordinating Social Beings in Motion," in *People Watching: Social, Perceptual, and Neurophysiological Studies of Body Perception,* ed. Kerri Johnson and Maggie Shiffrar (New York: Oxford University Press, 2013), 234–55.

In every culture: Barbara Ehrenreich, *Dancing in the Streets: A History of Collective Joy* (New York: Metropolitan Books, 2006). See also William Hardy McNeill, *Keeping Together in Time: Dance and Drill in Human History* (Cambridge: Harvard University Press, 1995).

"biotechnology of group formation": Walter Freeman, "A Neurobiological Role of Music in Social Bonding," in *The Origins of Music,* ed. Nils Lennart Wallin, Björn Merker, and Steven Brown (Cambridge: MIT Press, 2000), 411–24.

"human nature is 90 percent chimp": Jonathan Haidt, *The Righteous Mind: Why Good People Are Divided by Politics and Religion* (New York: Pantheon Books, 2012), 223.

219 *from "I" mode to "we" mode:* Mattia Gallotti and Chris D. Frith, "Social Cognition in the We-Mode," *Trends in Cognitive Science* 17 (April 2013): 160–65.

"muscular bonding": McNeill, *Keeping Together in Time,* 2.

220 *"to simulate conditions":* Joshua Conrad Jackson et al., "Synchrony and Physiological Arousal Increase Cohesion and Cooperation in Large Naturalistic Groups," *Scientific Reports* 8 (January 2018).

an experience of heightened emotion: Dmitry Smirnov et al., "Emotions Amplify

Speaker-Listener Neural Alignment," *Human Brain Mapping* 40 (November 2019): 4777–88. See also Lauri Nummenmaa et al., "Emotional Speech Synchronizes Brains Across Listeners and Engages Large-Scale Dynamic Brain Networks," *Neuroimage* 102 (November 2014): 498–509.

a group of individuals to cohere: Gün R. Semin, "Grounding Communication: Synchrony," in *Social Psychology: Handbook of Basic Principles,* ed. Arie W. Kruglanski and E. Tory Higgins (New York: Guilford Press, 2007), 630–49.

221 *generate greater cognitive synchrony:* Yong Ditch, Joseph Reilly, and Betrand Schneider, "Using Physiological Synchrony as an Indicator of Collaboration Quality, Task Performance and Learning," paper presented at the 19th International Conference on Artificial Intelligence in Education, June 2018. See also Dan Mønster et al., "Physiological Evidence of Interpersonal Dynamics in a Cooperative Production Task," *Physiology & Behavior* 156 (March 2016): 24–34.

"neural synchrony": Frank A. Fishburn et al., "Putting Our Heads Together: Interpersonal Neural Synchronization as a Biological Mechanism for Shared Intentionality," *Social Cognitive and Affective Neuroscience* 13 (August 2018): 841–49.

a certain kind of group experience: Kyongsik Yun, "On the Same Wavelength: Face-to-Face Communication Increases Interpersonal Neural Synchronization," *Journal of Neuroscience* 33 (March 2013): 5081–82. See also Jing Jiang et al., "Neural Synchronization During Face-to-Face Communication," *Journal of Neuroscience* 32 (November 2012): 16064–69.

"shared attention": Garriy Shteynberg, "Shared Attention," *Perspectives on Psychological Science* 10 (August 2015): 579–90.

along with other people: Garriy Shteynberg, "A Social Host in the Machine? The Case of Group Attention," *Journal of Applied Research in Memory and Cognition* 3 (December 2014): 307–11. See also Garriy Shteynberg, "A Silent Emergence of Culture: The Social Tuning Effect," *Journal of Personality and Social Psychology* 99 (October 2010): 683–89.

"cognitive prioritization": Garriy Shteynberg, "A Collective Perspective: Shared Attention and the Mind," *Current Opinion in Psychology* 23 (October 2018): 93–97.

we use shared attention: Garriy Shteynberg and Evan P. Apfelbaum, "The Power of Shared Experience: Simultaneous Observation with Similar Others Facilitates Social Learning," *Social Psychological and Personality Science* 4 (November 2013): 738–44.

222 *we learn things better:* Shteynberg and Apfelbaum, "The Power of Shared Experience."

We remember things better: Samantha E. A. Gregory and Margaret C. Jackson, "Joint Attention Enhances Visual Working Memory," *Journal of Experimental Psychology: Learning, Memory, and Cognition* 43 (February 2017): 237–49. See also Ullrich Wagner et al., "The Joint Action Effect on Memory as a Social Phenomenon: The Role of Cued Attention and Psychological Distance," *Frontiers in Psychology* 8 (October 2017).

more likely to act upon: Garriy Shteynberg et al., "Feeling More Together: Group Attention Intensifies Emotion," *Emotion* 14 (December 2014): 1102–14. See also Garriy Shteynberg and Adam D. Galinsky, "Implicit Coordination: Sharing Goals with Similar Others Intensifies Goal Pursuit," *Journal of Experimental Social Psychology* 47 (November 2011): 1291–94.

starts in infancy: Stefanie Hoehl et al., "What Are You Looking At? Infants' Neural Processing of an Adult's Object-Directed Eye Gaze," *Developmental Science* 11 (January 2008): 10–16. See also Tricia Striano, Vincent M. Reid, and Stefanie Hoehl, "Neural Mechanisms of Joint Attention in Infancy," *European Journal of Neuroscience* 23 (May 2006): 2819–23.

a baby begins to look: Rechele Brooks and Andrew N. Meltzoff, "The Development of Gaze Following and Its Relation to Language," *Developmental Science* 8 (November 2005): 535–43.

Infants will gaze longer: Catalina Suarez-Rivera, Linda B. Smith, and Chen Yu, "Multimodal Parent Behaviors Within Joint Attention Support Sustained Attention in Infants," *Developmental Psychology* 55 (January 2019): 96–109. See also Chen Yu and Linda B. Smith, "The Social Origins of Sustained Attention in One-Year-Old Human Infants," *Current Biology* 26 (May 2016): 1235–40.

more likely to recognize: Sebastian Wahl, Vesna Marinović, and Birgit Träuble, "Gaze Cues of Isolated Eyes Facilitate the Encoding and Further Processing of Objects in 4-Month-Old Infants," *Developmental Cognitive Neuroscience* 36 (April 2019). See also Vincent M. Reid et al., "Eye Gaze Cueing Facilitates Neural Processing of Objects in 4-Month-Old Infants," *NeuroReport* 15 (November 2004): 2553–55.

continually instructing their offspring: Louis J. Moses et al., "Evidence for Referential Understanding in the Emotions Domain at Twelve and Eighteen Months," *Child Development* 72 (May–June 2001): 718–35.

a baby will reliably look: Michael Tomasello et al., "Reliance on Head Versus Eyes in the Gaze Following of Great Apes and Human Infants: The Cooperative Eye Hypothesis," *Journal of Human Evolution* 52 (March 2007): 314–20.

visible whites of the eyes: Hiromi Kobayashi and Shiro Kohshima, "Unique Morphology of the Human Eye and Its Adaptive Meaning: Comparative Studies on External Morphology of the Primate Eye," *Journal of Human Evolution* 40 (May 2001): 419–35.

"cooperative eye hypothesis": Tomasello et al., "Reliance on Head Versus Eyes in the Gaze Following of Great Apes and Human Infants." See also Hiromi Kobayashi and Kazuhide Hashiya, "The Gaze That Grooms: Contribution of Social Factors to the Evolution of Primate Eye Morphology," *Evolution and Human Behavior* 32 (May 2011): 157–65.

"Our eyes see": Ker Than, "Why Eyes Are So Alluring," *LiveScience,* November 7, 2006.

makes all human achievements possible: Michael Tomasello, *The Cultural Origins of Human Cognition* (Cambridge: Harvard University Press, 1999).

remains important among adults: Shteynberg, "A Collective Perspective"; Shteynberg and Apfelbaum, "The Power of Shared Experience."

223   *our mental models of the world:* Garriy Shteynberg et al., "The Broadcast of Shared Attention and Its Impact on Political Persuasion," *Journal of Personality and Social Psychology* 111 (November 2016): 665–73.

laboring on a shared task: Bertrand Schneider, "Unpacking Collaborative Learning Processes During Hands-On Activities Using Mobile Eye-Trackers," paper presented at the 13th International Conference on Computer Supported Collaborative Learning, June 2019. See also Nasim Hajari et al., "Spatio-Temporal Eye Gaze Data Analysis to Better Understand Team Cognition," paper presented at the 1st International Conference on Smart Multimedia, August 2018.

the gaze of experienced surgeons: Ross Neitz, "U of A Lab Testing Technologies to Better Train Surgeons," *Folio,* April 25, 2019.

aren't always looking at the same place: Schneider, "Unpacking Collaborative Learning Processes During Hands-On Activities Using Mobile Eye-Trackers."

common conceptualizations of motivation: Duckworth, *Grit.*

a potent source of motivation: Gregory M. Walton and Geoffrey L. Cohen, "Sharing Motivation," in *Social Motivation,* ed. David Dunning (New York: Psychology Press, 2011), 79–102. See also Allison Master and Gregory M. Walton, "Minimal Groups Increase Young Children's Motivation and Learning on Group-Relevant Tasks," *Child Development* 84 (March–April 2013): 737–51.

a form of intrinsic *motivation:* Priyanka B. Carr and Gregory M. Walton, "Cues of Working Together Fuel Intrinsic Motivation," *Journal of Experimental Social Psychology* 53 (July 2014): 169–84.

and more easily maintained: Edward L. Deci and Richard M. Ryan, *Intrinsic Motivation and Self-Determination in Human Behavior* (New York: Plenum Press, 1985).

224 *"entitativity":* Brian Lickel et al., "Varieties of Groups and the Perception of Group Entitativity," *Journal of Personality and Social Psychology* 78 (February 2000): 223–46. See also Donald T. Campbell, "Common Fate, Similarity, and Other Indices of the Status of Aggregates of Persons as Social Entities," *Behavioral Science* 3 (January 1958): 14–25.

"They were pecking away": Paul Barnwell, "My Students Don't Know How to Have a Conversation," *The Atlantic,* April 22, 2014.

asynchronous communication: For an informative take on this topic, see Cal Newport, "Was E-Mail a Mistake?," *The New Yorker,* August 6, 2019.

teams that trained as a group: Richard L. Moreland and Larissa Myaskovsky, "Exploring the Performance Benefits of Group Training: Transactive Memory or Improved Communication?," *Organizational Behavior and Human Decision Processes* 82 (May 2000): 117–33. See also Diane Wei Liang, Richard Moreland, and Linda Argote, "Group Versus Individual Training and Group Performance: The Mediating Factor of Transactive Memory," *Personality and Social Psychology Bulletin* 21 (April 1995): 384–93.

225 *"silo effect":* Gillian Tett, *The Silo Effect: The Peril of Expertise and the Promise of Breaking Down Barriers* (New York: Simon & Schuster, 2015).

training together is not the norm: For a first-person perspective, see Dhruv Khullar, "Doctors and Nurses, Not Learning Together," *New York Times,* April 30, 2015.

group training across disciplinary lines: Maja Djukic, "Nurses and Physicians Need to Learn Together in Order to Work Together," post on the website of the Robert Wood Johnson Foundation, May 11, 2015, https://www.rwjf.org/en/blog/2015/05/nurses_and _physician.html. See also Andrew Schwartz, "Training Nurse Practitioners and Physicians for the Next Generation of Primary Care," post on the website of the University of California-San Francisco, January 2013, https://scienceofcaring.ucsf.edu/future -nursing/training-nurse-practitioners-and-physicians-next-generation-primary -care.

"escape room": Cheri Friedrich et al., "Escaping the Professional Silo: An Escape Room Implemented in an Interprofessional Education Curriculum," *Journal of Interprofessional Care* 33 (September–October 2019): 573–75. See also Cheri Friedrich et al., "Interprofessional Health Care Escape Room for Advanced Learners," *Journal of Nursing Education* 59 (January 2020): 46–50.

case study of a fictional patient: Hilary Teaford, "Escaping the Professional Silo: Implementing An Interprofessional Escape Room," post on the blog MinneSOTL, https:// wcispe.wordpress.com/2017/12/01/escaping-the-professional-silo-implementing-an -interprofessional-escape-room/.

"social glue": Brock Bastian, Jolanda Jetten, and Laura J. Ferris, "Pain as Social Glue:

Shared Pain Increases Cooperation," *Psychological Science* 25 (November 2014): 2079–85. See also Shawn Achor, "The Right Kind of Stress Can Bond Your Team Together," *Harvard Business Review,* December 14, 2015.

share their thoughts: Frederick G. Elias, Mark E. Johnson, and Jay B. Fortman, "Task-Focused Self-Disclosure: Effects on Group Cohesiveness, Commitment to Task, and Productivity," *Small Group Research* 20 (February 1989): 87–96.

226 *"That's a very different question":* Tony Schwartz, "How Are You Feeling?," *Harvard Business Review,* February 22, 2011.

searching or even wrenching: Tony Schwartz, "The Importance of Naming Your Emotions," *New York Times,* April 3, 2015.

"What's the most important thing": Tony Schwartz, "What If You Could Truly Be Yourself at Work?," *Harvard Business Review,* January 23, 2013.

engage in rituals *together:* Cristine H. Legare and Nicole Wen, "The Effects of Ritual on the Development of Social Group Cognition," *Bulletin of the International Society for the Study of Behavioral Development* 2 (2014): 9–12. See also Ronald Fischer et al., "How Do Rituals Affect Cooperation? An Experimental Field Study Comparing Nine Ritual Types," *Human Nature* 24 (June 2013): 115–25.

synchronized movement or shared physiological arousal: Kerry L. Marsh, Michael J. Richardson, and R. C. Schmidt, "Social Connection Through Joint Action and Interpersonal Coordination," *Topics in Cognitive Science* 1 (April 2009): 320–39.

"Morning Mile": Jenny Berg, "Students at Clearview Go the Extra (Morning) Mile Every Day Before School," *St. Cloud Times,* January 17, 2019. See also Sarah D. Sparks, "Why Lunch, Exercise, Sleep, and Air Quality Matter at School," *Education Week,* March 12, 2019.

match up their bodily movements: Christine E. Webb, Maya Rossignac-Milon, and E. Tory Higgins, "Stepping Forward Together: Could Walking Facilitate Interpersonal Conflict Resolution?," *American Psychologist* 72 (May–June 2017): 374–85. See also Ari Z. Zivotofsky and Jeffrey M. Hausdorff, "The Sensory Feedback Mechanisms Enabling Couples to Walk Synchronously: An Initial Investigation," *Journal of NeuroEngineering and Rehabilitation* 4 (August 2008).

role-play executives: Lakshmi Balachandra, "Should You Eat While You Negotiate?," *Harvard Business Review,* January 29, 2013.

227 *dine "family style":* Kaitlin Woolley and Ayelet Fishbach, "A Recipe for Friendship: Similar Food Consumption Promotes Trust and Cooperation," *Journal of Consumer Psychology* 27 (January 2017): 1–10.

consuming such food: Shinpei Kawakami et al., "The Brain Mechanisms Underlying the Perception of Pungent Taste of Capsaicin and the Subsequent Autonomic Responses," *Frontiers of Human Neuroscience* 9 (January 2016).

greater economic cooperation: Bastian, Jetten, and Ferris, "Pain as Social Glue."

"eating together is a more intimate act": Kevin Kniffin, quoted in Susan Kelley, "Groups That Eat Together Perform Better Together," *Cornell Chronicle,* November 19, 2015.

teams of firefighters: Kevin M. Kniffin et al., "Eating Together at the Firehouse: How Workplace Commensality Relates to the Performance of Firefighters," *Human Performance* 28 (August 2015): 281–306.

"Coworkers who eat together": Kevin Kniffin, "Upbeat Music Can Make Employees More Cooperative," *Harvard Business Review,* August 30, 2016.

228 *"rapport detection":* Philipp Müller, Michael Xuelin Huang, and Andreas Bulling, "De-

tecting Low Rapport During Natural Interactions in Small Groups from Non-Verbal Behaviour," paper presented at the 23rd International Conference on Intelligent User Interfaces, March 2018.

When rapport falls: Andrea Stevenson Won, Jeremy N. Bailenson, and Joris H. Janssen, "Automatic Detection of Nonverbal Behavior Predicts Learning in Dyadic Interactions," *IEEE Transactions on Affective Computing* 5 (April–June 2014): 112–25. See also Juan Lorenzo Hagad et al., "Predicting Levels of Rapport in Dyadic Interactions Through Automatic Detection of Posture and Posture Congruence," paper presented at the IEEE 3rd International Conference on Social Computing, October 2011.

"smart meeting rooms": Indrani Bhattacharya et al., "A Multimodal-Sensor-Enabled Room for Unobtrusive Group Meeting Analysis," paper presented at the 20th ACM International Conference on Multimodal Interaction, October 2018. See also Prerna Chikersal et al., "Deep Structures of Collaboration: Physiological Correlates of Collective Intelligence and Group Satisfaction," paper presented at the ACM Conference on Computer Supported Cooperative Work and Social Computing, February 2017.

"We set out to design": Katherine Isbister et al., "Yamove! A Movement Synchrony Game That Choreographs Social Interaction," *Human Technology* 12 (May 2016): 74–102.

"how being physically 'in sync'": Katherine Isbister, quoted in "The Future Is Now: Innovation Square Comes to NYU-Poly," article posted on the website of the NYU Tandon School of Engineering Polytechnic Institute, June 12, 2012, https://engineering .nyu.edu/news/future-now-innovation-square-comes-nyu-poly.

her game, called Yamove!: Isbister et al., "Yamove!" See also Katherine Isbister, Elena Márquez Segura, and Edward F. Melcer, "Social Affordances at Play: Game Design Toward Socio-Technical Innovation," paper presented at the Conference on Human Factors in Computing Systems, April 2018.

229 *"The more players look at each other":* Katherine Isbister, *How Games Move Us: Emotion by Design* (Cambridge: MIT Press, 2016), 96.

"engaging," "motivating," and even "fun": Jesús Sánchez-Martín, Mario Corrales-Serrano, Amalia Luque-Sendra, and Francisco Zamora-Polo, "Exit for Success. Gamifying Science and Technology for University Students Using Escape-Room. A Preliminary Approach," *Heliyon* 6 (July 2020).

"grouphate": Scott A. Myers and Alan K. Goodboy, "A Study of Grouphate in a Course on Small Group Communication," *Psychological Reports* 97 (October 2005): 381–86.

"Respected Sir": Satyendra Nath Bose, in a letter written to Albert Einstein, in *The Collected Papers of Albert Einstein,* vol. 14, *The Berlin Years: Writings & Correspondence, April 1923–May 1925,* ed. Diana Kormos Buchwald et al., (Princeton: Princeton University Press, 2015), 399.

The paper he was sending: A. Douglas Stone, *Einstein and the Quantum: The Quest of the Valiant Swabian* (Princeton: Princeton University Press, 2015), 215.

230 *"some combination of veneration and chutzpah":* Stone, *Einstein and the Quantum,* 215.

Bose had solved a problem: John Stachel, "Einstein and Bose," in *Satyendra Nath Bose: His Life and Times; Selected Works (with Commentary),* ed. Kameshwar C. Wali (Singapore: World Scientific, 2009), 422–41.

"a beautiful step forward": Albert Einstein, quoted in Stachel, "Einstein and Bose."

"I wanted to know how to grapple": Satyendra Nath Bose, quoted in Stachel, "Einstein and Bose."

the course of scientific history: Stone, *Einstein and the Quantum.* See also A. Douglas Stone, "What Is a Boson? Einstein Was the First to Know," *HuffPost,* October 1, 2012.

a newly precise measurement: Georges Aad et al., "Combined Measurement of the Higgs Boson Mass in pp Collisions at $\sqrt{s}$ = 7 and 8 TeV with the ATLAS and CMS Experiments," *Physical Review Letters* 114 (May 2015).

The shift is easiest to see: Stefan Wuchty, Benjamin F. Jones, and Brian Uzzi, "The Increasing Dominance of Teams in Production of Knowledge," *Science* 316 (May 2007): 1036–39. See also Brad Wible, "Science as Team Sport," *Kellogg Insight,* October 10, 2008.

fewer than 10 percent: National Research Council, *Enhancing the Effectiveness of Team Science* (Washington, DC: National Academies Press, 2015).

231 *"a sharp decline":* Truyken L. B. Ossenblok, Frederik T. Verleysen, and Tim C. E. Engels, "Coauthorship of Journal Articles and Book Chapters in the Social Sciences and Humanities (2000–2010)," *Journal of the Association for Information Science and Technology* 65 (January 2014): 882–97. See also Dorte Henriksen, "The Rise in Co-Authorship in the Social Sciences (1980–2013)," *Scientometrics* 107 (May 2016): 455–76.

solo-authored articles: Lukas Kuld and John O'Hagan, "Rise of Multi-Authored Papers in Economics: Demise of the 'Lone Star' and Why?," *Scientometrics* 114 (March 2018): 1207–25. See also Lukas Kuld and John O'Hagan, "The Trend of Increasing Co-Authorship in Economics: New Evidence," posted on the VOX CEPR Policy Portal, December 16, 2017, https://voxeu.org/article/growth-multi-authored-journal-articles -economics.

"team authors dominate": Christopher A. Cotropia and Lee Petherbridge, "The Dominance of Teams in the Production of Legal Knowledge," *Yale Law Journal* 124 (June 2014): 18–28.

each US patent application: Mark Roth, "Groups Produce Collective Intelligence, Study Says," *Pittsburgh Post-Gazette,* January 10, 2011.

"It suggests that the process of knowledge creation": Wuchty, Jones, and Uzzi, "The Increasing Dominance of Teams in Production of Knowledge."

"almost everything that human beings do today": Brian Uzzi, quoted in Roberta Kwok, "For Teams, What Matters More: Raw Talent or a History of Success Together?," *Kellogg Insight,* June 3, 2019.

"collective intelligence": Anita Williams Woolley et al., "Evidence for a Collective Intelligence Factor in the Performance of Human Groups," *Science* 330 (October 2010): 686–88.

232 *"uniquely-held information":* Garold Stasser and William Titus, "Pooling of Unshared Information in Group Decision Making: Biased Information Sampling During Discussion," *Journal of Personality and Social Psychology* 48 (June 1985): 1467–78.

"often hold back in meetings": Steven G. Rogelberg and Liana Kreamer, "The Case for More Silence in Meetings," *Harvard Business Review,* June 14, 2019.

"What do you all think?": Cass Sunstein, "What Makes Teams Smart (or Dumb)," *HBR IdeaCast,* Episode 440, December 18, 2014.

engage in "self-silencing": Cass R. Sunstein and Reid Hastie, *Wiser: Getting Beyond Groupthink to Make Groups Smarter* (Cambridge: Harvard Business Review Press, 2015), 33.

"more likely to silence themselves": Cass R. Sunstein, "Group Judgments: Deliberation,

Statistical Means, and Information Markets," University of Chicago Public Law and Legal Theory Working Paper, number 72, 2004.

"inquisitive and self-silencing": Sunstein and Hastie, *Wiser,* 105.

233 *"the mind is just less and less":* Clark, *Natural-Born Cyborgs,* 5.

a specific sequence of actions: Annelies Vredeveldt, Alieke Hildebrandt, and Peter J. van Koppen, "Acknowledge, Repeat, Rephrase, Elaborate: Witnesses Can Help Each Other Remember More," *Memory* 24 (2016): 669–82.

engaging in this kind of communication: Romain J. G. Clément et al., "Collective Cognition in Humans: Groups Outperform Their Best Members in a Sentence Reconstruction Task," *PLoS One* 8 (October 2013).

"error pruning": Suparna Rajaram and Luciane P. Pereira-Pasarin, "Collaborative Memory: Cognitive Research and Theory," *Perspectives on Psychological Science* 5 (November 2010): 649–63.

a study of airplane pilots: Michelle L. Meade, Timothy J. Nokes, and Daniel G. Morrow, "Expertise Promotes Facilitation on a Collaborative Memory Task," *Memory* 17 (January 2009): 39–48.

"shared artifacts": Gary M. Olson and Judith S. Olson, "Distance Matters," *Human-Computer Interaction* 15 (September 2000): 139–78.

234 *the best of all worlds:* Stephanie Teasley et al., "How Does Radical Collocation Help a Team Succeed?," paper presented at the Conference on Computer Supported Cooperative Work, December 2000. See also Lisa M. Covi et al., "A Room of Your Own: What Do We Learn About Support of Teamwork from Assessing Teams in Dedicated Project Rooms?," paper presented at the International Workshop on Cooperative Buildings, February 1998.

"As they discussed and agreed": Judith S. Olson et al., "The (Currently) Unique Advantages of Collocated Work," in *Distributed Work,* ed. Pamela Hinds and Sara B. Kiesler (Cambridge: MIT Press, 2002), 113–35.

gesture at large artifacts: Mathilde M. Bekker, Judith S. Olson, and Gary M. Olson, "Analysis of Gestures in Face-to-Face Design Teams Provides Guidance for How to Use Groupware in Design," *Proceedings of the 1st Conference on Designing Interactive Systems: Processes, Practices, Methods, & Techniques* (August 1995): 157–66.

"were often put up in the order": Olson et al., "The (Currently) Unique Advantages of Collocated Work," 122–23.

"inherent invisibility": Covi et al., "A Room of Your Own."

235 *"transactive memory":* Wegner, Giuliano, and Hertel, "Cognitive Interdependence in Close Relationships." See also Wegner, "Transactive Memory."

"Toni and I noticed": Daniel M. Wegner, "Don't Fear the Cybermind," *New York Times,* August 4, 2012.

"to understand the group mind": Wegner, "Don't Fear the Cybermind."

confirmed Wegner's claim: Gün R. Semin, Margarida V. Garrido, and Tomás A. Palma, "Socially Situated Cognition: Recasting Social Cognition as an Emergent Phenomenon," in *The SAGE Handbook of Social Cognition,* ed. Susan Fiske and Neil Macrae (Thousand Oaks, CA: Sage Publications, 2012), 138–64. See also Zhi-Xue Zhang et al., "Transactive Memory System Links Work Team Characteristics and Performance," *Journal of Applied Psychology* 92 (2007): 1722–30.

236 *transactive memory system:* John R. Austin, "Transactive Memory in Organizational

Groups: The Effects of Content, Consensus, Specialization, and Accuracy on Group Performance," *Journal of Applied Psychology* 88 (October 2003): 866–78.

much of their potential: Kyle J. Emich, "How Expectancy Motivation Influences Information Exchange in Small Groups," *Small Group Research* 43 (June 2012): 275–94.

thinking different *thoughts:* Babak Hemmatian and Steven A. Sloman, "Two Systems for Thinking with a Community: Outsourcing Versus Collaboration," in *Logic and Uncertainty in the Human Mind: A Tribute to David E. Over,* ed. Shira Elqayam et al. (New York: Routledge, 2020), 102–15. See also Roy F. Baumeister, Sarah E. Ainsworth, and Kathleen D. Vohs, "Are Groups More or Less Than the Sum of Their Members? The Moderating Role of Individual Identification," *Behavioral and Brain Sciences* 39 (January 2016).

sensitive and discriminating filters: Eoin Whelan and Robin Teigland, "Transactive Memory Systems as a Collective Filter for Mitigating Information Overload in Digitally Enabled Organizational Groups," *Information and Organization* 23 (July 2013): 177–97.

point us toward the people: Nathaniel Rabb, Philip M. Fernbach, and Steven A. Sloman, "Individual Representation in a Community of Knowledge," *Trends in Cognitive Sciences* 23 (October 2019): 891–902. See also Garold Stasser, Dennis D. Stewart, and Gwen M. Wittenbaum, "Expert Roles and Information Exchange During Discussion: The Importance of Knowing Who Knows What," *Journal of Experimental Social Psychology* 31 (May 1995): 244–65.

should be explicitly informed: Jim Steele, "Structured Reflection on Roles and Tasks Improves Team Performance, UAH Study Finds," article posted on the website of the University of Alabama in Huntsville, April 8, 2013.

237 *a meta-knowledge champion:* Julija N. Mell, Daan van Knippenberg, and Wendy P. van Ginkel, "The Catalyst Effect: The Impact of Transactive Memory System Structure on Team Performance," *Academy of Management Journal* 57 (August 2014): 1154–73. See also Alex Fradera, "Why Your Team Should Appoint a 'Meta-Knowledge' Champion — One Person Who's Aware of Everyone Else's Area of Expertise," *British Psychological Society Research Digest,* September 8, 2014.

"shared fate": Haidt, *The Righteous Mind,* 90.

public schools in Austin, Texas: My discussion of Elliot Aronson and his "jigsaw classroom" technique draws on the following sources: Elliot Aronson, *Not by Chance Alone: My Life as a Social Psychologist* (New York: Basic Books, 2010); Elliot Aronson, *The Social Animal,* 9th ed. (New York: Worth Publishers, 2004); Elliot Aronson, *Nobody Left to Hate: Teaching Compassion After Columbine* (New York: Henry Holt, 2000); Elliot Aronson, "Reducing Hostility and Building Compassion: Lessons from the Jigsaw Classroom," in *The Social Psychology of Good and Evil,* ed. Arthur G. Miller (New York: Guilford Press, 2004), 469–88; Elliot Aronson and Diane Bridgeman, "Jigsaw Groups and the Desegregated Classroom: In Pursuit of Common Goals," in *Readings About the Social Animal,* ed. Elliot Aronson (New York: Macmillan, 2003), 423–33; Elliot Aronson and Neal Osherow, "Cooperation, Prosocial Behavior, and Academic Performance: Experiments in the Desegregated Classroom," *Applied Social Psychology Annual* 1 (1980): 163–96; Susan Gilbert, "A Conversation with Elliot Aronson," *New York Times,* March 27, 2001; Ursula Vils, "'Jigsaw Method' Cuts Desegregation Strife," *Los Angeles Times,* August 10, 1978.

"The first step": Aronson, *Not by Chance Alone,* 201.

"The teacher stands": Aronson, *The Social Animal*, 281.

"students learn more": Aronson and Osherow, "Cooperation, Prosocial Behavior, and Academic Performance."

238 *"a situation where they needed to cooperate"*: Aronson, *Not by Chance Alone*, 203.

"Each student has possession": Aronson, "Reducing Hostility and Building Compassion."

"In this situation": Aronson and Osherow, "Cooperation, Prosocial Behavior, and Academic Performance."

CONCLUSION

241 *"I was totally intimidated"*: Author interview with Joshua Aronson.

"a modern classic": Susan T. Fiske, "The Discomfort Index: How to Spot a Really Good Idea Whose Time Has Come," *Psychological Inquiry* 14 (2003): 203–8.

described for the first time: Claude M. Steele and Joshua Aronson, "Stereotype Threat and the Intellectual Test Performance of African Americans," *Journal of Personality and Social Psychology* 69 (November 1995): 797–811.

242 *"a fixed lump"*: Joshua Aronson, quoted in Annie Murphy Paul, "It's Not Me, It's You," *New York Times*, October 6, 2012.

reinterpret bodily signals: Michael Johns, Michael Inzlicht, and Toni Schmader, "Stereotype Threat and Executive Resource Depletion: Examining the Influence of Emotion Regulation," *Journal of Experimental Psychology: General* 137 (November 2008): 691–705.

"cues of belonging": Sapna Cheryan et al., "Designing Classrooms to Maximize Student Achievement," Policy Insights from the Behavioral and Brain Sciences 1 (October 2014): 4–12.

expert feedback: Geoffrey L. Cohen, Claude M. Steele, and Lee D. Ross, "The Mentor's Dilemma: Providing Critical Feedback Across the Racial Divide," *Personality and Social Psychology Bulletin* 25 (October 1999): 1302–18.

"conditional stupidity": Joshua Aronson, quoted in Paul, "It's Not Me, It's You."

243 *we should offload information*: Risko and Gilbert, "Cognitive Offloading."

onto the world itself: Azadeh Jamalian, Valeria Giardino, and Barbara Tversky, "Gestures for Thinking," *Proceedings of the Annual Conference of the Cognitive Science Society* 35 (2013): 645–50.

244 *offloading may be* social: Sloman and Fernbach, *The Knowledge Illusion*.

246 *"spatializing the curriculum"*: Nora S. Newcombe, "Thinking Spatially in the Science Classroom," *Current Opinion in Behavioral Sciences* 10 (August 2016): 1–6.

247 *they don't design machines*: Andy Clark, "Extended You," talk given at TEDxLambeth, December 16, 2019.

248 *expands to encompass it*: Lucilla Cardinali et al., "Tool-Use Induces Morphological Updating of the Body Schema," *Current Biology* 19 (June 2009): R478–79. See also Angelo Maravita and Atsushi Iriki, "Tools for the Body (Schema)," *Trends in Cognitive Sciences* 8 (February 2004): 79–86.

250 *"to quantitatively assess"*: Bruno R. Bocanegra et al., "Intelligent Problem-Solvers Externalize Cognitive Operations," *Nature Human Behaviour* 3 (February 2019): 136–42.

251 *"Our study showed very clearly"*: Author interview with Bruno Bocanegra.

252 *"veil of ignorance"*: John Rawls, *A Theory of Justice* (Cambridge: Harvard University Press, 1971), 11.

Index